T0133106

Extending Moore's Law through Advanced Semiconductor Design and Processing Techniques

Extending Moore's Law through Advanced Semiconductor Design and Processing Techniques

By
Wynand Lambrechts
Saurabh Sinha
Jassem Abdallah
Jaco Prinsloo

CRC Press
Taylor & Francis Group
Boca Raton London New York

CRC Press is an imprint of the
Taylor & Francis Group, an **informa** business

CRC Press
Taylor & Francis Group
6000 Broken Sound Parkway NW, Suite 300
Boca Raton, FL 33487-2742

© 2019 by Taylor & Francis Group, LLC
CRC Press is an imprint of Taylor & Francis Group, an Informa business

No claim to original U.S. Government works

Printed on acid-free paper

International Standard Book Number-13: 978-0-8153-7074-1 (Hardback)

Library of Congress Cataloging-in-Publication Data

Names: Lambrechts, Wynand, author. | Sinha, S. (Saurabh), 1981- author. | Abdallah, Jassem Ahmed, author.
Title: Extending Moore's Law through advanced semiconductor design and processing techniques / edited by Wynand Lambrechts, Saurabh Sinha, Jassem Ahmed Abdallah.
Description: First edition. | Boca Raton, FL : CRC Press/Taylor & Francis Group, 2019. | "A CRC title, part of the Taylor & Francis imprint, a member of the Taylor & Francis Group, the academic division of T&F Informa plc." | Includes bibliographical references and index.
Identifiers: LCCN 2018015863 (print) | LCCN 2018016897 (ebook) | ISBN 9781351248679 (eBook) | ISBN 9781351248662 (Adobe PDF) | ISBN 9781351248655 (ePUB) | ISBN 9781351248648 (Mobipocket) | ISBN 9780815370741 (hardback : acid-free paper)
Subjects: LCSH: Integrated circuits--Design and construction--Technological innovations. | Semiconductors--Design and construction--Technological innovations. | Moore's law.
Classification: LCC TK7874 (ebook) | LCC TK7874 .E945 2018 (print) | DDC 621.39/5--dc23
LC record available at https://lccn.loc.gov/2018015863

Visit the Taylor & Francis Web site at
http://www.taylorandfrancis.com

and the CRC Press Web site at
http://www.crcpress.com

Contents

Authors

Wynand Lambrechts, SMIEEE, obtained his B.Eng., M.Eng., and Ph.D. degrees in Electronic Engineering from the University of Pretoria (UP), South Africa. He achieved his M.Eng. with distinction. He has authored 2 publications in peer-reviewed journals and has presented at various local and international conferences. Wynand is the lead author on two books; in the fields on sustainable energy and in microelectronic engineering, published by international publishers. He has co-authored four contributing chapters in other books in the fields of green energy and technology and the fourth industrial revolution. He previously held a position as an electronic engineer at Denel Dynamics, a state-owned company in South Africa. He is currently employed by SAAB Grintek Defence (SGD) and is also serving as a part-time research associate at the University of Johannesburg (UJ), South Africa.

Saurabh Sinha, Ph.D. (Eng.), Pr. Eng., SMIEEE, FSAIEE, FSAAE, obtained his B.Eng., M.Eng and Ph.D. degrees in Electronic Engineering from the University of Pretoria (UP). As a published researcher, he has authored or co-authored over 110 publications in peer-reviewed journals and at international conferences. Prof Sinha served UP for over a decade; his last service being as director of the Carl and Emily Fuchs Institute for Microelectronics, Department of Electrical, Electronic and Computer Engineering. On 1 October 2013, Sinha was appointed as Executive Dean of the Faculty of Engineering and the Built Environment (FEBE) at the University of Johannesburg (UJ). Sinha is currently the UJ Deputy Vice-Chancellor: Research and Internationalisation. Among other leading roles, Sinha also served the IEEE as a board of director and IEEE vice-president: educational activities.

Jassem Abdallah received his B.Sc. in Chemical Engineering from The University of Texas at Austin with honors (undergraduate research thesis), and both his M.Sc. and Ph.D. in Chemical Engineering from the Georgia Institute of Technology. He completed post-doctoral research assignments at the Georgia Institute of Technology, the IBM Corporation, and the University of Johannesburg. His research background spans both Materials Science Engineering and Process Development Engineering, with a focus on semiconductor technology and organic electronic applications. He has several years of research experience working in both academic-scale, 100 mm class 100 cleanroom facilities as well as state-of-the-art 300 mm class 10 semiconductor R&D facilities. In addition to co-authoring several journals articles and conference papers, Abdallah holds several patents, and serves as a peer reviewer for two scientific journals: The *Journal of the Electrochemical Society* and *Electrochemical and Solid-State Letters*.

Jaco Prinsloo received his B.Eng. in Electronic Engineering from the University of Pretoria (UP), South Africa, in 2015. He is currently employed as an embedded engineer and software developer, where his work focuses on the development of cryptographic technology. Apart from his full-time occupation, he has a keen interest in the latest advancements in semiconductor technologies and quantum physics, among other fields. He is also an active organist, classical musician and electronic systems developer in his free time.

1 The Driving Forces Behind Moore's Law and Its Impact on Technology

Wynand Lambrechts, Saurabh Sinha and Jassem Abdallah

INTRODUCTION

The rate of technological advancement as a function of time is governed by the correlation by which the number of components in very large-scale integration (VLSI) integrated circuits (ICs) and their total computing power double approximately every 18–24 months (Moore 1965; Hutcheson 2009). This phenomenon is known as Moore's law, and was predicted in a 1965 paper by Gordon Moore (Moore 1965), the co-founder of Fairchild Semiconductor and Intel. Moore predicted that the quantity of components integrated into ICs would increase twofold each year for the next decade (from 1965), and ten years later (during 1975), adjusted the prediction to the integrated components doubling every two years. This held true up until the current decade (five decades later), with deviations only occurring recently in response to several factors, primarily physical limitations (at atomic level) and changes in computing habits (such as multi-core and open-hardware systems). In 2009, IBM Fellow Carl Anderson noted a trend of semiconductor technology maturing and exponential growth slowing down, similar to that of the railroad, the automotive industry and aviation before it. Moore's law is related to the scaling law founded on the work of Dennard (1974), known as Dennard scaling. Dennard scaling is specifically related to power scaling in complementary metal-oxide semiconductor (CMOS) transistors. Essentially, the Dennard scaling law states that the overall IC power consumption for a set area – therefore, the power density – remains constant if the feature size (node) of the technology scales downward, typically by a decreasing gate length. Dennard (1974) provides scaling results for circuit performance, which have been adapted and are presented in Table 1.1.

This hypothesis, however, has started to derail since approximately 2005 (roughly 30 years after the work presented by Dennard (1974)), when 90/65 nm gate lengths were primarily manufactured by industry leaders such as Intel, TSMC, GlobalFoundries, Samsung and IBM, with leakage currents and thermal runaway becoming even larger, contributing to the power dissipation in smaller technology nodes (45, 32, 22 and 14 nm). Since increases in operating frequency have become largely incremental (compare the Intel Pentium II at 300 MHz by 1998 to the Intel Pentium 4 at 1.5 GHz by 2000, and again to the current generation, with Intel

TABLE 1.1

Scaling Results for Circuit Performance

Component or IC Parameter	Scaling Factor
Physical device dimensions (t_{ox}, L, W)	$1/\kappa$
Doping concentration (N_A)	κ
Operating voltage (V)	$1/\kappa$
Circuit current (I)	$1/\kappa$
Capacitance (C)	$1/\kappa$
Delay time per circuit (VC/I)	$1/\kappa$
Power dissipation per circuit (VI)	$1/\kappa^2$
Power density (VI/A)	κ

Source: Adapted from Dennard, R. H., et al., *IEEE Journal of Solid-State Circuits*, 9(5), 256–268.

processors averaging around 3-GHz operating frequency in 2016) and the transistor count has rapidly increased during this time, focusing on the scaling of transistors is essentially aimed at

- Achieving higher performance per unit area
- Lowering power requirements by using physically smaller components
- Decreasing the manufacturing cost through lowering the cost per transistor

In order to understand the mechanisms responsible for the demise of the Dennard scaling law, which assisted in the variance in Moore's law, the power consumption of CMOS transistors must be technically reviewed; such a review is presented in the following section.

The Inevitable Demise of Dennard's Scaling Law

The well-known power equation for a CMOS transistor is given by the relationship

$$P = fCV^2 \tag{1.1a}$$

where,
  f is the operating frequency in Hz
  C is the component capacitance and V is the operating (bias) voltage

On a CMOS IC consisting of multiple negative-channel metal-oxide semiconductor (NMOS) and positive-channel metal-oxide semiconductor (PMOS) transistors, the power equation can simply be adapted to

$$P = QfCV^2 \tag{1.1b}$$

where Q is the total number of active components (transistors) per unit area. Power dissipation in CMOS circuits is attributed to two primary categories, static (steady-state) and dynamic (transient) dissipation. Static power dissipation results from mechanisms such as

- Subthreshold conduction
- Tunneling current
- Leakage through reverse-biased diodes

where these mechanisms are generally attributed to losses, or the unexpected flow of current, when the transistor should ideally be non-conducting. These mechanisms are becoming more prevalent in technology nodes of 130 nm and smaller for various reasons, which include the sheer quantity of transistors per unit area, as well as the thin oxide layer of these devices, and are specified by current per gate length, with typical units of nA/µm. The second contributing category, dynamic power dissipation, results from mechanisms such as

- The charge and subsequent discharge of the capacitance seen by the circuit output (the load), including intrinsic capacitances of the technology
- Short-circuit current while both the PMOS and NMOS transistor networks are partially on during transitions.

These dynamic losses can largely be attributed to the frequency of operation of these circuits; with more transitions per second, there are more *opportunities* for these mechanisms to occur. To define Dennard's scaling law accurately, all components of power consumption should be taken into account and related to technology node size: dynamic, short-circuit and static biasing power dissipation. Static biasing power dissipation is dependent on the leakage current ($I_{leakage}$) through the active components in their steady state, which should ideally be zero for reverse-biased devices. The power dissipation of the CMOS transistor adds to the total power dissipation (P_{total}) of the IC in such a way that

$$P_{total} = P_{ds} + P_{sc} + P_{sb} \tag{1.2}$$

where,

P_{ds} is the dynamic switching power dissipation as a function of the operating frequency and charging and discharging through capacitances

P_{sc} is the short-circuit power dissipation

P_{sb} is the static biasing power dissipation due to the direct current (DC) consumed by the active device under its DC operating conditions, which include $P_{leakage}$

For the purpose of this introductory chapter and to thoroughly review the origins of Dennard's scaling law, stating that power consumption for a given area remains constant if the technology node scales, the large contributing power dissipations, P_{ds} and P_{sc}, are briefly reviewed. The power dissipation from leakage current has also become a larger contributor since approximately 2005, as high-performance modern ICs typically have in excess of a billion transistors on-chip, each contributing a small amount of leakage current and adding to the total power dissipation in steady state. The CMOS inverter is a popular and simplistic circuit to quantify semiconductor technology process characteristics such as

- Manufacturing cost, expressed by the complexity of a circuit and the area occupied on-chip
- Integrity and robustness of a circuit, expressed by its static behavior
- Circuit performance, determined by the dynamic response to input waveforms at a specific frequency
- Energy efficiency related to the energy supplied to the circuit and the power consumption to achieve the required performance

In this discussion, therefore, the CMOS inverter circuit is used to define and describe the power consumption in CMOS ICs, for both the dynamic and static power losses.

POWER CONSUMPTION IN CMOS INVERTERS

The dynamic switching power (a dominant power dissipation component for technologies of approximately 180 nm and larger) of a simple circuit, the CMOS inverter (adapted from Butzen and Ribas 2007), can be described by the schematic representation given in Figure 1.1.

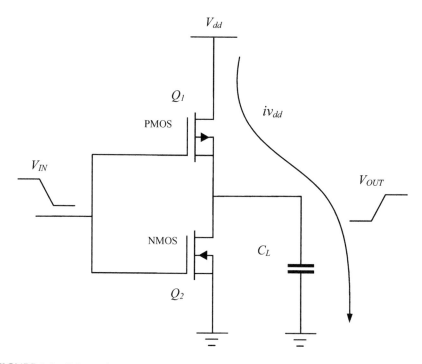

FIGURE 1.1 Schematic representation of a CMOS inverter to describe dynamic switching power consumption.

For Figure 1.1, Q_1 is a PMOS transistor and Q_2 is an NMOS transistor, with the load capacitor indicated as C_L. The dynamic power dissipation is not dependent on the design of the inverter, apart from the contributions of the parasitic output capacitances (Veendrick 1984). Considering an input waveform V_{IN}, as shown in Figure 1.1, with theoretically zero rise and fall times (where the signal in Figure 1.1 shows nonzero rise and fall times to accentuate its existence in practice), the output waveform is also shown in Figure 1.1 by V_{OUT}.

To determine the energy consumption E of the output waveform during a low-to-high change in signal amplitude, the integral of the instantaneous power over a defined time must be determined. This is mathematically described by

$$E_{V_{dd}} = \int_0^\infty i_{v_{dd}}(t) V_{dd} dt \tag{1.3}$$

where,

$i_{v_{dd}}$ is the small-signal current flowing through the PMOS transistor and through the load capacitor C_L toward ground

V_{dd} is the operating voltage

The current through a capacitor is defined as the rate of change of its voltage multiplied by the capacitance. This well-known relationship between the voltage and current of a capacitor is mathematically described by

$$i(t) = C \frac{dv}{dt} \tag{1.4}$$

and, therefore, the energy equation given in (1.3) can be rewritten as

$$E_{V_{dd}} = V_{dd} \int_0^\infty C_L \frac{dv_{out}}{dt} dt \tag{1.5}$$

where v_{out} is the small-signal output voltage of the CMOS inverter circuit, as shown in Figure 1.1. By separating the constants and the time-varying components in (1.5), the energy equation becomes

$$E_{V_{dd}} = C_L V_{dd} \int_0^{V_{dd}} dv_{out} \tag{1.6}$$

which, if solving the integral, results in

$$E_{V_{dd}} = C_L V_{dd}^2 \tag{1.7}$$

representing the energy, in joules, consumed by the output capacitor C_L. To determine the charge stored in the capacitor during one half-cycle, the energy equation given in (1.3) changes, with the operating voltage V_{dd} replaced by the observed output voltage, v_{out}, such that

$$E_{C_L} = \int_0^\infty i_{v_{dd}}(t) v_{out} dt \tag{1.8}$$

and by following a similar principle used to determine the energy from the voltage supply, the energy in the capacitor, and therefore its charge, is equated as

$$E_{C_L} = \frac{C_L V_{dd}^2}{2} \tag{1.9}$$

By comparing (1.7) and (1.9), it is noticeable that only half of the energy provided by the circuit supply, V_{dd}, is stowed in the capacitor as a charge. The remaining energy is consumed by the PMOS and NMOS transistors, depending on the direction of the input transition (low-to-high transition enables the PMOS transistor and high-to-low transition enables the NMOS transistor). The total amount of power coming from the supply therefore depends on the number of transitions of the circuit, known as its operating frequency, f, as observed in (1.1a and 1.1b). The average power of the circuit is determined by dividing the energy calculated in (1.7) and (1.9) by the period-time of the input signal, T, such that

$$E = \frac{1}{T} \int_0^\infty i(t) V dt \tag{1.10}$$

which leads to a result for the dynamic supply power consumption in (1.7) of

$$E_{V_{dd}} = \frac{C_L V_{dd}^2}{T} \tag{1.11}$$

where T can also be defined by

$$T = \frac{1}{f} = t_{pLH} + t_{pHL} = 2t_p \tag{1.12}$$

and where t_{pLH} and t_{pHL} are the propagation delay between low-to-high and high-to-low transitions, respectively. Assuming these propagation delays are equal for both transition types, the period-time of the signal T is defined as twice the propagation delay time (t_p). The propagation delay time can be determined by integrating the load capacitance charge or discharge current, such that

$$t_p = \int_{v_1}^{v_2} \frac{C_L(v)}{i(v)} dv \tag{1.13}$$

where v_1 and v_2 are the initial and final voltages defining the integration limits. Essentially, determining the propagation delay time using (1.13) is a complex task, since both $C_L(v)$ and $i(v)$ are non-linear functions of v. Determining the exact value for t_p from alternative methods, such as the voltage dependencies of the on-resistance of the transistors, falls outside the scope of this book. Thus, by combining (1.7) and the number of transistors in the IC, Q, the power consumption in the IC as defined by (1.1a and 1.1b) is achieved.

Leakage current in transistors is especially noticeable, albeit not dominant, in small node technologies such as those of 65 nm and below (Märtin 2014), where the oxide layer becomes extremely thin and electrons are able to tunnel through it. This phenomenon is more dominant in NMOS transistors than in PMOS transistors (Märtin 2014).

A significant contributor to the total power consumption in (1.2) is the short-circuit power consumption P_{sc}. The short-circuit power dissipation for an inverter circuit without a load is a function of the supply voltage and the mean current flowing through the transistor, such that

$$P_{sc} = VI_{mean} \tag{1.14}$$

and this component strongly depends on the design of the inverter (Veendrick 1984) and the symmetry of the active components. If, for simplicity, an inverter with zero load capacitance is considered and it is symmetrical, the gains of the PMOS and NMOS transistors are therefore identical, as well as the threshold voltages of both; it follows (Veendrick 1984) that the mean current through the circuit is given by

$$I_{mean} = \frac{1}{12} \frac{\beta}{V_{dd}} \left(V_{dd} - 2V_{th} \right)^3 \frac{\tau}{T} \tag{1.15}$$

where,

β is the gain factor of the MOS transistor
V_{th} is the threshold voltage of the MOS transistor
T is, again, the period-time of the input signal

The power consumption of the short-circuited node (Veendrick 1984) "is a function of the threshold voltage of the transistor (V_{th}), the rise or fall time of the input signal (τ) and the period-time of the signal," such that

$$P_{sc} = K \left(V_{dd} - 2V_{th} \right)^3 \frac{\tau}{T} \tag{1.16}$$

where K is therefore (from (1.15)) a technology-constant, depending on the physical transistor size, and thus the width and length ratio, which determines the gain, β, of the transistor. In (1.15) and (1.16), the period-time, T, is simply the inverse of the frequency f, specified in seconds. Therefore, as concluded in (1.16), V_{dd}, the rise and fall times of the input signal, τ, as well as T, are the only variables that are not technology/process-determined but are factors external to the circuit, which determine the short-circuit power consumption of the inverter circuit. V_{th} is the gate-source voltage (V_{GS}) of the MOS transistor required to produce an inversion layer and depends on several physical parameters of the transistor, including the oxide capacitance resulting from the thickness of the oxide layer of the technology. The threshold voltage V_{th}, as derived in Gray et al. (2001), is determined by

$$V_{th} = V_{t0} + \gamma\left(\sqrt{2\phi_f + V_{SB}} - \sqrt{2\phi_f}\right) \qquad (1.17)$$

where,

V_{t0} is the threshold voltage with $V_{SB}=0$
ϕ_f is the built-in Fermi level

given by

$$\phi_f = \frac{kT}{q}\ln\frac{N_A}{n_i} \qquad (1.18)$$

where,

k is Boltzmann's constant
T is the temperature in kelvin
q is the elementary electron charge
N_A is the acceptor-concentration
n_i is the intrinsic carrier concentration of the material (typically, silicon)

In (1.17), the body-effect parameter γ is defined as

$$\gamma = \frac{\sqrt{2q\varepsilon N_A}}{C_{ox}} \qquad (1.19)$$

where,

ε is the permittivity of the material
C_{ox} is the oxide capacitance related to the thickness of the oxide layer

The significance of (1.17) in the context of the discussion on the Dennard scaling law is based on the fact that the threshold voltage of transistors has not changed significantly since approximately 2005 because of the limitations in oxide thickness (resulting in high leakage currents if this layer is extraordinarily thin and allows electrons to cross this junction) and limitations on reducing the power supply voltage of smaller nodes. The band gap of silicon (discussed in subsequent chapters in this book) is 1.11 eV at 300 K, and this means that approximately 1 V is required to transition silicon between its insulating and conducting states, assuming that no intrinsic doping of additional materials, for example germanium, is introduced into the silicon. Since the operating voltage of ICs has not dropped significantly below the 1 V mark (for pure silicon technologies), the threshold voltage of the transistors has remained relatively constant (Kanter 2012).

In (1.9) and (1.16), power consumption is therefore determined by primarily external factors such as the power supply voltage, operating frequency, load capacitance, required gain and an intrinsic parameter, the threshold voltage, which has remained relatively constant since approximately 2005. Relating these attributes back to (1.9) and (1.16), dynamic power dissipation in a relatively simple IC circuit, the crux is that the Dennard scaling law is (was) an apt description to accompany Moore's law in stating that power consumption remains constant if technology scales. Recent technology advances (from 2005 to the present) have therefore limited the possibility of scaling threshold voltage (thus operating voltage) and "it is no longer possible to keep the power envelope constant" between advancing generations while simultaneously reaching potential performance improvements (Märtin 2014).

RETURNING TO MOORE'S LAW

Moore's law has evolved from an observation of trends within the semiconductor industry to the driving factor of technological advances in cost and capability (Chien and Karamcheti 2013; Holt 2016). It is the aim of this book to

- Introduce the reader to the science behind Moore's law and how it is complemented by the Dennard scaling law.
- Explain the importance of Moore's law over the past 50 years from a technical and economic perspective
- Highlight the driving forces that contribute to achieving the trend.
- Detail the physical and economic obstacles in the semiconductor industry that threaten to derail Moore's law.
- Present emerging strategies aimed to extend Moore's law if, arguably, it is necessary to ensure growth in the industry at the same rapid pace as seen in the previous half-century.

Electronic devices made possible through the transistor have become the cornerstone of modern society, with computers and other electronic devices becoming ubiquitous not only in our homes and businesses, but increasingly in the form of portable personal communication devices such as cellular phones, smart watches, wearables, tablets and development applications (Tummala 2000). The number of components in both microprocessors and memory devices has consistently increased at a relatively steady rate from the 1960s to modern electronics (Mack 2015, 2011).

During the 1950s, the notion of using ICs started to emerge and Werner Jacobi patented an amplifier consisting of five transistors and additional passive supporting circuitry (Guarnieri 2016 and Jacobi 1952). As seen in Figure 1.2, by 1970, five years after Gordon Moore's prediction about the density of transistors on-chip, the number of transistors per die had grown to over 1000 (10^3). In 2016, the predominantly graphics processing unit manufacturer, Nvidia, introduced its Titan X graphics processing unit with 12 billion transistors on its 471 mm^2 die, manufactured using 16-nanometer fin field-effect transistor (FinFET) technology. This growth is unprecedented and there is a general consensus that the traditional planar CMOS transistor will gradually become obsolete and that new nanotechnology, such as the FinFET technology described later in this chapter, will replace it for the foreseeable future (Bishop 2005).

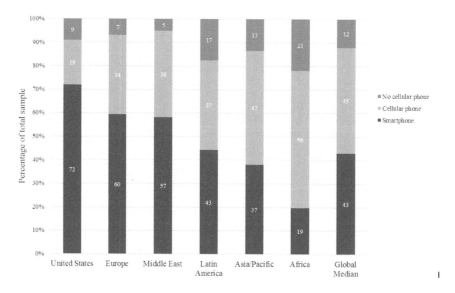

FIGURE 1.2 Regional medians of adults who report owning a smartphone, owning a cellular phone but not a smartphone or not owning either a cellular phone or a smartphone. (Adapted from Poushter, J., *PewResearchCenter*, February 2016.)

Moore's law is also evidently slowing down (Huang 2015) and has been adapted since its first manifestation, so that the doubling of components on ICs is occurring only every 24–36 months. This is not an indication of slowing progress or innovation, but is a factor of limitations resulting from, in part, the demise of Dennard's scaling law and other technical factors, as well as economic factors driving the types of changes in technology advancements, where manufacturing cost has become a very significant limitation, as discussed in Chapter 2 of this book.

There is a disparity between the linear trend in cumulative semiconductor revenue and the number of transistors per chip seen between 1971 and approximately 1998, as shown by Mack (2011). From 1998 onward, the number of transistors has increased rapidly; however, cumulative semiconductor revenue has slowed down. This is primarily due to the cost of starting and maintaining a semiconductor foundry as the limits of technology and physics are being pushed. As a consequence of semiconductor companies sharing the goal of maintaining the pace with Moore's law, the industry made a concerted effort to work synergistically to provide guidance to equipment suppliers, materials suppliers, engineers and researchers on the technical requirements of future technology nodes (Tummala 2000; Mack 2011; Liu and Chang 2009). This was done by the formation of consortia composed of representatives from key players in the semiconductor industry who would collaborate to identify technical challenges and estimate the performance requirements of future devices. These outcomes were then reported in freely available publications as the International Technology Roadmap for Semiconductors (ITRS) (Tummala 2000; ITRS 2014; Arden et al. 2010). These roadmaps are published periodically and, when necessary, revisions are made to account for inaccuracies or overly aggressive predictions (ITRS 2014). Thus, Moore's law has evolved from being a predictive model to being the fundamental business model by which the semiconductor industry judges its progress (Moore 1965; Hutcheson 2009).

The consistency of Moore's law over the five decades since it was first described shows the impressive power of the electronics industry to overcome technical challenges to this law when working in collaboration (Mack 2011, 2015; Moore 1995). As a result of the steady progress of electronic technology, the sizes of electronic devices and computers have reduced dramatically, while the devices have increased in both computing power and affordability. The first computers were both bulky and extremely expensive; they cost millions of US dollars and had footprints that were as large as rooms. Computers such as the UNIVAC could only be afforded and accommodated by large institutions, such as government departments, banks, large industries and educational facilities such as universities (Endres 2013). By the late 1970s, computers had become modular enough to be used in homes as desktops, for example the Apple II computer.

By the 1990s, desktop computers had vastly wider functionality, with the ability to be used as word processors, multimedia devices, telecommunication devices, internet portals and security systems (Tummala 2000). By the turn of the century, the feature sizes of ICs had scaled down to such an extent that electronic devices comparable in computing power to desktops from the 1980s were only a few centimeters in length (Tummala 2000; Moore 1995; Vardi 2014; Cho et al. 2003). Portable devices have become so versatile that billions of people around the world (including residents of under-developed countries) use cell phones not only to communicate with other people, but also to check their stock values in real-time, to play multimedia, to take impromptu photos or videos, to set up daily reminders of tasks, to write grocery lists, to browse the internet and to read electronic books. Smartphones and traditional cellular phones have become powerful tools for billions of users globally. Although smartphone and traditional cellular phone user data differ significantly between various sources, Poushter (2016) published data based on a 2015 global survey using a methodology that explicitly stated that "Results for the survey are based on telephone and face-to-face interviews conducted under the direction of Princeton Survey Research Associates International. The results are based on national samples, unless otherwise noted." A relatively concise summary of global cellular phone users can therefore be approximated from the data of Poushter (2016).

As seen in Figure 1.2, which is adapted from Poushter (2016), traditional cellular phone and smartphone usage, even in emerging markets such as Latin America, Asia/Pacific and Africa, is widespread, with respectively only 17%, 13% and 21% of adults in the survey reporting not owning one of these devices. Users not owning any of these devices in the United States, Europe and the Middle East only account for respectively 9%, 7% and 5% of the surveyed population. The global median for smartphone users (43%) and traditional cellular phone users (45%) therefore accounts for approximately 88% globally. If this is translated to the total global population in 2016, roughly 6.6 billion people worldwide have access to computers in their pockets, a testimony, in a sense, to the significance of Moore's law.

Not only have electronic systems undergone rapid growth in complexity, computing power, functionality and reliability, but these systems have also become more affordable and more ubiquitous (Tummala 2000). In fact, electronic devices have become so pervasive that they have an impact on almost every aspect of our daily lives. Therefore, over the past 50 years, mankind has moved from a paradigm where electronics were both too costly and unwieldy for all but a limited number of applications, to an era where our daily lives are nearly completely dependent on electronic devices (Tummala 2000; Mack 2011, 2015).

The success of Moore's law over the past five decades has been dependent on the ability of researchers and engineers working in semiconductor areas to overcome economic and technical challenges in each technology generation. However, since Moore's law results in circuitry of increasing complexity, there have always been trade-offs between the complexity of the ICs and their manufacturing yield (i.e. the fraction of manufactured ICs operating within pre-defined specifications) (Tummala 2000; Moore 1995; Vardi 2014). In addition, Moore's law predicts IC component sizes of ever-reducing dimensions for the foreseeable future. Since feature sizes must always remain finite, this trend cannot continue indefinitely (Mack 2011, 2015). Similar limitations apply to the traditional correlation between transistor scaling and overall device performance; beyond a threshold level of feature density, continued scaling of IC feature sizes no longer guarantees improvements in device performance, depending on the figure of merit (Tummala 2000; Vardi 2014; Lui and Chang 2009; ITRS 2014; Cho et al. 2003; Koo et al. 2005). The following section details some of the major factors that pose increasingly difficult challenges to Moore's law from a technical and economic perspective. Many of these challenges have begun to affect the scaling of feature sizes, the manufacturing yield, the profitability at the manufacturing level and even the succession time between the miniaturization of technology nodes (ITRS 2014). The general technological factors are discussed in the following section, followed by economic factors attributed to the slowing down of Moore's law.

TECHNICAL AND ECONOMIC FACTORS DRIVING MOORE'S LAW

The technical factors attributed to the slowing down of Moore's law are discussed in the following section.

TECHNICAL FACTORS DRIVING MOORE'S LAW

The early years of the doubling of processing power and memory capacity of computer semiconductors every 18 months, and, later, every 24–36 months, has induced substantial technology advances across many disciplines (Walter 2005). Memory, specifically, has drastically changed from estimations provided by Esener et al. (1999), where optical data storage, magnetic hard disk technology, magnetic tape and emerging (at the time) data storage technologies such as microelectromechanical structures (MEMS) were driving Moore's law. Optical and magnetic disk technologies are still used today, but trends show that solid-state memory has been driving Moore's law to achieve comparable capacity and speed advances in data storage.

In 2015, scientists Martinez-Blanco et al. (2015) created a transistor made up of a single molecule surrounded by only 12 atoms, possibly the smallest size that a transistor can reach and the hard limit of Moore's law. These transistors are made of a single planar π-conjugated molecule of either phthalocyanine or copper phthalocyanine circularly surrounded by 12 positively charged indium atoms of 167 picometers in diameter (0.167 nm), in an indium arsenide (InAs) crystal. These transistors represent a significant drive and progress toward realizing quantum computing, arguably considered the next phase in the evolution of computing devices. This research shows the extreme ends scientists must explore in order to maintain the scalability of transistors in future generations without compromising cost, yield, performance and variability. One of the factors that drives Moore's law is the scalability of semiconductor feature sizes. Feature sizes can be progressively reduced with time, thus allowing manufacturers to increase feature density with each generation, up to a point. The scalability of IC feature sizes is due to the fundamentals of photolithography, which is the patterning technique that is used to produce the IC features on semiconductor substrates.

During photolithography, photosensitive films are coated onto semiconductor substrates and selectively irradiated using an aerial image of the circuit design features. Depending on the photochemistry of the coatings, the irradiated regions form either positive or negative relief patterns when the wafer is immersed in a developing solution. The relief patterns are then etched onto the underlying substrates to produce permanent design features corresponding to the desired IC designs (Morita 2016). Since photolithography is a radiation-based technique, its imaging tools obey the laws of optics; hence, the beams of light that form the aerial image above the patterned wafers can be focused onto smaller areas using optical lenses. Thus, feature sizes can be progressively reduced at each technology node by increasing the focusing power of the imaging tools (Mack 2011, 2015). Since a reduction in feature size allows more IC components to be packed into a unit area of each chip, the steady advancement of photolithography materials and equipment results in a consistent increase in feature density at each technology node. According to Matt Paggi, the vice-president of advanced technology development at GlobalFoundries, photolithography is one of the most difficult challenges in technology nodes of 7 nm and smaller. Foundries initially assumed that extreme ultraviolet (EUV) lithography would be the way forward for sub-7 nm nodes; however, EUV is not yet ready to fully support high-volume manufacturing requirements at every patterning level (Lapedus 2016). Current implementations use variations of multi-patterning 193 nm lithography and EUV techniques rather than relying on EUV completely. Lithographic techniques such as EUV and immersion/multi-patterning will probably be used in a complementary fashion to achieve reliable patterning of these nodes.

The speed at which a transistor can be switched on or off is inversely related to the dimension of the gate (its length) between the source and drain (Tummala 2000; Huff and Gilmer 2005; Brock 2006). Hence, when the critical dimensions of transistors and other IC components shrink with each technology node, the overall speed of electronic devices increases correspondingly, but not linearly. Therefore, not only does the overall computing power of ICs increase with each technology node (because of the number of components and overall complexity of the ICs as well as increased feature density), but also the maximum operating speed of the ICs can potentially increase. This provides strong incentives for all members of the semiconductor industry to ensure that they do not fall behind their competitors in the pursuit of metrics predicted by Moore's law.

Holt (2016) lists several significant processing innovations in CMOS technology which contributed to achieving Moore's law over the past five decades. These innovations include, adapted from Holt (2016),

- Tungsten plugs
- Deep-trench isolation
- Chemical-mechanical polishing
- Copper interconnects
- Strained silicon
- High-dielectric metal gates
- FinFETs

This book provides technical reviews on these innovations and their significance at specific technology nodes, as well as their limitations and reasons why the technologies were superseded, improved or replaced. The following section briefly summarizes the most important economic factors that are considered when looking back at Moore's law, and for predicting the feasibility of Moore's law in the foreseeable future.

Economic Factors Driving Moore's Law

Another driving force for Moore's law is derived from economics (Hutcheson 2009): the *financial viability* of shrinking the size of the transistor gate length or introducing new materials. For a given area of silicon substrate, as feature density increases, the total value of the IC components and therefore the revenue generated by the sale of each microchip increase proportionally – a relatively historical trend since modern equipment upgrades are becoming extremely expensive. As the number of IC components (for a simple circuit) increases per unit area of silicon at each technology node, the manufacturing cost per component decreases inversely proportionally. Hence, not only does revenue increase at each technology node, owing to the higher value of IC components on the chip, but also the manufacturing cost per component decreases inversely proportionally – leading to greater profit margins per component at each technology node (Hutcheson 2009; Vardi 2014; Sullivan 2007). This seems like a win-win situation for semiconductor manufacturers, but it does not take into account crucial considerations such as

- More complex processing techniques
- Requirements for more advanced equipment
- An increase in the number of process steps
- The deviation from standard, relatively low-cost wafers to complex and expensive epitaxial-layered geometries
- The incorporation of new materials
- The need for skilled workers to operate and maintain complex foundries

These listed items contribute significantly to an increase in semiconductor manufacturing cost, which can undermine savings in cost per unit area of semiconductor materials. The financial benefits of being at the forefront of technological progress in the electronics industry provided the motivation for the semiconductor industry to formalize the predictions from the Moore's law model as objectives that the industry should strive to meet. These objectives, or nodes, were the common denominators used as competition guidelines between industry leaders, while they continually attempted to keep up with Moore's law. Thus, the actions of key players in the semiconductor industry made Moore's law a self-fulfilling prophecy. The critical economic aspect contained in Moore's law is that, as technology evolves and advances to new heights, a larger amount of functionality can be offered at the same cost associated with a particular component count and cost (Hutcheson 2009; Sullivan 2007). The complexity of modern electronic circuitry has created significant technical and economic challenges at the manufacturing level, such that the economic factors of modern-day production of ICs are no longer as attractive as in the past (Mack 2011). Expensive additional steps and equipment are required to reliably manufacture recent nodes, such as 22 nm or 14 nm nodes, with additional costs for processes such as package-on-package and through-silicon via packaging requiring costly and specialized equipment and operator skills. These issues have caused a slowing down in transistor scaling, superseded in many ways by the development of new and innovative techniques to increase yield and reliability and bring down the cost of production. It is also a common notion for foundries to purchase or transfer older technologies, especially 350 nm and 180 nm nodes, and to provide specialized services such as low minimum-order quantity options for research institutions or smaller companies aiming to develop ICs with higher complexity but with lower transistor count through innovation.

Another challenge in the quest to increase transistor density is its packaging and cooling. The ITRS guidelines for 2014 predicts that manufacturers and IC designers will be moving away from FinFET transistors by 2019 and migrating to gate-all-around transistor designs. Eventually, transistors will become vertical, using nanowire techniques, which would increase transistor density to similar orders compared with how three-dimensional (3D) vertical NAND (not and) memory technology improved upon traditional planar NAND flash memory, both in performance and capacity per unit area. Figure 1.3 shows a predicted evolution of transistor technology from the 22 nm node to 5 nm and 3 nm gate lengths.

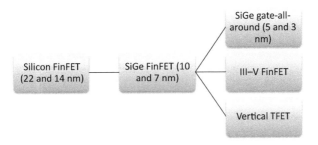

FIGURE 1.3   Predicted transistor evolution from the 22 nm node to achieving 5 and 3 nm gate lengths. (From Anthony, S., Transistors will stop shrinking in 2021, but Moore's law will live on. Retrieved 22 November 2016 from http://www.arstechnica.com.)

As shown in Figure 1.3, silicon FinFET technology used for the development of the current generation of 22 nm and 14 nm nodes will be replaced by introducing germanium into the gate of the silicon (SiGe) FinFET to achieve higher mobility and practical threshold voltages in the 10 nm and 7 nm nodes. In order to achieve 5 nm and 3 nm nodes, various advances are required. These are discussed in this book, but essentially the current predictions point toward a combination of SiGe gate-all-around transistors, III–V material FinFETs such as indium-gallium-arsenide (InGaAs) on silicon-on-insulator (SOI) substrates and vertical tunnel FET (TFET) technology. These technological advances are specifically aimed at improving and shrinking transistors, but importantly, the enabling technology does not rely only on transistor scaling to be feasible. The ITRS 2014 lists challenges such as the following as additional key challenges in future technologies:

- Electrostatic discharge
- Chip-package interaction reliability (process and design)
- Tensile and compressive stress management,
- Thermal management
- Passive device integration (performance and cost)
- Electro-optical interconnects

Cooling of these high-density components becomes a significant problem and transferring heat from the chip requires, again, costly, innovative and sometimes unreliable techniques. The deceleration of Moore's law provides some advantages, though. With the underlying technology changing rapidly, manufacturers need to introduce the latest hardware technology together with proprietary software, within turnaround times of six months or less. To mitigate these challenges, standardization of platforms with interchangeable components (Huang 2015) has become a de facto approach for manufacturers to stay relevant without the need to periodically upgrade and update costly manufacturing equipment. There is a stronger drive toward open platforms, which allow smaller businesses to focus on innovating incremental improvements, refining product features and improving software, as opposed to pushing toward higher transistor-count ICs. Achieving optimal performance from a multicore system, therefore fully utilizing the hardware, can give businesses and corporations the leading edge over competitors, essentially using similar hardware and underlying technology.

There are, however, certain limitations to technology that cannot be overcome easily, if at all. These limitations may challenge the future existence of Moore's law and progressively slow it down until innovative techniques to overcome these limitations are implemented. These inevitable limitations are briefly discussed in the following section.

THE LIMITATIONS OF MOORE'S LAW

Although the semiconductor industry and electronics industry have managed to keep up with Moore's law for several decades, the features have shrunk to dimensions that have introduced significant obstacles to continuing the historical trends of scaling (Tummala 2000; Moore 1995; Vardi 2014). These barriers have manifested themselves in areas such as

- Manufacturing
- Reliability
- Yield
- Power supply and threshold voltage
- Thermal management
- Performance

For example, photolithography, as discussed, one of the technical factors driving Moore's law, refers to the processing method by which electronic features are patterned; it is fundamental to the advancement of Moore's law, and yet the continual reduction in feature sizes is restricted by the resolution of the lithography process, a difficult process in semiconductor manufacturing to scale nodes below 14 nm (Shalf and Leland 2015; Bret et al. 2014; Mesawich et al. 2009; Lin et al. 1999).

PHOTOLITHOGRAPHY

The fidelity of photolithography patterns is limited by the wavelength of the radiation used in the imaging system as well as the properties of the photoresist. In fact, the minimum feature size that can be reliably patterned via photolithography is directly proportional to the wavelength of the light (Mack 2011, 2015). If the feature size is very small compared with the wavelength of the imaging source, Rayleigh scattering is observed, which can lead to lower pattern fidelity, such as corner rounding, tip-to-tip merging of contacts and incomplete development of features. Another patterning problem has been that the features have become so small and the aspect ratio of the features correspondingly so large that it has become difficult to avoid pattern collapse due to overwhelming capillary forces during the pattern development phase of photolithography. Thus, the process window of photoresists has become smaller, which has made it difficult for process engineers to reliably produce features with minimal defects (Moore 1995).

Immersion optical photolithography using an imaging wavelength of 193 nm, which has been the state-of-the-art method for patterning the smallest IC features onto a microchip for the past decade, is not capable of reliably producing features below 14 nm, which is required for the next generation of chips, according to Moore's law. To address the optical limitations of optical photolithography, the semiconductor industry has turned to extreme EUV to pattern feature sizes beyond the limitations of 193 nm lithography (Baksi et al. 2007; Zhang et al. 2008). The wavelength emitted by EUV scanners is 13 nm, which allows EUV lithography to pattern features much smaller than those of an optical scanner without any detrimental Rayleigh scattering (Zhang et al. 2008; Bodermann et al. 2009). However, EUV resists still have low sensitivity, resulting in a low throughput of wafers even though EUV equipment manufacturers try to compensate for the low sensitivity by increasing radiation power from the EUV scanners (Bakshi et al. 2007; Zhang et al. 2008;

Bodermann et al. 2009). Lastly, the soft X-rays emitted by EUV are, to a great extent, absorbed by the materials that are used to shield workers from high-energy radiation. This results in relatively low EUV output intensities from the imaging tools, thus compounding the problem of low wafer throughput. As alluded to in the previous statement, the soft X-ray radiation emitted by EUV tools is harmful to people, thus requiring extra measures to ensure safety. This results in significantly higher operating costs compared with optical lithography, as discussed in Chapter 2 of this book. In addition, the photochemistry of EUV resists results in unacceptably large amounts of volatile organic by-products during lithography imaging, a phenomenon referred to by engineers as outgassing (Denbeaux et al. 2007; Hada et al. 2004; Santillan et al. 2007; Watanabe et al. 2003). The volatile outgasses are condensable (i.e. they form a solid film when they meet cooler surfaces) and can cause very costly damage to manufacturing equipment, leading to increased production downtime and higher operating costs.

Another patterning method that has become a popular means of patterning sub-optical features is directed self-assembly (DSA) (Lui and Chang 2009; Jeong et al. 2013; Ruiz et al. 2012; Son et al. 2013; Tsai et al. 2012; Peters et al. 2015). As described in detail in Chapter 2, DSA involves the combination of photolithography with thermodynamics to create dense patterns of block-copolymer (BCP) domains that can be selectively removed to form CMOS-relevant features on a substrate. DSA guiding patterns are patterned at a relaxed pitch, which allows optical lithography to be used and avoids Rayleigh scattering, while the thermodynamics of the self-assembly avoids any limitations from optics (Jeong et al. 2013). However, at the present time, DSA suffers from relatively high line edge roughness as well as poor dry etch contrast, since the different domains of the BCPs have a fairly similar etch resistance when exposed to reactive ion etch (RIE) plasmas (Jeong et al. 2013; Kim 2014). In addition, a relatively large number of processing steps involved in DSA patterning compared with photolithography and EUV lithography results in DSA being susceptible to a large number of defect types and relatively poor pattern placement compared with standard photolithographic methods (Somervell et al. 2015). However, DSA is not the only process with high defect densities; in fact, defects become more prevalent and more critical at every step of manufacturing because feature sizes have grown so small that defects are no longer insignificant in size compared with feature sizes (Bret et al. 2014; Mesawich et al. 2009; Yoshikawa et al. 2015; DeHon 2009; Kawahira et al. 1999). Thus, almost any defect is now large enough to cause a critical part of a device to fail, which is why these types of defects are referred to as *device killers* or *killer defects* (Bret et al. 2014; Mesawich et al. 2009; Yoshikawa et al. 2015). Defects are unavoidable because, during manufacturing, there are multiple sources of contaminants or processing errors that may be introduced, for example

- Particulates buried during coating of layers
- Undissolved/residual material from photoresist films
- Dust particles or other foreign material landing on a wafer
- Imaging defects caused by patterning errors
- Etch residue and various other sources (Somervell et al. 2015)

Because of the increasing prevalence of *killer defects* as IC features grow smaller, an unwanted trend that follows Moore's law is that, as transistors scale down in size, failure rates increase, product yields decrease, operating costs increase (because of the increased time devoted to performing root cause failure analyses) and the rate of production of complete devices decreases (due to production downtime from low yields). Cumulative semiconductor revenue has followed a relatively linear path when graphed against feature size. Since approximately 1998, as feature size dropped below the 0.2 μm node (specifically, the 180 nm node), the cumulative semiconductor revenue slowed down significantly as the cost of upgrading and updating equipment, the requirement for skilled staff and the sheer size of foundry equipment, such as for photolithography, drastically increased.

INTERCONNECTS AND HEAT GENERATION

As feature sizes have become smaller and the density of transistors has increased with each technology node, interconnects have become smaller in cross-section. Their total length has become longer owing to the increased complexity of the circuitry, and interconnect density (i.e. the wiring density) has increased (Tammala 2000; Liu and Chang 2009; Maex et al. 2003). These factors have led to higher impedance and thus greater heat dissipation, since more electricity is converted into heat. The higher wiring density has exacerbated the heat dissipation problem to the extent that current chips that are optimally designed for high computing power emit heat with flux values approaching 100 mW/cm^2 (Tammal 2000; Cho et al. 2003; Koo et al. 2005). Since semiconductors must avoid high operating temperatures to prevent thermally induced failures (ITRS 2014 recommends that chips be kept below 80°C), high heat dissipation rates place a greater burden on packaging engineers to produce packages that can effectively remove the heat produced by the circuitry while maintaining a thin form factor for modern electronic devices – especially for portable electronics. Although cooling fans were the traditional means of cooling ICs for several decades, the thermal environment on some modern microchips compelled the industry to begin using more innovative techniques such as phase-change liquid cooling to improve the cooling efficiency of the thermal solutions (Cho et al. 2003). Efforts to improve thermal management efficiency continue, and will be discussed in detail in a later chapter.

ENGINEERING CHALLENGES

As described above, the recent challenges facing the electronics industry have led to increased costs at the research stage, as well as at the scale-up and production stages (Hutcheson 2009; Vardi 2014; Sullivan 2007). In addition, the time taken for a technology node to move from the research phase to the manufacturing stage has increased because of the additional time spent on "learning," which refers to the efforts required for engineers to develop solutions to technical problems experienced when advancing into new areas (Vardi 2014). The extra time taken for learning has forced the semiconductor industry to increase the time between technology nodes, while the cost of research and development has risen exponentially to meet the new challenges posed by sub-optical feature sizes. As the demand for more advanced technologies continues to drive the boundaries of technological and manufacturing capabilities, the associated costs involved are always a key determining factor in the feasibility of any state-of-the-art technological innovation. As technologies advance in complexity, the manufacturing processes involved also increase in complexity, which has a direct impact on manufacturing costs. From the late 1990s onward, the operating costs of semiconductor production have increased exponentially, leading to lower profit margins and the withdrawal of a few semiconductor companies from manufacturing (or merging of companies) because of excessive production costs (Mack 2011; Vardi 2014). The compound annual growth rate (CAGR) up to approximately 1990 was estimated at about 16.2%. The CAGR is typically calculated by

$$\text{CAGR} = \left(\frac{V_E}{V_B}\right)^{\frac{1}{n}} - 1 \tag{1.20}$$

where,
 V_E is the end-value of the investment
 V_B is the beginning-value of the investment
 n is the number of periods of the investment

CAGR is therefore an effective measure of return on investment. The CAGR (1.20) of the semiconductor industry had dropped to below 5% by the late 1990s and has followed a similar trend up to modern investments. The electronics industry is hence currently in a state of flux, and many assumptions used to extrapolate Moore's law into the future will consequently need to be re-examined going forward. The economic factors limiting the continued advancement of computing performance at the rates predicted by Moore's law are discussed in greater detail in a later chapter.

Several next-generation technologies have been identified to effectively fast-track the expected exponential growth within the semiconductor industry, taking into account the costs involved in developing new technologies as well as the feasibility of new technologies in terms of performance, yield, value and longevity. These next-generation technologies for the short-to-mid-term future are summarized in the following section.

NEXT-GENERATION TECHNOLOGIES

The semiconductor industry relies on specific enabling technologies to ensure rapid growth in the industry. The two primary specific technologies, among a variety of underlying improvements, are optoelectronic technology and 3D design of not only transistors, but entire system-on-chip (SoC) interfaces. The following section briefly describes how optoelectronic technology, specifically optical interconnects, aims to fast-track Moore's law in the future.

OPTICAL INTERCONNECTS

The underlying technology of optical interconnects uses modulated light signals to send and receive data between points. IBM Fellow Carl Anderson, mentioned in the introductory paragraph of this chapter, pointed out in 2009 that the exponential growth of the optical interconnect technology is generally divided into three main categories, based on the availability and short-term implementation potential of each category:

- Rack-to-rack interconnects use fiber as the primary technology to connect the back-end of analog and digital equipment.
- Chip-to-chip optical interconnects involve a technology used to connect the front-end of analog and digital equipment, where the output on chip level is modulated to optical signals and transmission between chips is done using optical interconnects only. This technology is more complex than rack-to-rack interconnect technology and requires optoelectronic circuit integration at the microelectronic scale.
- The more complex challenge, which also offers the largest benefits in bandwidth and interference of signals, is the use of on-chip optical signaling optoelectronic signals through waveguides. This is realized with MEMS, replacing traditional metallization techniques with thin nanotubes able to transmit signals using light at component level. An active research field today, but still in its infancy, this technology could allow for rapid growth in performance of future-generation semiconductor systems.

Optoelectronic circuitry uses a hybrid of optical and electrical signaling to combine the advantages of optical and electrical systems (Miller 2009; Orcutt et al. 2013). Optical systems have the advantage of using the fastest traveling signals (light), which present the biggest advantage in bandwidth gain and are less prone to heat dissipation and electromagnetic interference compared with traditional copper, resulting in relatively simple cooling and shielding solutions in comparison with modern-day CMOS ICs. Optoelectronic ICs are also well established because they have been investigated and developed since the 1970s, and fully functional optoelectronic ICs have even been demonstrated (Miller 2009; Orcutt et al. 2013). However, optoelectronic systems suffer from having relatively high packaging costs compared with CMOS ICs because some of the optical components cannot be manufactured using standard CMOS processes (i.e. monolithic integration is not yet possible). This forces manufacturers to pick and place many optical elements individually onto a motherboard, thus leading to high packaging costs and low production throughput (Tammala 2000).

An optoelectronic system, or optic communication network, essentially converts digital bits in the form of 1s and 0s to optical light signals, to a state where the light wave is either on, representing a 1, or off, representing a 0, or the other way round, depending on the application. A fiber optic communication network consists of transmitting and receiving electronic circuitry, a modulated light source, an optical link and a photodetector able to convert light back to electrical signals. These typical components of an optic communication network are shown in Figure 1.4.

The transmitter and receiver of a high-level optoelectronic system operate using electronic signals, typically in the digitized form of 1s and 0s, to represent binary-coded data (Figure 1.4). The optoelectronic system requires signals of light, and

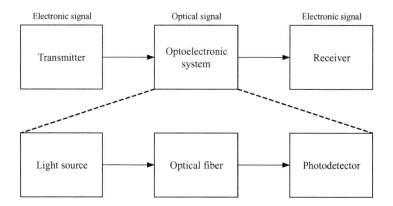

FIGURE 1.4 The basic functional diagram of an optoelectronic communication system. The transmitter, optoelectronic system and receiver in the top row represent the high-level description of such a system, whereas the light source, optical fiber and photodetector represent the high-level requirements to achieve optical communication.

therefore includes processes to convert electrical signals to light signals. This is typically achieved by modulating the output of a laser (the light source). Optical signals are then transmitted through optical fibers in the case of rack-to-rack or chip-to-chip networks, or can be transmitted on-chip using specialized semiconductor processing techniques to manufacture cavity-like 3D structures on-chip. Optical fibers can operate in single mode or in multi-mode. Single-mode fibers have a small diameter and are used to transmit one signal per fiber; whereas multi-mode fibers are thicker but allow for multiple signals to be transmitted through the fiber. Single- and multi-mode transmission depends on the v-number of the optical fiber, mathematically described by

$$v = \frac{2\pi d}{2\lambda} \sqrt{n_1^2 - n_2^2} \qquad (1.21)$$

where,

λ is the wavelength of the light signal in m
d is the diameter of the fiber core, also in m
n_1 and n_2 are the refractive indices of the core and the cladding, respectively

For v-numbers below approximately 2.4, a fiber only supports single-mode transmission, whereas for v-numbers larger than approximately 2.4, multi-mode signals are possible. From (1.21), it can be seen that a larger wavelength of light or smaller diameter of the fiber core translates to a smaller v-number, typically used to achieve single-mode transmission, depending on the application. An in-depth review of optic fiber light transmission is presented in a later chapter of this book. In the optoelectronic system, technology advancements (and, in a sense, a drive toward realizing Moore's law of performance and component density) strive toward compact light sources such as light-emitting diodes (LEDs) and laser diodes, low-loss cylindrical waveguide optical fibers and highly sensitive p-n photodiode or avalanche photodetectors.

The second primary enabling technology to ensure the pace of Moore's law is 3D design strategies for SoC and transistors, briefly discussed in the following section.

THREE-DIMENSIONAL DESIGN

The ITRS 2014 predicts that two-dimensional (2D) transistor scaling is likely to reach its limits by 2021, requiring another *dimension* in scaling. 3D scaling of transistors will replace conventional scaling techniques and breathe new life into Moore's law, 55 years after its first appearance. 3D scaling refers to the physical construction of transistors, a manipulation of the transistor processing to produce 3D transistors, such as FinFETs, as well as the 3D stacking of ICs, which provides the shortest possible interconnects between ICs, advantageous in high-frequency systems, as well as significant savings in the total area of integrated solutions. FinFET technology extends Moore's law beyond sub-20 nm process nodes and plays a significant role in technological advances in mobile computing such as smartphones, tablets and the Internet-of-things (IoT). The significance of the IoT to drive technology markets is discussed in a later chapter in this book, where devices such as the Raspberry Pi have boosted the development of sensors and actuators combined to create smart systems.

Heat is still a primary deterrent to 3D stacking becoming commonplace, predominantly in central processing units and, to a lesser extent, in NAND flash memory, which is already available and being manufactured for applications such as solid-state hard drives. The focus within the semiconductor industry has steadily been shifting from shrinking individual chips to a system-on-chip or device-centric model, which requires innovative integration and solutions reduction in power consumption that generates heat. Figure 1.5 represents a simplified diagram showing the reduction in chip area usage between traditional two-dimensional IC connections, combining technologies such as bipolar and CMOS technologies (BiCMOS) to allow interconnecting systems at chip level as well as the proposed 3D stacking of ICs to maximize area savings while allowing high heat generation modules to disperse heat effectively throughout the system.

Figure 1.5 therefore represents the potential of stacking ICs and designing interconnects such that the interface between chips effectively replaces traditional bond wires or on-chip metallization.

On transistor level, inventions such as the FinFET, essentially a 3D version of the traditional planar CMOS transistor, have proven to be a crucial enabling technology for future-generation technologies. The simplified construction of 3D transistors, such as the FinFET, is presented in Figure 1.6.

A FinFET is a type of multi-gate MOSFET where a thin silicon film is enfolded over the conducting channel of the substrate. The structure looks like a set of vertical fins, thus the name FinFET. The thickness of the device determines its channel length. In Huang et al. (2001), during the early days of FinFET technology, which was invented in the late 1990s, the FinFET was described as

- A transistor channel, which is formed on the vertical surfaces of an ultra-thin silicon fin and controlled by gate electrodes formed on both sides of the fin.

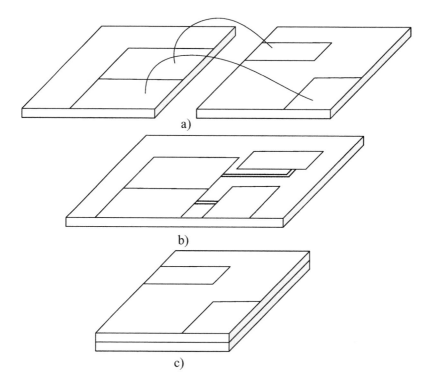

FIGURE 1.5 The fundamentals of 3D stacking of semiconductor ICs. Traditional bond wire connections in (a) have increased the total chip area and require larger packaging and additional integrity protection. Combining technologies, such as BiCMOS, allow (b) on-chip interconnects between technologies to reduce area. (c) 3D stacking of heat-generating processing semiconductor technologies and flash memory provide area benefits to re-imagine Moore's law.

- Dual gates, self-aligned both in conjunction with each other as well as with respect to the source and the drain sections to reduce the parasitic capacitance and resistance, as well as to control the absolute channel length.
- The physical construction of raised source/drain regions.
- A short fin of silicon to sustain quasi-planar geometry and topology (since the source/drain and gate are considerably thicker/taller) in order not to make the construction overly complex.

The gate material typically used in FinFETs is boron-doped $Si_{0.4}Ge_{0.6}$ to realize practical threshold voltages of these very thin gated devices (Hisamoto et al. 2000). The critical dimensions of a FinFET, as described by Huang et al. (2001), are the gate length, device width and body thickness. In advanced technologies, such as 20 nm planar FET technologies, the source and drain of the transistor intrude into its channel, and this means that a significant leakage current flows and that it is difficult to turn off the device completely. Since the FinFET's gate wraps around the fin structure, as seen in Figure 1.6, more control over the channel construction and

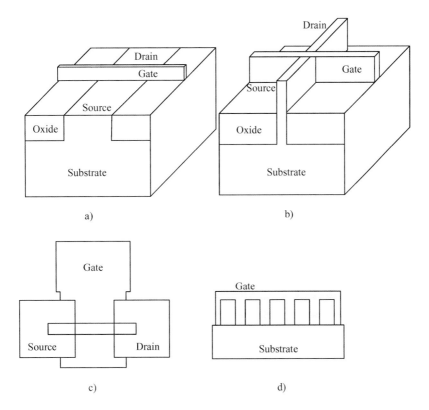

FIGURE 1.6   (a) Simplified construction of the conventional planar transistor and (b) the 3D FinFET transistor and (c) the FinFET as viewed from the top and (d) an example of a five-fin device. (From Hisamoto, D., et al., *IEEE Transactions on Electron Devices*, 47(12), 2320–2325, 2000.)

less leakage current through the body of the transistor are achieved, which enables the use of lower threshold voltages, leading to an improvement in performance and lower average power consumption.

FinFETs are already implemented, or close to being mass-produced, by global leaders in the semiconductor manufacturing of core devices, input/output devices and static random access memory; Samsung, Intel, Apple and TSMC ship 14 nm technologies, such as the Samsung Exynos7 chip, for their latest high-end devices. In September 2016, GlobalFoundries announced plans to deliver high-performance 7 nm FinFET semiconductor technology for future generations of computing applications to surpass current generation 16/14 nm FinFET offerings. FinFET technology has shown, since its invention, that it is an important enabling technology to ensure progress in the development of future applications.

Apart from the enabling technologies on which researchers, manufacturers and scientists rely to ensure that Moore's law continues, several future-generation technologies are also in development stages. Depending on their progress, these technologies are also aimed to fast-track Moore's law in the (relatively) far future. The dominant future-generation technologies are discussed in the following section.

FUTURE-GENERATION TECHNOLOGIES

The semiconductor industry and academia are also exploring exotic technologies such as molecular electronics, spintronics and quantum computing as alternatives to conventional CMOS devices. These techniques are discussed in further detail in a later chapter of this book.

MOLECULAR ELECTRONICS

Molecular electronics seeks to shrink the size of transistors down to the molecular level and to build circuits using the smallest possible feature sizes (Stan et al. 2009; Cui et al. 2015; Choi and Mody 2009). The motivation for this field stems from the fact that a minimum physical limit exists to the extent that features can be reduced in size. Therefore, by building ICs using molecular devices, the feature density would be as high as possible. However, research in this field is still at the nascent stage; researchers are still working out the fundamentals of this field and are only able to create single molecular devices at a time. Processing methods for mass-producing these devices and integrating them into complete circuits are still many years in the future.

GRAPHENE

New devices also involve the use of exotic materials such as graphene (Mohanram and Yang 2010; Xiying et al. 2012; Singh et al. 2015; Service 2015). Graphene is a two-dimensional form of carbon with a thickness of a single monolayer of carbon atoms covalently linked in a hexagonal comb structure similar to that of graphite. Its monolayer thickness makes it both flexible and transparent, thus providing the potential for future use in making flexible, transparent electronics or devices with ultrathin form factors. Graphene has semiconducting properties with zero energy band gap and hole and electron mobility values much higher than those of doped silicon substrates (Mohanram and Yang 2010). Although graphene has excellent conducting properties, its lack of a band gap precludes natural graphene from being switched off. Hence, researchers are actively investigating methods of introducing a band gap (e.g. by doping graphene sheets with other atoms) while maintaining high charge carrier mobility values. Thus, graphene transistors are still at the research stage but the electronic properties of graphene are so impressive that there is a great deal of interest from both academia and industry, hence funding for research is extensively supported (Mohanram and Yang 2010; Xiying et al. 2012; Singh et al. 2015; Service 2015).

SPINTRONICS

Spintronics refers to the use of the spin states of electrons as well as their charge to store and send information (Zahn 2007). Since spintronics would make use of both the spin and the charge of an electron, it is believed that an integrated spintronic circuit would be able to transmit more information and carry out more computations than a standard electronic device (Shalf and Leland 2015; Zahn 2007; Banerjee 2007; Kang et al. 2016). An advantage of spintronics over molecular electronics is that it can make use of standard CMOS processing methods to build devices, but this field is also fairly new and spintronic device theory is still being uncovered.

QUANTUM COMPUTING

One of the most exciting alternatives to CMOS technology is quantum computing. Unlike a standard CMOS bit, which can only exist in two binary logic states (0 or 1), a quantum bit (qubit) can exist in more than two logic states (0, 1, as well as 0 and 1 simultaneously); thus, according to device theory, quantum computers should be able to carry out more computations for the same number of logic components than conventional computers (DiVincenzo 2009). Hence, the realization of quantum computing would bring about a significant advancement in computing power that could support stronger encryption for more secure communications, as well as more complex simulations for scientific purposes, to name just a few applications.

The following section concludes this review on the fundamental principles of Moore's law and the inevitable limitations of this prediction in the five decades of its existence.

CONCLUSION

Given the importance of technology in our daily lives and the significance of Moore's law in guiding the development of each technology node, it should come as no surprise that any obstacles to the perpetuation of Moore's law cause great concern in the electronics industry. Subsequent chapters will go into further detail regarding the hindrances to Moore's law and the tactics the semiconductor industry has undertaken to overcome these limitations and the threats to derailing Moore's law.

This chapter highlights the introductory considerations of technical and economic factors that allowed Moore's law to emanate from its inception in 1965 to what it has established for modern technology. The aim of this book is to provide detailed and technical reviews of all factors that have influenced Moore's law and all factors that play a role in determining its longevity for future technologies. The limitations of Moore's law are not necessarily dependent on a shrinking economy or lack of innovation, but are due to physical limitations of scaling electronic components to levels beyond those of ordinary electronic and engineering principles, approaching the limitations of physics.

Semiconductor processing is expensive. Foundries that are at the forefront of technological innovation are paying the price to drive technology scaling to its limits, but are not necessarily reaping the financial rewards. Several technologies have been identified that will drive Moore's law in the near future; however, these technologies can only go as far as physical limitations permit. Future-generation technologies, aimed at sustaining Moore's law in the long term, are being researched and developed. This book aims to provide a relevant and detailed background on these enabling technologies.

REFERENCES

Anthony, S. (2016). Transistors will stop shrinking in 2021, but Moore's law will live on. Retrieved 22 November 2016 from http://www.arstechnica.com.

Arden, W., Brillouët, M., Cogez, P., Graef, M., Huizing, B., Mahnkopf, R. (2010). *More-than-Moore*. White Paper: International Technology Roadmap for Semiconductors, ITRS. Version 2, October 2010.

Bakshi, V., Lebert, R., Jaegle, B., Wies, C., Stamm, U., Kleinschmidt, J., Schriever, G., Ziener, C., Corthout, M., Pankert, J., Bergmann, K., Neff, W., Egbert, A., Gustafson, D. (2007). Status report on EUV source development and EUV source applications in EUVL. *23rd European Mask and Lithography Conference (EMLC)*, 1–11.

Banerjee, S. (2007). New materials and structures for transistors based on spin, charge and wavefunction phase control. *AIP Conference Proceedings*, 931, 445–448.

Bishop, D. (2005). Nanotechnology and the end of Moore's law? *Bell Labs Technical Journal*, 10(3), 23–28.

Bodermann, B., Wurm, M., Diener, A., Scholze, F., Gross, H. (2009). EUV and DUV scatterometry for CD and edge profile metrology on EUV masks. *25th European Mask and Lithography Conference (EMLC)*, 1–12.

Bret, T., Hofmann, T., Edinger, K. (2014). Industrial perspective on focused electron beam-induced processes. *Applied Physics*, 117(12), 1607–1614.

Brock, D. (2006). *Understanding Moore's Law: Four Decades of Innovation*. Philadelphia: Chemical Heritage Foundation, ISBN 978-094-1901413.

Butzen, P. F., Ribas, R. P. (2006). Leakage current in sub-micrometer CMOS gates. *Universidade Federal do Rio Grande do Sul*. 1–28.

Chien, A. A., Karamcheti, V. (2013). Moore's law: The first ending and a new beginning. *Computer*, 46(12), 48–53.

Cho, E. S., Koo, L., Jiang, L., Prasher, R. S., Kim, M. S., Santiago, J. G., Kenny, T. W., Goodson, K. E. (2003). Experimental study on two-phase heat transfer in microchannel heat sinks with hotspots. *19th Annual IEEE Semiconductor Thermal Measurement and Management Symposium*, 11–13 March 2003, 242–246.

Choi, H., Mody, C. C. M. (2009). The long history of molecular electronics: Microelectronics origins of nanotechnology. *Social Studies of Science*, 39(02), 11–50.

Cui, A., Dong, H., Hu, W. (2015). Nanogap electrodes towards solid state single-molecule transistors. *Small*, 11, 6115–6141.

DeHon, A. (2009). Sublithographic architecture: Shifting the responsibility for perfection. *Into the Nano Era*, 1st ed., H. Huff, Ed. Berlin, Heidelberg: Springer, 281–296.

Denbeaux, G., Garg, R., Waterman, J., Mbanaso, C., Netten, J., Brainard, R., Fan, Y. J., Yankulin, L., Antohe, A., DeMarco, K., Jaffem, M., Waldron, M., Dean, K. (2007). Quantitative measurement of EUV resist outgassing. *23rd European Mask and Lithography Conference (EMLC)*, 1–5.

Dennard, R. H., Gaensslen, F. H., Rideout, V. L., Bassous, E., Leblanc, A. R. (1974). Design of ion-implanted MOSFET's with very small physical dimensions. *IEEE Journal of Solid-State Circuits*, 9(5), 256–268.

DiVincenzo, D. (2009). Quantum computing. *Into the Nano Era*, 1st ed., H. Huff Ed. Berlin Heidelberg: Springer, 297–313.

Endres, A. (2013). Early language and compiler developments at IBM Europe: A personal retrospection. *IEEE Annals of the History of Computing*, 35, 18–30, October 2013.

Esener, S. C., Kryder, M. H., Doyle, W. D., Keshner, M., Mansuripur, M., Thompson, D. A. (1999). WTEC panel report on the future of data storage technologies. *International Technology Research Institute*. Retrieved 4 November 2016 from http://www.wtec.org.

Guarnieri, M. (2016). The unreasonable accuracy of Moore's law. *IEEE Industrial Electronics Magazine*, 21 March 2016.

Hada, H., Hirayama, T., Shiono, D., Onodera, J., Watanabe, T., Lee, Y. Kinoshita, H. (2004). Outgassing characteristics of low molecular weight resist for EUVL. *Microprocesses and Nanotechnology Conference*, 2004. Digest of Papers, 234–235.

Hisamoto, D., Lee, W-C., Kedzierski, J., Takeuchi, H., Asano, K., Kuo, C., Anderson, E., King, T-J., Bokor, J., Hu, C. (2000). FinFET – A self-aligned double-gate MOSFET scalable to 20 nm. *IEEE Transactions on Electron Devices*, 47(12), 2320–2325.

Holt, W. M. (2016). Moore's law: A path going forward. *IEEE International Solid-State Circuits Conference*. San Francisco, CA, 31 January–4 February 2016.

Huang, A. (2015). Moore's law is dying (and that could be good). *spectrum.ieee.org*. Posted 15 March 2015.

Huang, X., Lee, W.-C., Kuo, C., Hisamoto, D., Chang, L., Kedzierski, J., Anderson, E., Takeuchi, H., Choi, Y.-K., Asano, K., Subramanian, V., King, T.-J., Bokor, J., Hu, C. (2001). Sub-50 nm P-channel FinFET. *IEEE Transactions on Electron Devices*, 48(5), 880–886.

Huff, H. R., Gilmer, D. C. (2005). *High Dielectric Constant Materials: VLSI MOSFET Applications*. Berlin Heidelberg: Springer, ISBN 978-3-540-21081-8.

Hutcheson, G. (2009). The economic implications of Moore's law. *Into the Nano Era*, 1st ed., H. Huff, Ed. Berlin Heidelberg: Springer, ISBN 978-3-540-74558-7, 11–38.

ITRS. (2014). ITRS Reports – International Technology Roadmap for Semiconductors. Retrieved 20 March 2016 from http://www.itrs2.net/itrs-reports.html.

Jacobi, W. (1952). Halbleiterverstärker. Patent DE833366 (C), 15 May 1952.

Jeong, S., Kim, J. Y., Kim, B. H., Moon, H., Kim, S. O. (2013). Directed self-assembly of block copolymers for next generation nanolithography. *Materials Today*, 16, 468–476.

Johnson, R. C. (2009). IBM Fellow: Moore's Law defunct. Retrieved 22 November 2016 from www.eetimes.com.

Kang, W., Wang, Z., Zhang, Y., Klein, J., Lu, W., Zhao, W. (2016). Spintronic logic design methodology based on spin Hall effect-driven magnetic tunnel junctions. *Journal of Physics D-Applied Physics*, 49, 065008.

Kanter, D. (2012). Intel's near-threshold voltage computing and applications. Retrieved 13 November 2016 from http://www.realworldtech.com.

Kawahira, H., Hayashi, N., Hamada, H. (1999). PMJ' 99 panel discussion review: OPC mask technology for KrF lithography. *Proceedings of SPIE*, 3873, 19th Annual Symposium on Photomask Technology, 318.

Kim, S. (2014). Stochastic simulation studies of line-edge roughness in block copolymer lithography. *Journal of Nanoscience and Nanotechnology*, 14, 6143–6145.

Koo, J., Im, S., Jiang, L., Goodson, K. E. (2005). Integrated microchannel cooling for three-dimensional electronic circuit architectures. *Journal of Heat Transfer*, 127, 49–58.

Lapedus, M. (2016). 7 nm fab challenges. Retrieved 20 November 2016 from http://www.semiengineering.com.

Lin, H., Lin, J. C., Chiu, C. S., Wang, Y., Yen, A. (1999). Sub-0.18-μm line/space lithography using 248-nm scanners and assisting feature OPC masks. *Proceedings of SPIE*, 3873, 19th Annual Symposium on Photomask Technology, 307.

Liu, T., Chang, L. (2009). Transistor scaling to the limit. *Into the Nano Era*, 1st ed., H. Huff, Ed. Berlin Heidelberg: Springer, 191–223.

Mack, C. A. (2011). Fifty years of Moore's law. *IEEE Transactions on Semiconductor Manufacturing*, 24(2), 202–207.

Mack, C. A. (2015). The multiple lives of Moore's law. *IEEE Spectrum*, 52(5), 67–67.

Maex, K., Baklanov, M. R., Shamiryan, D., Iacopi, F., Brongersma, S. H., Yanovitskaya, Z. S. (2003). Low dielectric constant materials for microelectronics. *Journal of Applied Physics*, 93, 8793–8841.

Märtin, C. (2014). Post-Dennard scaling and the final years of Moore's law. *Informatik und Interactive Systeme*. Technical Report, Hochschule Augsburg University of Applied Sciences, September 2014.

Martinez-Blanco, J., Nacci, C., Erwin, S. C., Kanisawa, K., Locane, E., Thomas, M., von Oppen, F., Brouwer, P. W., Fölsch, S. (2015). Gating a single-molecule transistor with individual atoms. *Nature Physics*, 11, 640–645.

Mesawich, M., Sevegney, M., Gotlinsky, B., Reyes, S., Abbott, P., Marzani, J., Rivera, M. (2009). Microbridge and e-test opens defectivity reduction via improved filtration of photolithography fluids. *Proceedings of SPIE*, 7273, Advances in Resist Materials and Processing Technology, XXVI, 72730O.

Miller, D. A. B. (2009). Device requirements for optical interconnects to silicon chips. *Proceedings of IEEE*, 97, 1166–1185.

Mohanram, K., Yang, X. (2010). Graphene transistors and circuits. *Nanoelectronic Circuit Design*, N. K. Jha, and D. Chen, Eds. New York: Springer Science & Business Media, 349–376.

Moore, G. E. (1965). Cramming more components onto integrated circuits. *Electronics*, 38(8), 1–6.

Moore, G. E. (1995). Lithography and the future of Moore's law. *Proceedings of SPIE*, 2438(2), Optical/Laser Microlithography VIII, Santa Clara, CA.

Morita, H. (2016). Lithography process simulation studies using coarse-grained polymer models. *Polymer Journal*, 48, 45–50.

Orcutt, J., Ram, R., Stojanović, V. (2013). Chapter 12 – CMOS photonics for high performance interconnects. *Optical Fiber Telecommunications*, 6th ed., I. P. Kaminow, T. Li, and A. E. Willner Eds. Boston: Academic Press, 419–460.

Peters, A. J., Lawson, R. A., Nation, B. D., Ludovice, P. J., Henderson, C. L. (2015). Simulation study of the effect of molar mass dispersity on domain interfacial roughness in lamellae forming block copolymers for directed self-assembly. *Nanotechnology*, 26, 385301.

Poushter, J. (2016). Smartphone ownership and internet usage continues to climb in emerging economies. *PewResearchCenter*, February 2016.

Ruiz, R., Wan, L., Lille, J., Patel, K. C., Dobisz, E., Johnston, D. E., Kisslinger, K., Black, C. T. (2012). Image quality and pattern transfer in directed self-assembly with block-selective atomic layer deposition. *Journal of Vacuum Science & Technology*, B, 30, 06F202.

Santillan, J. J., Kobayashi, S., Itani, T. (2007). Outgas quantification analysis of EUV resists. *Microprocesses and Nanotechnology Conference*, 2007. Digest of Papers, 436–437.

Service, R. F. (2015). Beyond graphene. *Science*, 348, 490–492.

Shalf, J. M., Leland, R. (2015). Computing beyond Moore's law. *Computer*, 48(12), 14–23.

Singh, R., Kumar, D., Tripathi, C. C. (2015). Graphene: Potential material for nanoelectronics applications. *Indian Journal of Pure & Applied Physics*, 53, 501–513.

Somervell, M., Yamauchi, T., Okada, S., Tomita, T., Nishi, T., Kawakami, S., Muramatsu, M., Iijima, E., Rastogi, V., Nakano, T., Iwao, F., Nagahara, S., Iwaki, H., Dojun, M., Yatsuda, K., Tobana, T., Romo Negreira, A., Parnell, D., Rathsack, B., Nafus, K., Peyre, J., Kitano, T. (2015). Driving DSA into volume manufacturing. *Proceedings of SPIE*, 9425, Advances in Patterning Materials and Processes XXXI, 94250Q.

Son, J. G., Son, M., Moon, K., Lee, B. H., Myoung, J., Strano, M. S., Ham, M., Ross, C. A. (2013). Sub-10 nm graphene nanoribbon array field-effect transistors fabricated by block copolymer lithography. *Advanced Matter*, 25, 4723–4728.

Stan, M., Rose, G., Ziegler, M. (2009). Hybrid CMOS/molecular integrated circuits. *Into the Nano Era*, 1st ed., H. Huff, Ed. Berlin Heidelberg: Springer, 257–280.

Sullivan, R. F. (2007). The impact of Moore's law on the total cost of computing and how inefficiencies in the data center increase these costs. *ASHRAE Transactions*, 113, 457–461.

Tsai, H., Miyazoe, H., Engelmann, S., To, B., Sikorski, E., Bucchignano, J., Klaus, D., Liu, C., Cheng, J., Sanders, D., Fuller, N., Guillorn, M. (2012). Sub-30 nm pitch line-space patterning of semiconductor and dielectric materials using directed self-assembly. *Journal of Vacuum Science & Technology*, B, 30, 06F205.

Tummala, R. R. Ed. (2000). *Fundamentals of Microsystems Packaging.* New York: McGraw-Hill, ISBN 978-007-1371698.

Vardi, M. Y. (2014). Moore's law and the sand-heap paradox. *Communication ACM*, 57(5), 5.

Veendrick, H. J. M. (1984). Short-circuit dissipation of static CMOS circuitry and its impact on the design of buffer circuits. *IEEE Journal of Solid-State Circuits*, SC19(4), 468–473.

Walter, C. (2005). Kryder's law. *Scientific American*, 32–33.

Watanabe, T., Hamamoto, K., Kinoshita, H., Hada, H., Komano, H. (2003). Resist outgassing characteristics in EUVL. *Microprocesses and Nanotechnology Conference*, 2003. Digest of Papers, 288–289.

Xiying, M., Weixia, G., Jiaoyan, S., Yunhai, T. (2012). Investigation of electronic properties of graphene/Si field-effect transistor. *Nano Express: Nanoscale Research Letters*, 7, 677.

Yoshikawa, S., Fujii, N., Kanno, K., Imai, H., Hayano, K., Miyashita, H., Shida, S., Murakawa, T., Kuribara, M., Matsumoto, J., Nakamura, T., Matsushita, S., Hara, D., Pang, L. (2015). Study of defect verification based on lithography simulation with a SEM system. *Proceedings of SPIE*, 9658, Photomask Japan 2015: Photomask and Next-Generation Lithography Mask Technology XXII, 96580V.

Zahn, P. (2007). Spintronics: Transport phenomena in magnetic nanostructures. *Materials for Tomorrow*, 1st ed., Gemming, S., Schreiber, M. and Suck, J., Eds. Berlin Heidelberg: Springer, 59–89.

Zhang, C., Katsuki, S., Horta, H., Imamura, H., Akiyama, H. (2008). High-power EUV source for lithography using tin target. *Industry Applications Society Annual Meeting*, 1–4.

2 The Economics of Semiconductor Scaling

*Wynand Lambrechts, Saurabh Sinha,
Jaco Prinsloo, and Jassem Abdallah*

INTRODUCTION

At the modern core of Moore's law and the economic benefits thereof is a different way of thinking about semiconductor scaling; the historical path that Moore's law has followed is becoming difficult to realize. The change in thinking involves considering hardware and software co-design elements that complement each other at every iteration of node size improvement or change in wafer diameter. These changes are partly due to two main factors:

- The price of upgrading semiconductor manufacturing facilities today.
- The physical limitation reached at sub-10 nm technology nodes, especially in photolithography.

It is becoming more difficult to argue, in the case of wafer sizes, that bigger is better, or in the case of node sizes, that smaller is better. Many foundries and semiconductor wafer manufacturers are having to change their approach to be more viable for all parties along the supply line. All the way from wafer production to integrated circuit (IC) design, the sheer cost of producing sub-10 nm chips is becoming a critical factor; the parties involved realize this and have started to work in conjunction with one another, depending on the demand, supply and other economic factors that influence the volume of IC production. For example, Advanced RISC Machines (ARM) has recently (end of 2016) announced that it is working with Taiwan Semiconductor Manufacturing Company (TSMC) to make future-generation 7 nm chips. ARM shared its intellectual property for its 7 nm designs with TSMC, allowing TSMC to prepare and configure its foundry for the upcoming products. The manufacture of the 7 nm chips is expected to commence in 2018, according to the vice-president of ARM, and an estimated 15%–20% improvement in speed and power consumption compared with its 16 nm process is expected. This shows how design companies and semiconductor manufacturers are working together in preparation for next-generation products, a relatively new phenomenon, as opposed to designers only being able to use foundry services as-is, which ultimately slows progress. Current-generation applications, which are still in their infancy compared with traditional computing, include virtual reality and machine learning and these applications demand much higher performance from chips; this is an inevitable driver of Moore's law.

This chapter aims to highlight the economic perspective of improving semiconductor engineering on two main fronts:

• Technology node reduction and the challenges associated with it.
• Wafer diameter increases with their accompanied challenges.

Technology node size reduction is arguably the main driver toward achieving Moore's law in current- and future-generation chips, but a fair argument for improving yield and managing the overall cost of processing is made by corporations to adopt larger-diameter wafers. Both these perspectives are discussed in this chapter.

THE ECONOMICS OF GEOMETRIC SCALING OF SUBSTRATES AND DIES

As semiconductor manufacturing technology continually develops with time, feature sizes on ICs become increasingly smaller, in line with the predictions of Moore's law, as discussed in Chapter 1 of this book (Moore 1965). As a result of the shrinking feature sizes, more components are able to fit in a unit area of a semiconducting substrate, which leads to increases in feature density and device capability. Historically, as circuit design complexity increases, so does the footprint of the integrated circuitry, necessitating an increase in the average size of a chip (Doering and Nishi 2007) if the technology node is kept constant. Had the sizes of substrates remained constant as the sizes of chips increased, manufacturers would not have been able to fit as many ICs per wafer and this would have resulted in the reduction of overall revenue generated per wafer. Therefore, the semiconductor industry had both an economic incentive and a technological/engineering incentive to increase the sizes of wafers in order to accommodate the larger chip sizes (Doering and Nishi 2007). When silicon-based circuitry was first introduced in the 1960s, the diameter of silicon wafers used at the manufacturing level was only 1 inch (approximately 25 mm), but by the 1970s, their diameter had doubled to 2 inches (approximately 50 mm) (Doering and Nishi 2007). Production-level silicon wafers continued to grow exponentially, such that by the beginning of the 1990s, they had reached 200 mm in size, as shown in Figure 2.1 (Doering and Nishi 2007).

As also shown in Figure 2.1, a wafer size of 11.8 inches (300 mm, commonly referred to as 12-inch wafers) has been used in production in various foundries since the late 1990s and early 2000s. Currently, there are approximately 100 foundries implementing the 300 mm wafer size, whereas only 15 foundries that were using the 300 mm wafer size, were actively producing ICs in 2002 (IC Insights 2016a). An estimated 117 foundries using the 300 mm wafer size will be operating by 2020, reaching a projected peak of approximately 125 foundries before declining again as 450 mm foundries start to become more commonplace. A similar trend is seen with 200 mm foundries, with as many as 200 of these foundries closing in recent history. Refurbishment of foundries with older-generation equipment and relocation of machinery in areas where larger wafer handling is not necessary is commonplace, and a trend that has been followed throughout the evolution of wafer technology. The

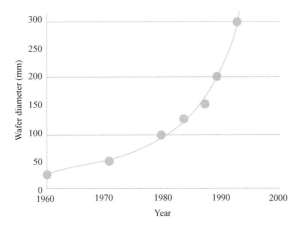

FIGURE 2.1 The evolution of wafer diameters over time. (Adapted from Doering and Nishi, *Handbook of Semiconductor Manufacturing Technology*, CRC Press, 2007.)

estimates include pilot and volume production facilities, but exclude research and development facilities. Current 300 mm foundries may have capacities to produce 25,000–40,000 wafers per month, translating to higher profit margins from increased wafer capacity. Figure 2.2 shows a relative comparison between wafer sizes, ranging from initial 1 inch (25 mm) wafers during the 1960s up to the 17.7 inch (450 mm, commonly referred to as 18 inch wafers) wafers, which will become a preferred solution within the next decade or even sooner.

As shown in Figure 2.2, the increase from 1 inch (25 mm) to the current 11.8 inch (300 mm) wafers shows a substantial increase in area available for dies. For a typical 5 mm × 8 mm die size, a 11.8 inch (300 mm) wafer-fab can produce approximately 1450 dies per wafer, whereas a 17.7 inch (450 mm) wafer-fab can produce approximately 3400 dies per wafer, not taking into account the yield of each wafer, scribe line width and wafer edge exclusion. An example is the Intel Ivy Bridge wafer shown in Figure 2.3 (Hruska 2012).

The increase in size has certain geometric advantages due to its round shape. Let us consider the effects of the geometric scaling of wafers and devices to elucidate the technical and economic benefits of increasing the sizes of silicon wafers. Since we know that wafers are circular in shape, the laws of basic geometry tell us that the expression shown in (2.1) holds true. The area (A) is calculated by

$$A = \pi r^2 = \pi \frac{d^2}{4} \tag{2.1}$$

where,
    r is the radius of the wafer
    d is the diameter of the wafer

From (2.1), we observe that the area of the wafer is proportional to the square of the diameter of the wafer; hence, when the sizes of wafers doubled from 25 mm to

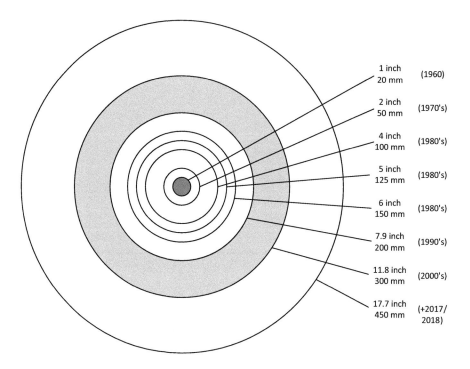

FIGURE 2.2 Relative sizing of wafers between the 1960s and current-generation technologies. The dark-filled wafer size shows the initial 1 inch (25 mm) wafers and the light-filled wafer size shows the current-generation 11.8 inch (300 mm) wafers.

50 mm between the early 1960s and the early 1970s, the available area quadrupled. Mathematically, this is shown through

$$\text{if} \quad d_2 = 2 \times d_1. \tag{2.2}$$

It then follows that

$$A_2 = \pi \frac{(d_2)^2}{4} = \pi \frac{(2d_1)^2}{4}; \tag{2.3}$$

therefore

$$A_2 = \pi (d_1)^2 = 4 \times A_1. \tag{2.4}$$

Since larger wafers can accommodate more complex and more powerful ICs, the total production revenue per wafer is approximately proportional to the wafer area available, hence production revenue is approximately proportional to the square of the wafer diameter (Hutcheson 2009; Mack 2011; Doering and Nishi 2007; Chien et al. 2007; Vardi 2014; Sullivan 2007). Total revenue is, of course, also dependent on the capacity of the foundry, typically measured in kilo-wafers per month (thousands

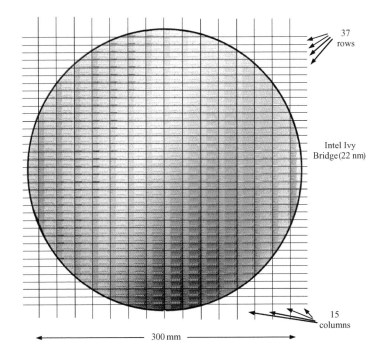

FIGURE 2.3   The 2011 Intel Ivy Bridge wafer: a 300 mm wafer, using a 22 nm process node and die size of approximately 8 mm×20 mm. (Adapted from Hruska, J., Deliberate excellence: Why Intel leads the world in semiconductor manufacturing. Retrieved 9 December 2016 from http://www.extremetech.com, 2012.)

of wafers manufactured each month). Table 2.1 summarizes the capacity of high-volume foundries operating in 2015 (AnySilicon 2016). These capacities refer to the installed capacity of each corporation, therefore the number takes into account all the foundries owned by the specific company.

The capacity of a foundry/company is important, since it translates to potential sales. The capacity results given in Table 2.1 are for both integrated device manufacturers (IDMs) such as Intel, Samsung and Texas Instruments, which design, manufacture and sell their own solutions, and fabless manufacturers such as Qualcomm, TSMC, Nvidia and Advanced Micro Devices (AMD), which design and sell solutions but outsource manufacturing to foundries. Over the last decade or less, it has become more difficult to distinguish between manufacturing companies that qualify as pure IDM and pure fabless foundries. Mergers and outsourcing have become commonplace for most large semiconductor manufacturers, depending on the demand for primarily personal computing, mobile phones and mobile computers or tablets. Merchant foundries, for example, have also become more prevalent in recent years; these are foundries that only manufacture devices under contract to other corporations and do not design any of the devices. From an economic perspective, the high cost of building and maintaining a semiconductor foundry can be mitigated by maintaining full utilization of the equipment through careful planning and scheduling process runs on contract. This model makes merchant foundries profitable and

TABLE 2.1

High-Volume Production Foundry Installed Capacities as of December 2015, in terms of 200 mm Wafer Size Equivalent and Measured in Wafers Processed per Month

Foundry	200 mm Equivalent Capacity (kilo-wafers/month) as of December 2015	Headquarters/Share of Global Total Capacity
Samsung	2534	South Korea (15.5%)
TSMC	1891	Taiwan (11.6%)
Micron	1601	North America (9.8%)
Toshiba/San Disk	1344	Japan (8.2%)
SK Hynix	1316	South Korea (8.1%)
GlobalFoundries	762	North America (4.7%)
Intel	714	North America (4.4%)
UMC	564	Taiwan (3.4%)
Texas Instruments	553	North America (3.4%)
STMicorelectronics	458	Europe (2.8%)

Source: AnySilicon, Top semiconductor foundries capacity 2015–2016 (Infographic). Retrieved 6 December 2016 from http://www.anysilicon.com, 2016.

allows fabless semiconductor manufacturers to use multiple foundries to process their unique designs and production at high volumes. In summary, the three major categories of foundries are

- IDMs such as Intel, which design and manufacture their own units.
- Fabless semiconductor companies such as Qualcomm, which design products and outsources design to.
- Pure-play/merchant semiconductor foundries such as TSMC and GlobalFoundries, which only manufacture designs from contracted third-party companies.

As shown in Table 2.1, according to the 2015 capacity capabilities provided by AnySilicon (2016), Samsung had the highest capacity at 2534 kilo-wafers per month, followed by TSMC, Micron, Toshiba/San Disk and SK Hynix, which were all able to produce more than 1 million wafers per month. The type of product, or IC, also affects the revenue of a foundry. Intel (4.4% of global share in capacity), for example, is the largest semiconductor brand globally, although its capacity is relatively low by comparison. The types of products Intel manufactures differs from those of the other foundries; Intel produces complete systems-on-chip and sells these solutions to markets that can accept higher margins, leading to a higher gross margin per product and the ability to generate more sales – along with the fact that Intel is an IDM. Samsung, for example, having a 15.5% global share in total capacity and capable of producing upward of 2.5 million wafers a month, primarily manufactures lower-cost memory devices for computers, laptops and highly competitive mobile devices, leading to a much higher capacity requirement; however, this does not translate to a leading position in the semiconductor market share.

In terms of sales revenue, companies such as TSMC, GlobalFoundries and United Microelectronics Corporation (UMC) are generating the highest revenue, based on their capacity abilities as well as their foundry model. Table 2.2 ranks the worldwide semiconductor foundries by revenue for the year 2015 (Gartner 2016).

According to Table 2.2, TSMC, GlobalFoundries and UMC dominated global semiconductor revenue in 2015. These three companies collectively hold approximately 73% of the global market in terms of revenue generated. The success of these companies is, again, attributed to the demand for semiconductor solutions for mobile smart phones and tablets, the foundry model implemented by these companies (pure-play manufacturers) and, in the case of TSMC, the success of 20 nm planar and 16 nm FinFET technologies. Following these companies on the list of highest revenue, with substantial contributions of 5.3% and 4.6% respectively, are Samsung Electronics and Semiconductor Manufacturing International Corporation (SMIC). Both Samsung Electronics and SMIC reported revenue upward of US $2 billion for 2015.

In terms of total sales of global leaders in the semiconductor industry, Table 2.3 lists the top 20 companies that reported the highest sales for the first quarter of 2016. These companies include IDMs, fabless and merchant foundries.

From Table 2.3, it is seen that Intel led global sales for the first quarter of 2016 by a substantial margin. Samsung Electronics, TSMC, Broadcom Ltd. and Qualcomm followed with sales of between US $9.34 billion and US $3.34 billion. The global sales of semiconductor devices declined from the first quarter of 2015 to the first

---

TABLE 2.2
Revenue of Global Top 10 Fabless Semiconductor Foundries in 2015, Measured in US Dollars

Rank	Foundry	Revenue in Millions of US $ (2015)	Market Share in % (2015)
1	TSMC	26,566	54.3
2	GlobalFoundries	4673	9.6
3	UMC	4564	9.3
4	Samsung Electronics	2607	5.3
5	SMIC	2229	4.6
6	Powerchip Technologies	985	2.0
7	TowerJazz	961	2.0
8	Fujitsu Semiconductor	845	1.7
9	Vanguard International	736	1.5
10	Shanghai Huahong Grace Semiconductor	651	1.3
	Top 10 for 2015	44,814	91.7
	Others	4077	8.3
	Total	48,891	100.0

TABLE 2.3

Sales of the Global Top 20 Semiconductor Suppliers, Which Include IDMs, Fabless Companies and Foundries, in US Dollars

Rank	Foundry	Sales in Millions of US Dollars (1st quarter of 2016)
1	Intel	13,115
2	Samsung Electronics	9340
3	TSMC	6122
4	Broadcom Ltd.	3550
5	Qualcomm	3337
6	SK Hynix	3063
7	Micron	2930
8	Texas Instruments	2804
9	Toshiba	2446
10	NXP	2224
11	Infineon	1776
12	MediaTek	1691
13	ST Microelectronics	1601
14	Renesas	1415
15	Apple	1390
16	GlobalFoundries	1360
17	Nvidia	1285
18	Sony	1125
19	UMC	1034
20	AMD	832
	Top 20 for 2016	62,440

Source: Adapted from ICInsights 2016b.

quarter of 2016, from US $66.745 billion to US $62.440 billion, a reduction of 6% in total sales. The only companies that saw an increase in total sales for this period were Intel (+9%), Infineon (+7%), MediaTek (+12%), Apple (+10%) and Nvidia (+15%). Companies such as Qualcomm (−25%), SK Hynix (−30%) and Micron (−28%) were among those that showed the largest decline in sales over this period. The semiconductor industry is likely to undergo a significant amount of shuffling over the next few years as mergers and foundry models change and the industry reaches maturity by the early 2020s. These mergers (such as the merger between Microchip and Atmel in 2016) significantly change the capacity and models of individual companies and typically shift the category quantifying the production capabilities.

There are three primary categories of wafer fabrication facilities: MiniFab (a monthly wafer production of 10,000–30,000 wafers), Megafab (monthly production of between 30,000 and 100,000 wafers) and GigaFab (monthly production of more than 100,000 wafers per month). These wafer starts per month (WSPM) are determined for single, integrated foundries, not for total company capacity as per Table 2.1. A common trend among large-volume manufacturers is to have numerous large wafer fabrication facilities in concentrated cluster parks, similar to science parks. Typically, analogue foundries fall into the MiniFab or MegaFab categories, producing somewhere around 10,000–50,000 WSPM, attributed mostly to an increase in testing and integration time due to the complexity of the manufacturing (proportional to the operating frequency of these devices). Logic foundries typically have around 80,000 WSPM, whereas a typical memory fab exceeds 120,000 WSPM.

The cost of building a wafer fabrication facility has steadily become more expensive for new generations of technology as well as for larger silicon wafers. Larger wafers effectively decrease the cost per chip, but they also substantially increase the cost of building a facility, as shown in Figure 2.4.

In Figure 2.4, the current-generation foundries used most frequently are the 11.8 inch (300 mm) technology foundries and can cost up to US $5 billion to build, not accounting for the operating cost, maintenance, staff and consumables. The 17.7 inch (450 mm) foundries cost in excess of US $10 billion, and a careful comparison between the expected revenue and the substantial initial capital must therefore be taken into account. As a calculated bet, presented by Hruska (2013), the advantage of 450 mm wafers is that long-term production costs are lower, assuming that the formidable initial cost can be covered. In the comparison presented by Hruska (2013), based on information provided by GlobalFoundries (which is planning to launch 450

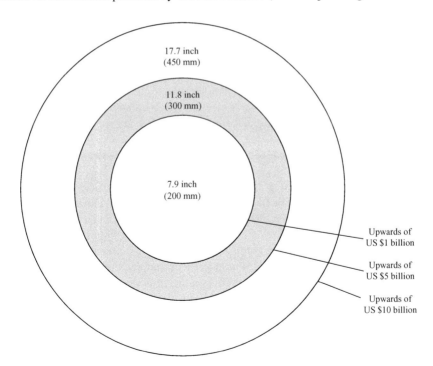

FIGURE 2.4 Simplified illustration of building a wafer production facility based on the wafer size only.

mm foundries in China, believing that its capability will reach 50% of the semiconductor market by 2028) the following arguments are provided:

- A 450 mm fab with a 40,000–45,000 WSPM can produce the same volume of die as a 300 mm fab with 100,000 WSPM.
- A 450 mm fab producing 40,000 WSPM requires approximately 25% less capital expense than a 300 mm fab with 100,000 WSPM with an equivalent number-of-die output.
- Models indicate approximately 10%–15% savings in cost per die for 32 nm processes and savings of between 20% and 25% for 22 nm nodes, for a 450 mm fab compared with a 300 mm fab.
- Total savings are more prominent for smaller nodes since the equipment share of overall costs per die is higher – therefore, the constant push for smaller nodes and larger wafers (currently, the combination of 14 nm on 450 mm wafers) to increase the profitability of foundries.

Historic semiconductor manufacturing data show that total manufacturing cost also increases with wafer sizes. This is because more IC components must be produced to efficiently use the extra area available on larger wafers (Mönch et al. 2013; Ibrahim et al. 2014). It is therefore in the best interest of any semiconductor device manufacturer to maximize the use of the available area so that total profits increase as the wafer footprint increases. One of the key parameters of the semiconductor manufacturing process is the gross die per wafer (GDW). This parameter refers to the number of dies that will fit onto a silicon wafer, where one die represents the footprint of an individual microchip (De Vries 2005), similar to the Intel Ivy Bridge example shown in Figure 2.3. Maximizing the number of dies of a specific surface area that can fit onto a single wafer is a critical factor in minimizing the cost per die. However, there is a geometric mismatch between wafers and dies in that silicon wafers are circular in shape and dies are square or rectangular in shape. Because of this mismatch, there are regions on every wafer that cannot accommodate a complete die – typically around the edges of the wafer, as shown in Figure 2.5.

For the example presented in Figure 2.5, the following parameters were used in the online calculator: die width and length of 10 mm, horizontal and vertical spacing of 80 µm (effectively, the scribe lines), edge clearance of 5 mm and a wafer-flat/notch height of 10 mm. The wafer sizes are 17.7 inch (450 mm) for Figure 2.5a and 4 inch (100 mm) for Figure 2.5b. From Figure 2.5a and b, it can be seen that if the size of each die is kept constant (10 mm × 10 mm in this example), the fraction of the total wafer area that is wasted by imperfect fitting of the chips decreases as the size of the wafer increases. As the wafer diameter increases, the fraction of the wafer area that is not used therefore decreases.

Various formulae for calculating GDW have been developed by semiconductor manufacturers. However, the shapes and sizes of dies are dependent on the complete IC designs, which are kept confidential by each manufacturer (De Vries 2005; Ferris-Prabhu 1989). There are, however, some basic principles on which these GDW formulae are based, which we will use to illustrate the advantages of the geometric

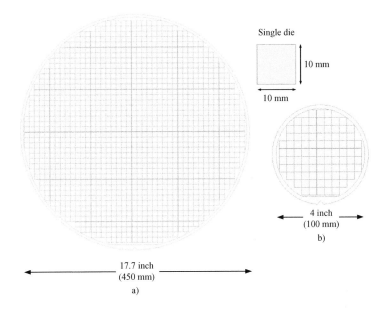

FIGURE 2.5 A comparison of the wasted wafer area (white areas) between two different-sized wafers. The gray squares represent complete dies. The images were generated using the online die-per-wafer calculator at http://www.silicon-edge.co.uk.

scaling of wafers (De Vries 2005). In addition to geometric mismatch between the shapes of the wafer and the die, wasted wafer area is also caused by kerf loss (Dross et al. 2008). The kerf regions are the safety margins around the outer portions of each die where circuitry is purposefully avoided because of the need to account for damage due to die-cutting by the dicing saws (Tummala 2001). Since no IC components exist in the kerf region, the total area of the kerf regions contributes to the total wasted space on a wafer. However, the total area of the kerf is itself dependent on the IC design (which is confidential) and the parameters of the die-cutting process (which will vary from manufacturer to manufacturer). The area used up by the kerf is therefore specific to a particular IC design. Hence, the values calculated from GDW formulae are not uniform across the industry, nor are they uniform across technology nodes within the same manufacturer; they are only used here to explain the motivation behind the geometric scaling of substrates. Several GDW formulas and algorithms have been proposed by researchers, each of which contains parameters to account for manufacturing and equipment-related defects. Typically, this is done by estimating the defect rate and excluding a fraction of the radius from the total wafer radius by using a term called an effective wafer radius, R_{eff} (De Vries 2005; Ferris-Prabhu 1989). The reduction in the wafer radii, or edge exclusion, is typically in the order of 1–5 mm (De Vries 2005). Ferris-Prabhu (1989) proposed the following formula for GDW (De Vries 2005; Ferris-Prabhu 1989):

$$\text{GDW} = \left(\frac{\pi R_{eff}^2}{A} \right) e^{\left(-L / R_{eff} \right)} \qquad (2.5)$$

where,

L   denotes the maximum die dimensions for a square die, making provision for variations in aspect ratio

A   denotes the area of each die destined for the wafer

R_{eff} denotes the effective wafer radius

Upon closer inspection of (2.5), it can be noted that the first bracketed term is a modified form of (2.1) that now represents the total effective wafer area that can consistently produce working ICs, while the exponential term in (2.5) is a correction factor. Using an unpublished algorithm for verification, the results of the algorithm seem always to result in an odd number of calculated dies (De Vries 2005; Ferris-Prabhu 1989). A variant on the GDW formulation, which does *not* make any provision for any variation in aspect ratio, is shown in De Vries (2005) by

$$\text{GDW} = \frac{\pi\left(R_{\text{eff}} - \sqrt{A}\right)^2}{A} = \left(\frac{\pi R_{\text{eff}}^2}{A}\right)\left(1 - \frac{\sqrt{A}}{R_{\text{eff}}}\right)^2. \tag{2.6}$$

It is noticeable that both (2.5) and (2.6) essentially comprise (2.1) multiplied by a correction factor. From the approximations given in (2.5) and (2.6), it is clear that the amount of edge exclusion on the wafer would have to be limited to a minimum in order to not decrease the GDW count. Considering a 300 mm diameter wafer (with a radius of 150 mm), for an edge exclusion of 1 mm, the ratio between the resulting effective diameter and total diameter is

$$\frac{R_{\text{eff}}}{R_{\text{total}}} = \frac{149 \text{ mm}}{150 \text{ mm}} = 0.9933 \tag{2.7}$$

and the effective area resulting from the edge exclusion is thus calculated as

$$A_{\text{eff}} = \pi \times \left(0.9933R\right)^2 = 0.9866\pi \times R^2. \tag{2.8}$$

Calculating the ratio between the total wafer area and effective wafer area due to the edge exclusion yields

$$\frac{A_{\text{eff}}}{A_{\text{total}}} = \frac{0.9866\pi \times R^2}{\pi \times R^2} = 0.9866 \Rightarrow 98.66\%. \tag{2.9}$$

Thus, for an edge exclusion of 1 mm for a 300 mm diameter wafer, a total of 98.66% of the total wafer area can be utilized. This corresponds to a loss of 1.34% of the total wafer area due to edge exclusion. Similarly, for an edge exclusion of 5 mm, the ratio between the resulting effective diameter and total diameter is

$$\frac{R_{\text{eff}}}{R_{\text{total}}} = \frac{145 \text{ mm}}{150 \text{ mm}} = 0.9667 \tag{2.10}$$

and the effective area resulting from the edge exclusion is thus calculated as

$$A_{\text{eff}} = \pi \times (0.9667R)^2 = 0.9344\pi \times R^2 \tag{2.11}$$

and, similarly, calculating the ratio between the total wafer area and effective wafer area due to the edge exclusion yields

$$\frac{A_{\text{eff}}}{A_{\text{total}}} = \frac{0.9344\pi \times R^2}{\pi \times R^2} = 0.9344 \Rightarrow 93.44\%. \tag{2.12}$$

Thus, for an edge exclusion of 5 mm for a 300 mm diameter wafer, a total of 93.44% of the total wafer area is utilized. This corresponds to a loss of 6.56% of the total wafer area due to edge exclusion.

While these losses may seem trivial at first glance, when taking into account that IC manufacturers generate product revenues in the order of billions of dollars or more, it becomes clear that any reduction in wasted substrate area can lead to substantial financial gains. It has been discussed how one of the main driving forces for the gradual increase in the size of silicon wafers was the ability to fit more chips per wafer; therefore, the GDW increases as wafer sizes increase (as shown in Figure 2.5), which results in higher revenue per wafer (Doering and Nishi 2007; Chien et al. 2007; Mönch et al. 2013; Ibrahim et al. 2014; Peters 1997; ITRS 2014). An additional motivation is that the manufacturing cost per die is inversely proportional to the number of complete dies per wafer. Since the entire area of a wafer is processed simultaneously, the manufacturing cost per unit area is inversely proportional to the wafer area. This implies that the manufacturing cost per die is also inversely proportional to the wafer area; hence it is cheaper per die to process a larger wafer than a smaller wafer. However, it should be noted that the cost of the actual bare silicon substrate increases proportionally with the silicon area, as indicated in Hutcheson (2009) by

$$\text{Price per unit area} = \frac{\text{Wafer cost}}{\text{Total wafer area}}. \tag{2.13}$$

Fortunately, when the cost of bare silicon substrates is converted into a per unit area basis, it remains fairly constant, and since we have seen that the percentage of wasted area decreases as the sizes of wafers increase, it is actually more economical for semiconductor manufacturers to purchase and use larger wafers rather than smaller ones (Hutcheson 2009; Mack 2011; Vardi 2014; Sullivan 2007). Therefore, if the primary factor dictating the manufacturing cost per unit area of a fully processed wafer is the processing cost, not the cost of the substrate, then the economic incentives for using large wafers at the manufacturing level become apparent. Geometric scaling of semiconductor substrates is not without technical and economic challenges; each transition from one wafer size to the next requires the development of new processing technologies and the purchase of new manufacturing equipment to accommodate the larger wafers (Hutcheson 2009). This partly explains why the semiconductor industry has been slow to transition from the 11.8 inch (300 mm) manufacturing technology to the 17.7 inch (450 mm) technology even though prior transitions occurred at a fairly steady rate, as shown in Figure 2.1 of this chapter.

THE WAFER MANUFACTURING INDUSTRY

While on the subject of wafers, the (silicon and other III–V, including epitaxially grown) wafer industry is a critical player in the IC supply chain and enables foundries to manufacture their components and devices. Without wafer suppliers, semiconductor companies would just be extremely expensive laboratories at a standstill. The current merger and acquisition activity in the semiconductor industry is influencing the number of suppliers (Lapedus 2016). Silicon wafer production flow is a complex process where polysilicon is melted in a quartz crucible, a silicon seed is lowered into the crucible and the resulting ingot is pulled out and diced into wafers of different sizes.

The profits of semiconductor wafer production facilities are relatively low and since the 1990s, only five major suppliers have survived from approximately 20 (Lapedus 2016). The average selling price of silicon wafers has declined from US $1.04 per square inch (2.54 cm^2) in 2009 to US $0.76 per square inch in 2015. A major contributing factor to wafer production companies not being profitable, especially between 2014 and 2016 (the fourth quarter of 2015 was the lowest point), is that wafer foundries suffered from a period of excess capacity of wafers, along with global price pressure and low profit margins. Foundries can only operate at 100% capacity if the demand is high, and various economic factors influence consumer demand; a chain reaction emanates from under-utilized foundry operations down to wafer production. By the end of 2016, the silicon wafer market was predicted to be around US $7 billion (down 1% from 2015) and shipments at over 10,800 million square inches (27,432 million cm^2) of wafer surface area were estimated in 2016. The 11.8 inch (300 mm) wafer size is still the driving force for this growth in demand for both logic and memory circuits, whereas the 7.9 inch (200 mm) wafers are considered the workhorse of industries such as automotive, consumer and Internet-of-things (IoT) – albeit changing in favor of the larger wafer size.

In 2016, 11.8 inch (300 mm) wafers accounted for 60% of wafers sold, whereas 7.9 inch (200 mm) wafers accounted for 31%; the remaining 9% were smaller sizes. The silicon wafer capacity for 11.8 inch (300 mm) wafers grew from 43 million wafers per year in 2009 to 76 million wafers per year in 2015, according to Sage and reported by Lapedus (2016). The issue is, however, that only 57 million of the 76 million wafers were processed by foundries, which now had excess capacity and were lowering orders at wafer production facilities, leading to further increases in liability for these manufacturers.

Since 17.7 inch (450 mm) wafers are considered the future of the semiconductor industry, at least in the near/foreseeable future, and are necessary to facilitate Moore's law and to allow more chips to be manufactured on current-generation technology nodes, the characteristics, limitations and advantages of 17.7 inch (450 mm) wafers are discussed in the following section.

THE MOVE TO 17.7 INCH (450 MM/18 INCH) WAFERS

The move to 450 mm wafers has been delayed several times in recent history, notably during the recession of 2008; the original plans were to start the move in 2012, have pilot lines in 2013/2014 and full production facilities in the 2016/2017 timeframe (Edwards 2012). In a paper published by Pettinato and Pillai (2005), technology decisions to minimize the risk of transitioning to 450 mm wafers were outlined. Three key elements identified by Pettinato and Pillai (2005) to take into account when moving to 450 mm wafers, while ensuring minimal risk in the long term, were

- Wafer attributes such as wafer thickness, which influences the wafer sag and equipment required to mount the wafers in equipment such as sputtering and photolithographic tools.
- Production equipment configuration and the cost-effectiveness of single versus batch wafer processing.
- Wafer carrier size and transport within foundries and between processing steps, again considering the cost-effectiveness and complexity of single versus batch transport equipment.

These elements have been key in many previous-generation transitions toward larger wafer diameters, but have become exponentially more complex with wafer sizes reaching the 300 mm and 450 mm diameters, especially considering the significance of wafer sag at these large diameters.

Larger-diameter wafers also suffer from a higher probability of bending, since their flexural rigidity is compromised by their larger diameter. The sag (δ) of a circular component (such as a wafer), if supported at its periphery (as is generally done with wafers to avoid touching and damaging the geometries/devices on the wafer), is determined (as given in Watanabe and Kramer 2006) by taking into account its stiffness (k), given by

$$k = \frac{F}{\delta} \tag{2.14}$$

where,

F   is the force applied to the body (or gravitational force in the case of a circular wafer supported around its edges)

δ is the displacement produced by the force along the same degree of freedom. The flexural rigidity of a plate is dependent on

- The thickness of the plate.
- The elastic properties of the plate (in this case, of silicon).
- The load force applied to the plate (such as gravity).

The flexural rigidity (D) is determined by

$$D = \frac{Et^3}{12\left(1-v^2\right)} \tag{2.15}$$

where,

E   is Young's modulus
t   is the thickness of the wafer
v   is Poisson's ratio Watanabe and Kramer (2006) deduced that the sag/deflection on a wafer surface due to film stress is given by

$$\delta = \frac{R^2}{t^2} \frac{6(1-v)\sigma}{E}$$ (2.16)

where,

R   is the radius of the wafer
σ is the film stress
R^2/t^2 is a scaling factor

Therefore, for a constant film stress, the sag of a 450 mm wafer is 2.25 times greater than that of an equally thick 300 mm wafer (Watanabe and Kramer 2006). This relationship is also evident in the work presented by Goldstein and Watanabe (2008), where the relationship between fracture strength and wafer thickness results in the equation

$$t_{450} = 2.25^{\frac{1}{2m}} t_{300}$$ (2.17)

where,

t_{450} and t_{300} are the thicknesses of the 450 mm and 300 mm wafers, respectively
m is the Weibull modulus of the silicon plate, which can range between 3.7 and 11.5 (unitless) depending on the size, crystal orientation and surface quality

Dunn et al. (2012) have reported the sag for various 300 mm and 450 mm wafers, as shown in Table 2.4 (adapted from Dunn et al. 2012).

According to Table 2.4, the silicon materials tested have parameters of 2.33 g/cm³ density, an elastic modulus of 141 GPa, Poisson's ratio of 0.22 and a three-point support structure (as opposed to typical foundry ring support structures). It is evident

TABLE 2.4

Wafer Sag Variability with Wafer Dimension and Material Properties

Material	Diameter (mm)	Thickness (μm)	Support Ring Radius (mm)	Sag (μm)
Silicon	50	400	22	0.35
Silicon	300	690	147	212
Silicon	300	700	147	206
Silicon	300	710	147	200
Silicon	450	700	222	1060

Source: Adapted from Dunn et al., Metrology for characterization of wafer thickness uniformity during 3DS-IC processing, *62nd Electronic Components and Technology Conference*, 1239–1244, 2012.

that the sag/deflection of the larger radius wafers is higher, as can be expected. The 2 inch (50 mm) wafer deflection is minimal, at around 350 nm, but for the 300 mm wafers the deflection is significant, averaging approximately 206 μm with a slight variation in wafer thickness. The 450 mm wafer, however, has a substantially higher sag value, at 1060 μm, more than four times that of the 300 mm wafers (supported by the Weibull parameter in (2.17). The flexural integrity of larger-diameter wafers introduces several challenges during manufacturing and processing, including

- Increased vibrational effects (varying natural frequency) of thin wafers, which can cause wafer breakage or cracks.
- Increased gravitational bending (sag).
- Compromised flatness over the entire area of the wafer.
- Compromised wafer process uniformity across large areas and the maintenance of uniformity along all process steps.
- Significantly heavier crystal ingots (up to 1 ton for a single ingot).
- Longer cooling times during processing with more stringent temperature control required to reduce thermal stresses.
- Typically longer processing times, which may include additional processing steps, increasing overall cost.

Among the challenges listed above, the migration to larger wafer sizes requires collaboration of the supply chain with upstream suppliers such as equipment manufacturers, wafer production companies, wafer carrier manufacturers and factory automation suppliers, as well as downstream suppliers such as IC designers, testing houses and chipmakers (Chien et al. 2007). It is a challenging and risky decision for most foundries to move to 450 mm wafer production; essentially, a single 450 mm fabrication facility can cost up to 2.25 times that of a 300 mm fab, but has additional advantages of fewer engineers required to operate the facility, hence decreasing the payroll; better automation equipment available; faster ramp-up to high-capacity manufacturing and overall higher productivity; and reduced operating overheads. This is, of course, as with any technology upgrade, only true in the current milieu and will become less cumbersome as more foundries and upstream and downstream suppliers support and are committed to the 450 mm technology. Global semiconductor manufacturing leaders such as Intel and TSMC are also seeing the implementation plans for 450 mm production slipping by several years, primarily because of lower demand in the personal computing market and the challenges of developing and integrating FinFET technology fully. In fact, all challenges are related to decreasing node size below 14 nm, allowing the 450 mm plans to be accorded lower priority until these issues are addressed. Moving to a new wafer size in a foundry can therefore have just as high ramifications as moving to a smaller technology node, require an equal amount of commitment and imply high cost related to research and development to ensure a qualified process on the new wafer size. In reality, the cost versus benefit of the 450 mm wafer is difficult to gauge, especially in view of the enormous initial capital investment and rapid changing of the semiconductor market that drives the need, all of this in order to bring down the cost of chips per die. Cost benefits are expected in etching and deposition steps, but photolithography steps will still dictate the overall benefits, and currently (in 2016/2017), these steps remain expensive and need to be completely controllable before they can be rolled out to mainstream processes. Mack (2015) presents a diagram of photolithography cost relative to wafer size, which is adapted and reproduced in Figure 2.6.

From Figure 2.6 it is evident that, according to Mack (2015), photolithography costs scale with wafer area; an increase in wafer size (such as from 150 mm to 200 mm to 300 mm, as shown in Figure 2.6a, b and c, respectively), results in an increase in the photolithography cost as a fraction of the wafer area.

Several other mechanical aspects are to be considered in the manufacturing process of ICs. These have particular influences on the total manufacturing cost of ICs. Certain silicon wafers used in the manufacturing processes can become extremely thin, in the order of 40–60 μm (Kashyap et al. 2015), if mechanically/chemically polished and thinned down from the standard 500–700 μm thickness (for example, as used in ultrathin single-crystal wafers for integration on flexible substrates). Because of the thin dimension of silicon wafers in general (less than 1 mm), they can become fragile and need to be handled with extreme care in order to prevent any form of mechanical damage or wafer sag. Furthermore, micro-defects such as miniature cracks originating from the thinning process could also cause localized stress concentrations at the defect locations, causing the wafer to fracture under an applied load (Kashyap et al. 2015) or deteriorate over time with variations in temperature causing mechanical stress. Fracturing of the wafer could also occur when the wafer is placed on an uneven surface and a small force (such as gravity, for example) is exerted on the wafer. These types of mechanical issues are only some of the reasons why such sensitive and expensive manufacturing equipment is required for the manufacturing of ICs (Balasinksi 2012). The introduction of 450 mm diameter wafers would, in some cases, require a complete redesign of some manufacturing equipment to accommodate the larger size. Furthermore, a larger series of quality and reliability testing considerations would be required, which adds to the total general manufacturing costs (Balasinksi 2012).

Although geometric scaling of semiconducting substrates is a key component of Moore's law, the true cornerstone of Moore's law is still primarily device scaling, which is discussed in the following section.

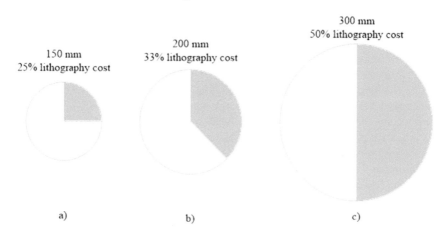

FIGURE 2.6   The cost of photolithography relative to wafer diameter. (Adapted from Mack, *IEEE Spectrum*, 52(5), 67–67, 2015.)

The Economics of Device Scaling/Feature Scaling

Device scaling, as discussed in Chapter 1 of this book, refers to the ability of the semiconductor industry to progressively reduce the feature sizes of silicon-based ICs at each technology node (Moore 1965,1995). Since a reduction in feature size allows more IC components to be packed into the unit area of each chip, the steady advancement of patterning and processing technology results in a consistent increase in feature density at each technology node (Hutcheson 2009; Mack 2011, 2015; Vardi 2014; Sullivan 2007; Morita 2016; Huff and Gilmer 2005). As shown in Figure 1.2 in Chapter 1, the number of components in both microprocessors and memory devices has consistently increased at a steady rate for several decades, as predicted by Moore's law (Huff and Gilmer 2005). Furthermore, the speed at which a transistor can be switched on or off is inversely proportional to the length of the gate between the source and the drain (Tummala 2001; Morita 2016; Huff and Gilmer 2005). Hence, when the critical dimensions of transistors and other IC components shrink, the overall speed of electronic devices correspondingly increases (Mack 2011, 2015; Moore 1995; Brock 2006). Therefore, not only does the overall computing power of ICs increase with each technology node (because of the number of components and overall complexity of the ICs as well as increased feature density), but the speed of the ICs increases as well. This provides strong economic incentives for all members of the semiconductor industry to ensure that they do not fall behind their competitors in the pursuit of metrics predicted by Moore's law (Hutcheson 2009; Vardi 2014; Sullivan 2007).

It is also observed from Figure 1.3 in Chapter 1 of this book that the revenue generated by the semiconductor industry is strongly correlated with the number of IC components that are produced (Mack 2011). By combining the data from these two figures, it can be deduced that the revenue generated by the semiconductor industry is directly tied to the ability of manufacturers to scale feature sizes, i.e. device scaling is the main driving force for generating revenue, and thus profits for the semiconductor industry, and takes precedence over increasing wafer size. Although device feature scaling has both economic and technical benefits, the technical requirements for continually reducing the sizes of features while correspondingly increasing their per unit area are becoming more challenging and more costly (Hutcheson 2009; Vardi 2014; Sullivan 2007). As the feature size decreases, manufacturing defects become much more significant and can no longer be regarded as insignificant in size compared with feature sizes (Mesawich et al. 2009; Yoshikawa et al. 2015; Bret et al. 2014; DeHon 2009; Kawahira et al. 1999; Lin et al. 1999). Defects have led to higher failure rates for finished devices, resulting in lower production yields and higher operating costs due to the increased time devoted to performing root cause failure analyses (Mesawich et al. 2009; Yoshikawa et al. 2015; Bret et al. 2014). The manufacturing cost is influenced by the production yield. The expression is actually an over-simplification, since it does not account for wasted space on the wafer, but serves as a reasonable first approximation of the influence of yield on manufacturing cost (Touma 1993), as shown by

$$C_{\text{manuf.}} = \frac{C_{\text{wafer}}}{Y_{\text{die}} \cdot N_{\text{dies}}} \tag{2.18}$$

where,

C_{manuf} is the manufacturing cost per wafer
C_{wafer} is the cost of the bare silicon wafer prior to processing
Y_{die} is the average yield per die after processing is completed
N_{dies} is the number of dies per wafer

It is clear from (2.14) that the manufacturing costs are inversely proportional to the yield; hence, in order for manufacturing to be profitable, the yield must be kept high. Therefore, higher failure rates lead to lower profit margins for device manufacturers. In addition to lowering production yields, higher defect counts and higher defect densities cause the rate of production of complete devices to decrease, since manufacturers have to shut down production to be able to determine the cause of low yields (Mesawich et al. 2009; Yoshikawa et al. 2015; Bret et al. 2014; DeHon 2009; Kawahira et al. 1999; Lin et al. 1999). Since decreased production rates and production yields both lead to lost revenue, they both represent de facto manufacturing costs. This is consistent with the trend shown in Figure 1.7 of Chapter 1 of this book, where the rate of revenue generation in the semiconductor industry has been slowing noticeably from the late 1990s onward (Mack 2011). Thus, historical data confirm that defects and failure rates and low production yields are becoming more prevalent, which makes device scaling less attractive than it used to be, i.e. before the late 1990s (Hutcheson 2009; Vardi 2014; Sullivan 2007).

However, this has not stopped technology scaling in any way; in fact, scaling has continued to push the limits of each technology. Table 2.5 lists the most significant years of technology node scaling in the semiconductor business. Each node is also accompanied by the most prevalent and notable featured products that implemented the node and were successfully sold as commercial products.

As shown in Table 2.5, there has been a reduction in technology node size from 10 μm (10,000 nm) during 1971 down to the 16/14 nm FinFET used currently, and 11 nm nodes expected to appear in the near future, with trials to qualify this technology undertaken by corporations such as Intel, TSMC and Samsung. Figure 2.8 displays the same data as presented in Table 2.5 visually on a logarithmic y-axis (node size in nm), omitting the featured products column for simplicity of the resulting figure. In addition, added to Figure 2.8 are vertical orange lines, which indicate the years where technology node size decreased by an order of magnitude. The orange-dashed line is the prediction for when the technology node size will decrease below the 10 nm barrier.

From Figure 2.8 (and Table 2.5), it is noticeable that there were steep reductions in node size in earlier years, for example the reduction from 10 μm to 3 μm between 1971 and 1975, and especially the reductions during the 1990s (from 800 nm to 130 nm in a span of approximately 11 years). According to the vertical lines shown in Figure 2.8, the first reduction of an order of magnitude (from 10,000 nm to 1000 nm) occurred between 1971 and 1985 (14 years), followed by the period 1985–2002 (17 years from 1000 nm to 100 nm). During these phases, manufacturers and engineering institutions saw relatively smaller iterations of physical limitations during each technology reduction, especially in photolithography. More recent advances have been implemented at a relatively constant pace, but with much smaller steps

TABLE 2.5

Process Geometry Scaling Since 1971 with Featured Products Used by Each Significant Node

Process Geometry	Year (Approximate)	Featured Products
10 μm	1971	Intel 4004, 8008
3 μm	1975	Intel 8085, 8086, 8088
1.5 μm	1982	Intel 80286
1 μm	1985	Intel 80386
800 nm	1989	Intel 80486
600 nm	1994	Intel Pentium, IBM/Motorola PowerPC 601
350 nm	1995	Intel Pentium Pro, Intel Pentium II, AMD K5
250 nm	1998	Intel Pentium III, Sony PlayStation 2
180 nm	1999	Intel Coppermine, AMD Athlon Thunderbird
130 nm	2000	Intel Pentium 4, AMD Athlon XP
90 nm	2002	Intel Pentium 4 Prescott, Nvidia GeForce 8800 GTS
65 nm	2006	Intel Core 2, Sony PlayStation 3, Microsoft Xbox 360
45 nm	2008	Intel Core i7, AMD Athlon II, Microsoft Xbox 360 S
32 nm	2010	Intel Core i7 980X Extreme Edition
28 nm	2011	Nvidia Tegra K1, Qualcomm Snapdragon
22/20 nm	2012/2014	Intel Core i7 and i5 Ivy Bridge
16/14 nm	2014/2015	Intel Core M, Samsung Galaxy S6/S6 Edge
11 nm	2017/2018 (est.)	Trials by Intel, TSMC and Samsung

in the decline in each technology node reduction; the drop from the 100 nm barrier to 14 nm occurred between 2002 and 2016, another 14-year period. In summary, it is evident from Figure 2.8 that the reduction in technology node size by an order of magnitude has historically been relatively constant, with only the number of steps in between increasing for more recent technology reductions.

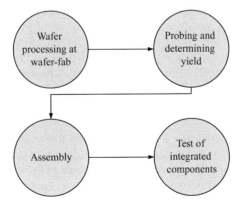

FIGURE 2.7 The various stages involved in the semiconductor manufacturing process. (Adapted from Mönch et al., *Production Planning and Control for Semiconductor Wafer Fabrication Facilities: Modeling, Analysis and Systems*, Springer, New York, 2013.)

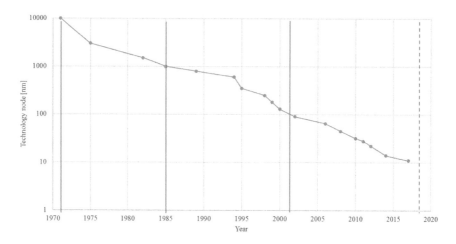

FIGURE 2.8   Visual representation of the technology significant node reduction from 1971 to the current implementation. Orange lines indicate the years when technology node size decreased by an order of magnitude.

Not only have physical limitations played their part in requiring more in-between steps of technology reduction to ensure that each node functions properly and can be qualified for high-volume production; general manufacturing operating cost has also increased significantly and forces manufacturers to reduce the step size in each advancement, primarily to reduce the cost of replacing most equipment, as is needed with large jumps in node reduction. These challenges of general manufacturing operating costs are discussed in the following section.

General Manufacturing Operating Costs

The largest component of semiconductor manufacturing costs, making up approximately 70% of the total manufacturing costs, is the capital costs associated with manufacturing machinery and equipment (Mönch et al. 2013). The price of manufacturing equipment has risen by nearly eight orders of magnitude during the past 50 years. For example, some of the earliest contact aligners used for photolithography in the 1960s cost approximately US $3000–US $5000; the price of these tools increased to US $10,000 by the 1970s (Hutcheson 2009). The first steppers (the next generation of lithography equipment) introduced in the 1980s cost about US $500,000, but by 2000 the cost of state-of-the-art steppers had risen to approximately US $10 million. The cost of modern-day steppers is now approaching US $100 million dollars. The steady increase in equipment costs has not been limited to patterning equipment; rather, it manifested itself in every processing module. The average price of evaporators, which are used in metallization, has risen from a few thousand dollars in the 1970s to over US $4 million today. Furthermore, some of the latest semiconductor manufacturing tools can cost in the order of US $50 million and even higher (Hutcheson 2009). These staggering increases in prices have been attributed to the continually advancing and developing manufacturing equipment and the associated technologies. Because of these high equipment prices, constructing and setting up a new semiconductor fabrication plant can cost in excess of US \$5 billion (Mönch et al. 2013), with 450 mm wafer production upgrades pushing this amount to the US \$10 billion mark. Therefore, optimal and efficient machine utilization is a crucial aspect in fabrication and production processes. As illustrated in Figure 2.7, four main steps are involved in the manufacturing process, each of which further involves a number of process steps.

A typical manufacturing process can consist of up to 700 individual process steps with a total manufacturing time of up to three months. While the main focus of semiconductor manufacturers in the past was on delivering a good product, the sheer scale and costs involved in modern manufacturing processes necessitated an approach of optimizing and utilizing manufacturing equipment in the best possible manner in order to keep the manufacturing processes cost-effective (Mönch et al. 2013). One aspect that is of particular concern to semiconductor manufacturers is the downtime of manufacturing equipment and machinery; downtime refers to a time period where equipment cannot participate in production. Downtime is typically due either to scheduled maintenance or to an automatic shutdown event caused by the equipment performance degrading beyond limits specified by the production engineers.

Downtime has a negative impact on a production cycle since the flow of production sequences is disrupted and production capacity is lost, which ultimately leads to decreased yield and lost revenue. The nature of these failures is not necessarily catastrophic in the conventional sense (such as breakages of equipment), but rather involves events where some form of maintenance or adjustment needs to be performed. Maintenance (and preventative maintenance) should typically be performed by skilled workers, often contracted by the equipment manufacturing corporation or as part of these corporations themselves. This again leads to a higher maintenance cost for complex machinery, especially in countries where these companies do not have an active presence, requiring that these skilled workers be flown in and contracted in foreign currency, generally leading to much higher costs. If maintenance staff are called out for issues on equipment, high-quality data analysis and capabilities are essential, which includes post-acquisition event triggers, recalling of all related status or process variables, multi-variate statistical process control and principal component analysis.

Instead of regularly replacing equipment components altogether, production cycles are planned to complete a batch or a number of batches before such components are scheduled for downtime in order to perform the necessary maintenance. For example, downtime for manufacturing equipment such as ion implanters could span up to 30%–40% of the production cycle time, which has a tremendous impact on overall productivity (Rose 2004). It is therefore clear that minimizing downtime is of the utmost importance to semiconductor manufacturers. Downtime can be divided into scheduled and unscheduled downtime (Rose 2004; Qi et al. 2002; Babbs and Gaskins 2008). Scheduled downtime occurs when equipment cannot perform its intended function because of planned events, such as preventative maintenance, change of consumables, equipment test runs and setup for performing a different manufacturing process. Owing to the large impact of downtime on production, scheduled downtimes are meticulously planned and factored into a production run cycle. There are two main approaches to scheduled downtime (Qi et al. 2002).

Downtime can either be scheduled more frequently for shorter durations, or be scheduled less frequently for longer durations. Regular short-duration downtime can result in more predictable downtimes and aid in streamlining fab scheduling. Alternatively, downtime can be scheduled in response to lower yields and inadequate performance (Babbs and Gaskins 2008). Montgomery (2000) gives examples of several issues that can lead to downtime due to a lack of preventative maintenance, including

- Afer transport or automatic alignment robots out of alignment.
- Load locks where wafers are not touching the edge of cassettes.
- Incorrect gas mixing.
- Low-quality incoming gas supplies.
- Clogged gas exhaust lines.
- Chamber contamination from particle build-up.
- Incorrect recipe of incoming wafers.
- Malfunctioning power supply capacitors.
- General automation failures.
- Low-quality or insufficient preventative maintenance (Montgomery 2000).

Unscheduled downtimes occur when equipment cannot perform its intended function because of unplanned events. Causes of unscheduled downtime include technical equipment failure and unplanned measures concerning safe and secure operation (Rose 2004; Qi et al. 2002; Babbs and Gaskins 2008). This is undoubtedly the type of downtime that is least desirable to semiconductor manufacturers. While unscheduled downtimes are often highly unpredictable in view of their intrinsic nature, preventative maintenance can greatly reduce the risk of encountering unscheduled downtime (Rose 2004; Babbs and Gaskins 2008). Downtime scheduling, where various modeling tools and methods are used to determine and plan downtimes that have the least negative impact on production efficiency, is an intensely studied field among semiconductor manufacturers. In general, downtimes have a direct cost implication for semiconductor manufacturers, regardless of the nature of the downtime, which ultimately reduces yield and revenue (Babbs and Gaskins 2008).

RUNNING A FOUNDRY OPTIMALLY

Since manufacturing plants are typically run continuously for 24 hours per day, they require constant skilled staff presence to cover all working shifts (Mönch et al. 2013; Peters 1997) without overworking the staff. Each fab must employ hundreds or even thousands of people to maximize production throughout the year – therefore, labor costs are one of the major contributors to the operating costs of a fab. Semiconductor manufacturing equipment consumes large amounts of power, since many of the processing modules involve high temperature (for example oxidation processes, annealing or curing and metal deposition processes) and/or high power (in the case of plasma equipment such as plasma etchers) and/or high vacuum environments, and these physical and chemical conditions must be maintained throughout the operating hours of the facility (Hutcheson 2009; Vardi 2014; Sullivan 2007; Mönch et al. 2013; Tummala 2001). The average amount of electricity consumed per square centimeter of silicon wafers being processed in the late 1990s was in the order of 1.44–1.52 kWh/cm^2 (Williams et al. 2002). The peak electricity consumption of an entire fab may be around 30–50 MW, which is enough to power a small city (Williams et al. 2002) of approximately 80,000 residents. According to Öchsner (2014), the power consumption of a semiconductor fabrication facility can be categorized into five main categories, these being

- Equipment responsible for up to 41% of the total energy consumption in the facility.
- Pumps and cooling towers, which make up an additional 24% of the total consumption.
- Gas and vacuum generation and distribution, which total up to 18.4% of the total consumption.
- Exhaust and cleanroom fans using about 8.4%.
- Backup energy systems and general usage such as administration, lighting and uninterruptable power supplies accounting for the remaining 8.2%.

Öchsner (2014) also lists several techniques and approaches with respect to tools and equipment in foundries to improve energy efficiency; these techniques are not discussed in detail in this book, as they fall outside its scope, but it remains important to highlight them briefly, as they can indirectly be traced back to allowing Moore's law to continue successfully through future generations, as cost savings are crucial to advancing technologies. Running a 300 mm wafer-fab implementing 14 nm nodes at a significantly lower operating cost greatly increases the potential and financial prospects of upgrading to 450 mm wafer sizes. The techniques to reduce energy usage with respect to tools, as given in Öchsner (2014), are

- Front-end of line and back-end of line improvements, which include mask-level reduction and reducing the number of processing steps.
- Potentially increasing wafer diameter to increase the yield and reduce the number of etching steps (photolithography might be more expensive for larger-diameter wafers).
- Increasing process-cooling water temperature requirements to potentially enable free cooling instead of using chillers.
- Replacing and maintaining chillers.
- Implementing vacuum pumps with certified and specified idle modes, which include smart idle mode for various tools within the facility.

There are obviously a multitude of changes in routine and awareness of energy consumption that can also be implemented to save energy, including gas-to-power initiatives, efficient energy storage, renewable energy and load shifts. Energy consumption is a key aspect of semiconductor manufacturing costs in addition to labor costs (ITRS 2014; Williams et al. 2002). Chen (2013) provides a breakdown of potential savings at a 300 mm fab in terms of energy reduction across such a facility, and this breakdown is adapted and presented here. According to Chen (2013), potential savings from energy cost reduction are most notable in

- Plant management (with potential savings of up to 50%).
- Efficiency in air conditioning systems, which can save up to 21%.
- Bulk gas efficient handling and distribution for savings of up to 10%.
- Higher process-cooling water temperature, which can save 8% of energy.
- Use of ultrapure water, requiring an initial investment, but energy savings can amount to 5%.
- Various other mechanisms that can lead to an additional 6% of energy savings (Chen 2013).

Plant management, where energy cost savings can amount to 50%, is divided into specific management processes by Chen (2013). These specific processes, along with their potential savings, include thin film (31%), diffusion (19%), etch (15%), implant (12%), lithography (10%), polish (8%) and clean technologies (5%). Actual wafer throughput falls under the general operating cost umbrella, as discussed in the following paragraph.

Because of stringent design rules and standards of modern IC designs, the manufacturing equipment used to create the silicon ingots from which the wafers are cut has become increasingly sophisticated with time. Although the development of sophisticated equipment incurred significant costs, it has led to increased yield and higher-quality wafers, permitting a decrease in the overall manufacturing cost of the wafers. One of the most prominent factors of a manufacturing cost model is the wafer throughput. Typically measured in wafers per hour, the wafer throughput T in wafers per hour is generally expressed as (Doering and Nishi 2007)

$$T[\text{wph}] = \frac{3600 \left[\dfrac{\text{s}}{\text{h}}\right]}{t_{\text{wafer}} \ [\text{s}]} \tag{2.19}$$

where,

t_{wafer} the denominator (wafer process time, t_{wafer}) of (2.15) is described by

$$t_{\text{wafer}} = t_{\text{woh}} + N_f \left(t_{\text{exp}} + t_{\text{foh}}\right) \tag{2.20}$$

where,

t_{woh} represents the overhead time per wafer
t_{exp} represents the exposure time
t_{foh} represents the overhead time per exposure field
N_f represents the number of exposure fields per wafer

These overhead and exposure times include, among others, the time taken by various intrinsic processes of the manufacturing process, such as wafer loading and unloading, wafer alignment, and other times occupied by the manufacturing stages and machine operations. Maintenance delays and other delays caused by unforeseen circumstances are generally not included in the wafer throughput calculation (Doering and Nishi 2007).

CONSIDERING DIE AND WAFER YIELD TO KEEP DRIVING MOORE'S LAW

Another important aspect in terms of ensuring the most cost-effective manufacturing solution is the die yield, which is determined indirectly from the wafer yield. Die yield is generally regarded as the single most significant factor that influences wafer processing costs. This is due to the fact that even a marginal increase in yield of about 1% could significantly reduce wafer costs and device manufacturing costs (Krueger 2011). Several formulations for die yield have been proposed, of which the approximation by Hennessy and Patterson (Touma 1993) will be considered first.

$$Y_{die} = Y_{wafer}\left(1+\frac{N_{defects} \cdot A_{die}}{N_{CM}}\right)^{-N_{CM}} \tag{2.21}$$

where,

Y_{wafer} denotes the wafer yield
N_{CM} denotes the number of critical masks
$N_{defects}$ denotes the number of defects per unit area
A_{die} denotes the total surface area of the die

The wafer yield is included in the equation because of the possibility of a defective die being caused by a wafer defect and not specifically a die defect. Wafer yields typically range between 80% and 90% (Touma 1993). The number of critical masks is high because during each photolithography step, there is the potential for wafer contamination either from foreign particulates (for example dust particles) or other sources, which may ultimately result in device failure. Another yield model was proposed by Murphy (Krueger 2011), and takes into account the variations in defect densities between wafers and dies. The yield formulation proposed by Murphy (Krueger 2011) is given as

$$Y_{die} = \int_{0}^{\infty} e^{-DA} \cdot f(D)dD \tag{2.22}$$

where,

$f(D)$ refers to the normalized distribution function of defect densities in dies
A refers to the area of the device susceptible to defects

This is a bell-shaped distribution, but in order to simplify yield calculations, Murphy proposed a triangular distribution as an approximation to the actual bell-shaped distribution, asss given by (Krueger 2011):

$$Y_{die} = \left(\frac{1-e^{-D_0 A}}{D_0 A}\right)^2 \tag{2.23}$$

where,

D_0 refers to the mean defect density

However, Krueger (2011) later noted that this approximation had a limitation, since it assumed that only one type of spot defect occurred in the manufacturing process. A number of other yield models have been proposed in addition to the two models discussed here, but a discussion of each of these models is beyond the scope of this book. The key aspect in any of the proposed yield models is that device defects are often encountered owing to subtle manufacturing errors and contamination with impurities such as dust particles (Touma 1993). In some cases, yield loss is attributed to electrostatic discharge in the fabrication environment (Peters 1997). Nevertheless, each semiconductor manufacturer uses its own yield model, or a combination of yield models, in order to predict and optimize the yield for its product manufacturing. For example, Figure 2.9 shows three similar-sized wafers (300 mm) with different sizes of dies on each wafer (due to a smaller technology node or improved circuit design), and the yield of each wafer is calculated below Figure 2.9.

In all three examples given in Figure 2.9, the wafer sizes are (arbitrarily) 300 mm in diameter. Figures 2.9a, b and c have dies of sizes 10×10 mm, 20×20 mm and 40×40 mm, respectively. In all three examples, the gray dies are discarded dies, typically found around the edges of the wafer and not processed as full dies; therefore, these dies are not included in the yield criteria. The black dies in these examples are dies that have passed optical and electrical inspection and the white dies have failed optical or electrical inspection. The yield equation used in this example is simplified and only determines the true yield based on the total number of fully processed dies and the number of dies that passed optical and electrical inspection, such that

$$Y_{die} = \frac{Y_{pass}}{Y_{wafer}} \times 100 \tag{2.24}$$

where,

Y_{pass} is the number of dies that passed inspection
Y_{wafer} is the total wafer yield

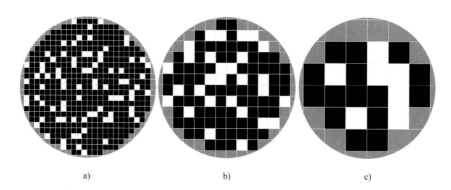

a)   b)   c)

FIGURE 2.9 Comparison of die yield on three identically sized wafers (300 mm), with different die sizes on each.

TABLE 2.6
Comparison of the Die Yield of Three 300 mm Wafers with Different Die Sizes on Each Wafer

Sample	Wafer Size	Die Size	Pass	Fail	Discard	Yield
(a)	300 mm	10×10 mm	388	84	90	82%
(b)	300 mm	20×20 mm	78	29	49	73%
(c)	300 mm	40×40 mm	15	7	25	68%

The wafer sizes and die sizes are given in the table.

To determine the overall yield, the number of dies that passed is divided by the sum of the dies that passed and failed, not considering the discarded dies. The results of the three examples in Figure 2.9 are given in Table 2.6.

From Table 2.6, it is evident that the number of pass and failed dies (and therefore the total number of usable dies on a wafer) increases significantly as the die size decreases. In Figure 2.9a there are a total of 472 usable (dies that are not on the edge of the wafer and automatically discarded) dies, in Figure 2.9b there are 107 usable dies and in Figure 2.9c there are only 22 usable dies. Also noticeable from Table 2.6 is the relatively low increase in discarded dies in each example. Figure 2.9a has 90 discarded dies, which is only 16% of all dies on the wafer (including the discarded/ unusable dies). In Figure 2.9b, there are 49 unusable dies (amounting to 31.4% of the total number of dies on the wafer) and in Figure 2.9c there are 25 unusable dies; 53% of the dies on the wafer are therefore not usable and must be discarded. From this information, it is evident that decreasing the die size, either by decreasing the technology node or through innovative electronic design techniques, dramatically increases wafer yield and reduces the ratio of dies that are automatically discarded, typically around the edges of the wafer.

INTEGRATED CIRCUITS: A COMPLEX BUSINESS

The manufacturing of integrated circuitry on silicon wafers is arguably one of the most complex processes in the manufacturing industry today. This complexity is attributed to the range of delicate and sensitive manufacturing steps, machinery and the controlled environment involved. The manufacturing of ICs involves four main processes, namely

- Wafer fabrication.
- Sorting.
- Assembly.
- Testing.

Among these four processes, the costliest and most time-consuming step is wafer fabrication. The machinery used in these manufacturing processes is expensive and can cost in excess of US $40 million (Mönch et al. 2013). Consequently, a modern 300 mm fabrication plant could typically cost in the order of US $7 billion. In fact, the GlobalFoundries fab in New York, known as Fab 8, allegedly cost more than US $10 billion to erect (Ibrahim et al. 2014).

CONCLUSION

Given the importance of technology in our daily lives and the significance of Moore's law in guiding the development of each technology node, it should come as no surprise that any obstacles to the continuation of Moore's law cause great concern to the electronics industry. Subsequent chapters will go into further detail regarding the hindrances to Moore's law and the strategies that the semiconductor industry has adopted to overcome these limitations and threats to derailing Moore's law.

As technological demands continue to push the boundaries of semiconductor manufacturability, constant research and development (R&D) in semiconductor manufacturing processes is required. Although higher component densities following the demands of advancing technologies have decreased the end-consumer prices, the cost of continual R&D in semiconductor manufacturing technology actually continues to rise. The trend of increasing development costs for semiconductor manufacturing has grown exponentially over time. This characteristic is sometimes referred to as Moore's Second Law, or Rock's Law (Mack 2011; Ross 2003; Rupp and Selberherr 2011). The two main factors that essentially determine fabrication costs are process complexity and capacity. Process complexity is mainly the outcome of R&D aimed at determining a trade-off between manufacturing process complexity and manufacturing cost. However, R&D also bears a significant price tag for semiconductor manufacturers in general. The R&D involved in semiconductor manufacturing technology typically addresses aspects such as wafer handling, production process flow and yield maximization. Owing to the extent of the complete manufacturing process, it could consist of up to 700 single processes spanning a production time of up to three months (Mönch et al. 2013). Although some semiconductor processes produce a near-100% yield, this is not the case for all manufacturing processes. Low production yield is largely caused by impurities encountered on the silicon wafers during manufacturing. This is mainly caused by one of two possible situations: uneven wafer manufacturing processes, or an unclean manufacturing environment (Doering and Nishi 2007).

In order to ensure high-quality and highly reliable end-products, extensive quality testing is required. If not properly managed and optimized, the cost of testing could match the actual manufacturing costs. This would mainly be due to the actual time taken for testing to be completed, since the final product's release in the industry will be delayed by the time it takes to complete the testing. Another factor involved here is equipment complexity. Since testing equipment would have to be highly automated and optimized, additional costs could be incurred through this test equipment (Balasinski 2012). Design and development costs are also important factors to consider when performing a cost analysis of an IC manufacturing process. The concept of design for manufacturability (DfM) is particularly important in the design and development of a new product, since a great deal of design, process and production costs could be saved by applying DfM principles from the very start of a new product's development and design (Balasinski 2012).

The economics of semiconductor scaling are therefore dependent on many external and internal factors, and the deciding factor in upgrading and improving a qualified process is not necessarily a straightforward decision. This chapter highlights these factors and provides a brief introduction to the most critical and typically the most expensive transactions involved in semiconductor advances.

REFERENCES

AnySilicon. (2016). Top semiconductor foundries capacity 2015–2016 (Infographic). Retrieved 6 December 2016 from http://www.anysilicon.com.

Babbs, D., Gaskins, R. (2008). Effect of reduced equipment downtime variability on cycle time in a conventional 300 mm fab. *2008 IEEE/SEMI Advanced Semiconductor Manufacturing Conference*, 237–242.

Balasinski, A. (2012). *Semiconductors – Integrated Circuit Design for Manufacturability*. Florida, USA: CRC Press.

Bret, T., Hofmann, T., Edinger, K. (2014). Industrial perspective on focused electron beam-induced processes. *Applied Physics*, 117(12), 1607–1614.

Brock, D. (2006). *Understanding Moore's Law: Four Decades of Innovation*. Philadelphia: Chemical Heritage Foundation, ISBN 978-094-1901413.

Chen, S. (2013). Bringing energy efficiency to the fab. Retrieved 16 December 2016 from http://www.mckinsey.com

Chien, C., Wang, J. K., Chang, T., Wu, W. (2007). Economic analysis of 450 mm wafer migration. *Semiconductor Manufacturing, 2007. ISSM 2007. International Symposium on*, 1–4.

De Vries, D. K. (2005). Investigation of gross die per wafer formulas. *IEEE Transactions on Semiconductor Manufacturing*, 18, 136–139.

DeHon, A. (2009). Sublithographic architecture: Shifting the responsibility for perfection. *Into the Nano Era*, 1st ed., H. Huff, Ed. Berlin Heidelberg: Springer .

Doering, R., Nishi, Y. (2007). *Handbook of Semiconductor Manufacturing Technology,* 2nd ed.Boca Raton: CRC Press, 2007.

Dross, F., Milhe, A., Robbelein, J., Gordon, I., Bouchard, P. O., Beaucarne, G., Poortmans, J. (2008). Stress-induced lift-off method for kerf-loss-free wafering of ultra-thin (~50 µm) crystalline Si wafers. *33rd IEEE Photovoltaic Specialists Conference PVSC '08*, 1–5.

Dunn, T., Lee, C., Tronolone, M., Shorey, A. (2012). Metrology for characterization of wafer thickness uniformity during 3DS-IC processing. *62nd Electronic Components and Technology Conference*, 1239–1244.

Edwards, C. (2012). 450 mm or bust? *Engineering & Technology,* 6(12), 76–79.

Ferris-Prabhu, A. V. (1989). An algebraic expression to count the number of chips on a wafer. *IEEE Circuits and Devices Magazine*, 5, 37–39.

Gartner. (2016). Worldwide semiconductor foundry market grew 4.4 percent in 2015, according to final results by Gartner. Retrieved 11 December 2016 from http://www.gartner.com

Hruska, J. (2012). Deliberate excellence: Why Intel leads the world in semiconductor manufacturing. Retrieved 9 December 2016 from http://www.extremetech.com

Hruska, J. (2013). Atom everywhere: Intel breaks ground on first 450 mm fab. Retrieved 9 December 2016 from http://www.extremetech.com

Huff, H. R., Gilmer, D. C. (2005). *High Dielectric Constant Materials: VLSI MOSFET Applications*. Berlin Heidelberg: Springer, ISBN 978-3-540-21081-8.

Hutcheson, G. (2009). The economic implications of Moore's law. *Into the Nano Era*, 1st ed., H. Huff, Ed. Berlin Heidelberg: Springer, ISBN 978-3-540-74558-7, 11–38.

Ibrahim, K., Chik, M. A., Hashim, U. (2014). Horrendous capacity cost of semiconductor wafer manufacturing. *2014 IEEE International Conference on Semiconductor Electronics (ICSE)*, 329–331.

IC Insights. (2016a). Research Bulletin: Number of 300 mm IC wafer fabs expected to reach 100 in 2016. Retrieved 6 December 2016 from http://www.icinsights.com

IC Insights. (2016b). Seven top-20 1Q16 semiconductor suppliers show double-digit declines. Retrieved 10 December 2016 from http://www.icinsights.com

ITRS. (2014). ITRS Reports - International Technology Roadmap for Semiconductors. Retrieved 20 March 2016 from http://www.itrs2.net/itrs-reports.html

Kashyap, K., Zheng, L. C., Lai, D. Y., Hou, M. T., Yeh, J. A. (2015). Rollable silicon IC wafers achieved by backside nanotexturing. *IEEE Electron Device Letters*, 36, 829–831.

Kawahira, H., Hayashi, N., Hamada, H. (1999). PMJ' 99 panel discussion review: OPC mask technology for KrF lithography. *Proceedings of SPIE*, 3873, 19th Annual Symposium on Photomask Technology, 318.

Krueger, D. C. (2011). Semiconductor yield modeling using generalized linear models. *A Dissertation Presented in Partial Fulfillment of the Requirements for the Degree Doctor of Philosophy*, University of Arizona, May 2011.

Lapedus, M. (2016). Will there be enough silicon wafers? Retrieved 13 December 2016 from http://www.semiengineering.com

Lin, H., Lin, J. C., Chiu, C. S., Wang, Y., Yen, A. (1999). Sub-0.18-μm line/space lithography using 248-nm scanners and assisting feature OPC masks. *Proceedings of SPIE*, 3873, 19th Annual Symposium on Photomask Technology, 307.

Mack, C. A. (2011). Fifty years of Moore's law. *IEEE Transactions on Semiconductor Manufacturing*, 24(2), 202–207.

Mack, C, A. (2015). The multiple lives of Moore's law. *IEEE Spectrum*, 52(5), 67–67, April 2015.

Mesawich, M., Sevegney, M., Gotlinsky, B., Reyes, S., Abbott, P., Marzani, J., Rivera, M. (2009). Microbridge and e-test opens defectivity reduction via improved filtration of photolithography fluids. *Proceedings of SPIE*, 7273, Advances in Resist Materials and Processing Technology, XXVI, 727300.

Mönch, L., Fowler, J. W., Mason, S. J. (2013). *Production Planning and Control for Semiconductor Wafer Fabrication Facilities: Modeling, Analysis and Systems*. New York: Springer.

Montgomery, S. R. (2000). Higher profits from intelligent semiconductor-equipment maintenance. *Future Fab International*, 8, 63–67.

Moore, G. E. (1965). Cramming more components onto integrated circuits. *Electronics*, 38, 114–117.

Moore, G. E. (1995). Lithography and the future of Moore's law. *Proceedings of SPIE*, 2438(2), Optical/Laser Microlithography VIII, Santa Clara, CA.

Morita, H. (2016). Lithography process simulation studies using coarse-grained polymer models. *Polymer Journal*, 48, 45–50.

Öchsner, R. (2014). Energy efficiency in semiconductor manufacturing – tool and fab aspects. *SEMICON Europa 2014*.

Peters, L. (1997). *Cost Effective IC Manufacturing 1998–1999*. Scottsdale, AZ: Integrated Circuit Engineering Corporation.

Pettinato, J. S., Pillai, D. (2005). Technology decisions to minimize 450-mm wafer size transition risk. *IEEE Transactions on Semiconductor Manufacturing*, 18(4), 501–509.

Qi, C., Tang, T., Sivakumar, A. I. (2002). Simulation based cause and effect analysis of cycle time and WIP in semiconductor wafer fabrication. *Proceedings of the 2002 Winter Simulation Conference*, 2, 1423–1430.

Rose, O. (2004). Modeling tool failures in semiconductor fab simulation. *Proceedings of the 2004 Winter Simulation Conference*, 2, 1910–1914.

Ross, P. E. (2003). Five Commandments [technology laws and rules of thumb]. *IEEE Spectrum*, 40, 30–35.

Rupp, K., Selberherr, S. (2011). The economic limit to Moore's Law. *IEEE Transactions on Semiconductor Manufacturing*, 24, 1–4.

Sullivan, R. F. (2007). The impact of Moore's law on the total cost of computing and how inefficiencies in the data center increase these costs. *ASHRAE Transactions*, 113, 457–461.

Tolia, A. (2015). Linde Electronics: Addressing challenges in gas supply to fabs. *Silicon Semiconductor*, 37(III).

Touma, W. R. (1993). *The Dynamics of the Computer Industry: Modeling the Supply of Workstations and their Components*. The Netherlands: Springer Science & Business Media.

Tummala, R. (2001). *Fundamentals of Microsystems Packaging*. New York: McGraw-Hill.

Vardi, M. Y. (2014). Moore's law and the sand-heap paradox. *Communication ACM*, 57(5).

Watanabe, M., Kramer, S. (2006). 450 mm silicon: An opportunity and wafer scaling. *The Electrochemical Society, Interface*, Winter 2006.

Williams, E. D., Ayres, R. U., Heller, M. (2002). The 1.7 kilogram microchip: Energy and material use in the production of semiconductor devices. *Environmental Science and Technology*, 36(12/01), 5504–5510.

Yoshikawa, S., Fujii, N., Kanno, K., Imai, H., Hayano, K., Miyashita, H., Shida, S., Murakawa, T., Kuribara, M., Matsumoto, J., Nakamura, T., Matsushita, S., Hara, D., Pang, L. (2015). Study of defect verification based on lithography simulation with a SEM system. *Proceedings of SPIE*, 9658, Photomask Japan 2015: Photomask and Next-Generation Lithography Mask Technology XXII, 96580V.

3 The Importance of Photolithography for Moore's Law

Wynand Lambrechts, Saurabh Sinha and Jassem Abdallah

INTRODUCTION

Moore's law remains a key driver of advances in lithography to reduce the sizes of semiconductor features. Photolithography (photo-litho-graphy, which is translated from Latin as light-stone-writing) is the technology of patterning and enables complex integrated circuits to be manufactured on wafers. The wavelength of light determines the feature size limitation of the process and with a decrease in feature size, a smaller wavelength of light is also required to pattern these geometries at high resolution. In the early 2000s, optical lithography at 180 nm using step-and-repeat technology (steppers are discussed in subsequent sections of this chapter) with illumination at a wavelength of 248 nm had already become routine (Schellenberg 1999). Recently developed 45 nm and 32 nm technology nodes use optical lithography at typically 193 nm. These radiation wavelengths are rapidly becoming comparable to feature size and this poses new challenges for semiconductor manufacturers. The photolithographic process step(s) in semiconductor manufacturing, which is repeated multiple times for each wafer, can roughly be categorized into ten steps. Each step is crucial to ensure the accuracy and compatibility of the subsequent step and with decreasing feature size, acceptable tolerances are becoming much smaller. The ten steps are listed, in order, as

- Surface preparation, typically through surface etching to remove unwanted organic oxides and defects from the surface of the wafer before processing further.
- Coating the surface of the wafer with photoresist (spin casting).
- Pre-baking or soft-baking the wafer to remove moisture from the photoresist.
- Alignment of the photomask.
- UV exposure from a light source with specific wavelength.
- Development of the photoresist.

- Post-baking or hard-baking the wafer.
- Processing of the wafer using the photoresist as a masking film.
- Etching away the top layer on the wafer.
- Post-processing, which includes cleaning and inspection of the wafer.

In semiconductor manufacturing, photolithography plays a prominent role in the scaling of feature size and is arguably the most important process step in determining the future longevity of Moore's law. Reducing feature sizes entails shorter wavelength radiation. A shorter wavelength translates to a larger depth of focus, discussed in further detail in this chapter, as well as a reduction in the critical dimension (CD) of geometries. The progress of shortening the wavelength of light for these processes is relatively slow and arguably threatens the perpetuation of Moore's law. In this chapter, the terms photolithography, optical lithography and lithography are used interchangeably. All three terms refer to the process of transferring geometries from a mask to a semiconductor wafer by UV exposure on light-sensitive resist (photoresist); this is also described in further detail in this chapter.

Over the last four decades, lithography wavelengths have decreased from the blue spectral region at 436 nm, 405 nm and near-UV light at 365 nm produced by vaporized mercury (Hg) arc gas discharge lamps to deep-UV (248 nm) generated by an excited dimer (excimer) krypton fluoride (KrF) laser, also referred to as an exciplex laser. Argon fluoride (ArF) light sources emitting light at a wavelength of 193 nm are typically used for modern 45 nm and 32 nm IC technologies. As feature sizes and feature pitches continue to shrink below 20 nm and 40 nm, respectively, financial pressures and technical challenges force FABs to move away from multiple patterning photolithography using 193 nm wavelengths towards EUV. Developing light sources within this low-wavelength spectrum that are powerful, robust, and stable for 24 hours and 365 days of operation, each year, is a monumental challenge and its progress has been hindered by high cost and reliability issues.

Photolithography exposure sources in semiconductor manufacturing are essentially divided into three categories according to the mechanism responsible for inducing changes in the chemical properties of the light-sensitive material (photoresists). The three main mechanisms that define each category are

- *Photons*: Generated by sources such as white light, mercury-arc gas lamps, excimer lasers and X-rays.
- *Electrons*: Direct-write techniques generated through focused electron-beams.
- *Charged ions*: Also a direct-write technique through focused ions.

Photon generation, under which mercury-arc gas lamps are categorized, is most commonly used in the semiconductor industry; it has been replaced by excimer lasers for semiconductor geometries below the 180 nm node. Optical elements in a photolithographic system that influence its performance and limitations include practices of phase-shifting masks and the chemical, optical and quality attributes of photoresist, as well the characteristics of the optics – all contributing to the complexity and cost of developing low-wavelength and high-power light sources. Kapoor and Adner (2007) present a summary of semiconductor lithography technologies in accordance with the year they were first adopted as industry standards and the initial feature size resolution that could be achieved with each technology. The summary presented by Kapoor and Adner (2007) is adapted and presented in Table 3.1.

The first photolithography techniques, during the 1960s, involved contact lithography, a technique where the photomask makes hard contact with the photoresist on the wafer. The initial resolution that was obtainable using this technique was 7 μm (7000 nm), although improvements over the years have decreased this significantly. (This is generally true for most of the techniques listed in Table 3.1; therefore, only the initial resolution obtainable is recorded in this table.) Contact photolithography, along with its limitations (discussed in this chapter), was improved upon through proximity lithography. Resolutions of 3 μm (3000 nm) were recorded with the first uses of this technique by the early 1970s. Proximity lithography improves on contact lithography by reducing defects introduced on photomasks during hard contact. Further improvements, albeit at the cost of complex optics, are achieved by projection lithography that converges de-magnified light onto the wafer, a process initiated shortly after proximity lithography became industrialized. Improvements in various disciplines have ensured that photolithography enhancements remain constant, with recent 12 nm FinFET-based processors introduced in commercially available components. The near future of Moore's law seems to be secured by developments that

TABLE 3.1

The History of Semiconductor Lithography Technologies

Lithography Technology Generation	Initial Resolution (nm)	First Recorded Industry Use
Contact	7000	1962
Proximity	3000	1972
Projection	2000	1973
E-beam	500	1976
X-ray	300	1978
G-line	1250	1978
I-line	800	1985
Deep-UV (248 nm)	450	1986
Deep-UV (193 nm)	150	1996
Deep-UV (157 nm)	100	1998
Deep-UV (193 nm)	65	2007
Deep-UV immersion (193 nm)	45	2009
Deep-UV immersion (193 nm)	32	2010
Deep-UV immersion (193 nm)[a]	10	2016/2017
Extreme UV (13.5 nm)	7 (est.)	2018 (est.)
Extreme UV (13.5 nm)	5 (est.)	2020 (est.)

Source: Adapted from Kapoor, R., Adner, R., Technology interdependence and the evolution of semiconductor lithography, *Solid State Technology*, 2007.

[a] Modern deep-UV immersion lithography techniques adopt several additional techniques such as multiple-patterning and phase-shifting masks to achieve smaller feature size with 193 nm lithography.

enable 5 nm nodes through EUV, which may possibly be used for high-volume IC manufacturing as early as 2018 (Bourzac 2016). This key tool holds the promise of offsetting the deceleration of Moore's law and has led to billions of US dollars being invested by industry leaders such as Samsung and Intel. Taiwanese chip-making company TSMC predicts that it will start using EUV as an industry standard by 2020. Current 193 nm wavelength photolithography has been pushed to its limits by using multiple patterning steps for each layer, which adds complexity, time, cost and reliability issues, and by immersion lithography. Multiple patterning lithography splits each photomask into multiple, separate masks representing intermediate steps in the photolithography patterning process. Certain multiple patterning process technologies can introduce mask shift and potential misalignment, which can cause spacing variation in both the x- and y-directions that leads to variations in process batches. The move to EUV would bring relief and again decrease the number of process steps for semiconductor manufacturing. ASML, a company in which Intel invested US $4 billion in 2011 (Bourzac 2016), recently announced that it had overcome the largest challenge of EUV wavelength light sources – the lack of intensity (lumens) of light sources. Low-intensity light increases the time required to expose photoresist, which lowers the throughput of wafers, a limit that significantly decreases the profitability of manufacturers. ASML invested research and development into better understanding the materials involved in EUV lithography, and in incorporating advances in plasma and laser physics to increase the brightness of these light sources. According to Bourzac (2016), the power of light sources increased from 40 watts in 2015 to 200 watts in 2016, increasing the manufacturing capabilities from 400 wafers per day to 800 wafers per day. The industry status quo is approximately 3000 wafers per day, although this number will inevitably decrease with more patterning steps required to process higher-complexity wafers.

In 1968, when Intel was founded, its manufacturing facility was set up and a piece of equipment cost approximately US $12,000 (Moore 1995). Currently, a photolithography stepper capable of feature resolutions below 16 nm with numerical aperture of 0.33 costs in the range of US $90 million. The most significant enhancements to photolithography are discussed in the following chapter, whereas this chapter focuses on the enabling principles of semiconductor photolithography, reviewing the technologies of lithography advances since the 1960s.

PHOTOLITHOGRAPHY LIGHT SOURCES

Photolithography requires high resolution, high photoresist sensitivity, alignment with high precision and accuracy, exact process parameter control and low defect density. The light source performance of photolithography equipment requires dose, energy, wavelength, bandwidth and beam stability. The global photolithography equipment market, categorized by light source, is divided into

- Mercury lamps.
- Excimer lasers.
- Fluorine lasers.
- Laser plasmas.

This section reviews the photolithography market as described by the various light sources used to pattern semiconductor geometries on wafers.

MERCURY LAMPS

A mercury-arc gas discharge lamp generates light by channeling an electric arc through vaporized Hg. The Hg in the tube of the lamp must first be vaporized and ionized in order for it to conduct electricity and for the arc to start. This is achieved by a starter mechanism, similar to a balun used to inject high power over a short period into fluorescent lights. The tube of the lamp is filled with an inert starter gas such as argon (Ar) or xenon (Xe), which is ionized when a high enough voltage is supplied by the starter, leading to the Hg being heated and ionized and an arc being generated in the Hg between the two primary electrodes connected to the power supply of the lamp. The wavelengths emitted from the light range are between approximately 184 nm and 578 nm, depending on the pressure in the tube. The typical spectrum of a mercury-arc lamp is dominated by discrete lines that span the visible spectrum (violet, blue, green and yellow/orange), through the near-UV region and into the shortwave-UV region. Table 3.2 lists the wavelengths produced by Hg lamps under varied pressure.

In Table 3.2, the wavelengths of mercury-arc gas lamps under varied pressure of the inert starting gas are shown; these vary from approximately 184 nm to 578 nm. The energy E of each light wave is also given in Table 3.2, specified in electron-volt (eV) and determined by

$$E = \frac{hc}{\lambda} \tag{3.1}$$

where,

h is Planck's constant (6.626×10^{-34} m^{2}kg/s)
c is the speed of light (299,792,458 m/s)
λ is the wavelength of the light wave in meters

In previous-generation semiconductor photolithography, light-sensitive photoresist was primarily exposed through masks using 436, 405 and 365 nm wavelengths.

TABLE 3.2
Typical Wavelengths Produced by Mercury-Arc Gas Lamps under Varying Ar Pressure

Excimer Formula	Color	Emitted Wavelength (nm)	Energy (eV)
Deep-UV	Ultraviolet C	184	6.74
Deep-UV	Ultraviolet C	254	4.88
I-line	Ultraviolet A	365	3.40
H-line	Violet	405	3.06
G-line	Blue	436	2.84
E-line	Green	546	2.27
(Krypton)	Yellow-orange	578	2.15

EXCIMER LASERS

The most commonly used wavelength in modern semiconductor lithography is at 193 nm, generated primarily by excimer lasers and suitable for exposing geometries smaller than 100 nm. For geometries of 180 nm, an excimer laser with a wavelength generated at 248 nm is typically used. Excimer lasers contain a mixture of noble gases such as Ar, krypton (Kr) or Xe and a halogen gas such as fluorine (F) or chlorine (Cl) to produce UV light at wavelengths ranging from 157 nm to 351 nm. Table 3.3 lists the wavelengths produced by various excimer gases.

The noble gases used in the excimer lasers listed in Table 3.3 are used as the laser gain medium. The gain medium is pumped with pulses of current of typically nano-second duration in a high-voltage electric discharge circuit (electron beam), which generates molecules in the excited electronic state, representing the bound state of their elements. If the gain medium is stimulated and spontaneous emission occurs, the excimer rapidly dissociates to avoid reabsorption of the generated radiation. Excimer lasers also fall under the category of molecular lasers, since the gain medium consists of ionized molecules. The short wavelengths of excimer lasers in the UV spectrum are ideal for photolithography applications in semiconductor manufacturing. Additional applications of excimer lasers include material processing with laser ablation, laser marking of glass and plastic, manufacturing of fiber Bragg gratings, eye surgery, psoriasis treatment and pumping of certain dye lasers. Doolittle (2008) presents a table that summarizes optical lithography system parameters based on the years of use, the achieved numerical aperture (NA), k_1, wavelength, CD and depth of focus (DOF). The table presented in Doolittle (2008) is adapted and presented here as Table 3.4.

As shown in Table 3.4, optical lithography parameters improved significantly between 1990 and 2008, with further improvements up to 2017. The NA of optical lithography, for example, increased from 0.5 in 1990 to approximately 0.8 by 2008; developments up to 2015 led to apertures of 0.25 and lower. The k_1 parameter was reduced from 0.7 in 1990 to 0.35 by 2008 and wavelengths decreased from the 365 nm range in 1990 to 193 nm and 157 nm systems by 2008. Critical dimensions have

---

TABLE 3.3

Typical Types of Excimer Lasers Consisting of Noble Gases such as Ar, Kr and Xe Combined with Halogen Gases such as F and Cl; the Resulting Emitted Light Wavelengths Are Listed

Excimer Formula	Excimer Name	Emitted Wavelength (nm)/ Suitable Geometry (nm)	Energy (eV)
F_2	Fluorine	157	7.90
ArF	Argon fluoride	193/<100	6.42
KrF	Krypton fluoride	248/180	5.00
XeBr	Xenon bromide	282	4.40
XeCl	Xenon chloride	308	4.03
XeF	Xenon fluoride	351	3.53

TABLE 3.4

Optical Lithography System Parameters between 1990 and 2008

Year Parameter	1990	1995	1999	2002	2005	2008
NA	0.5	0.6	0.7	0.7	0.75	0.8
k_1	0.7	0.6	0.5	0.45	0.4	0.35
λ (nm)	365	248	248	248/193	248/193	193/157
CD (nm)	500	250	180	150/130	130/100	80/70
DOF (μm)	1.5	1.0	0.6	0.5	0.4	0.3

Source: Adapted from Doolittle, A., 2008, Lithography and pattern transfer. *Georgia Tech – ECE 6450*, Retrieved 21 January 2017.

since decreased dramatically, from 500 nm in 1990 to 70 nm in 2008, with current production of 12 nm FinFET transistors planned for high-volume production. The DOF is another parameter that has improved, decreasing from 1.5 μm in 1990 to 300 nm by 2008. Optical lithography techniques to improve minimum feature size in recent years, such as the use of laser-pulsed plasmas with wavelengths of 13.5 nm, further ensure that Moore's law of reducing feature size remains achievable.

LASER-PULSED PLASMA

Laser-pulsed plasmas producing 13.5 nm EUV light sources are under development and are planned for the high-volume production of semiconductor circuits by approximately 2020. EUV essentially enables the processing of nodes of 5 nm and below; however, significant improvements to various parameters listed in Table 3.1 are required to achieve this. EUV lithography development is discussed later in this chapter.

Photolithography light sources used in semiconductor manufacturing challenge Moore's law and decrease the wavelength of operation, while maintaining high output power from these sources is challenging. Technological advances in various disciplines are ensuring constant improvement in optical lithography. The following sections briefly review the most essential attributes and contributors to the photolithographic process in semiconductor manufacturing. These process step reviews aim to provide sufficient background on optical lithography, followed by the techniques available and currently being developed to improve photolithography and further scale down on feature size, as Moore's law predicts.

SIMPLIFIED PROCESS FLOW FOR PATTERNING SEMICONDUCTOR LEVELS

Semiconductor component processing consists of multiple levels of circuitry integrated with one another along both the lateral and vertical planes (Tummala 2001). Individual lateral planes are referred to as design levels, and each level is patterned separately via the combination of photolithography, metal deposition and wet

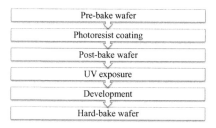

FIGURE 3.1   A process flow diagram depicting the typical processing sequence for the patterning of an IC using photolithography. The process is followed by an etching step to permanently transfer patterns to the wafer.

chemical or dry plasma etching, as shown in Figure 3.1 (Cui 2008; Guy and Burwell 2013; Huff and Gilmer 2005; Mack 2011,2015; Tummala 2001).

Prior to photolithography, wafers are coated with photosensitive films called photoresists. Upon exposure to UV light of appropriate wavelengths, the materials that make up these films undergo photochemical changes that result in a solubility switch (conversion of soluble materials into insoluble materials or vice versa, depending on the type of resist being irradiated) (Bartczak and Galeski 2014; Bulgakova et al. 2014; Guy and Burwell 2013; Ito 2005). Conventional photolithography projection tools use mirrors, windows, prisms and lenses to direct and focus beams of light through photomasks onto semiconductor wafers (Bartczack and Galeski 2014; Bulgakova et al. 2014; Cui 2008; Guy and Burwell 2013; Indykiewicz et al. 2012; Ito 2005; Nouralishahi et al. 2008; Rudnitsky and Serdyuk 2012). The opaque regions of the photomasks block photons from reaching the areas of photoresist directly below them, while the transparent areas allow photons to reach these regions (Ciu 2008; Guy and Burwell 2013). These transparent and opaque regions are composed of lateral patterns corresponding to the IC design features of the specific level being processed, but at a larger geometric scaling than what will be produced on the wafer. Each physical dimension on a photomask is typically about five times larger than the corresponding dimension that will be imaged on a wafer; hence, the total footprint of the aerial image is scaled down by a ratio of about 25:1 from the mask to the wafer (Ciu 2008; Guy and Burwell 2013). As shown in Figure 3.2, the geometric scaling of the aerial image that is transmitted through the mask is achieved with an objective lens, which focuses the beams of photons as they travel, thereby shrinking the aerial image down to the size of a chip as it arrives at wafer level (Ciu 2008; Guy and Burwell 2013).

Since a projection scanner only irradiates a small region on the wafer at any given time, multiple exposures are required in order to pattern the photoresist across an entire wafer surface (Cui 2008; Tummala 2001). Full wafer exposures are accomplished by stitching up individual chip exposures, for example if the wafer is laterally offset by a chip length between individual chip exposures. This chip-to-chip sequence of projection exposure and lateral offset is commonly referred to as step and repeat imaging or step and scan imaging (Cui 2008; Tummala 2001), hence the name stepper given to the photolithography equipment capable of these imaging techniques.

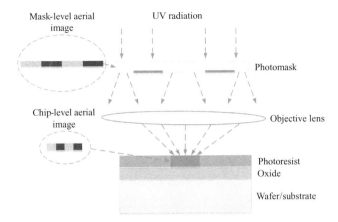

FIGURE 3.2   A figure showing the focusing principle by which a projection photolithography scanner works. The dark gray and light gray areas of each aerial image represent the dark and irradiated regions, respectively.

The following section provides a more in-depth review of the basic principles of lithographic imaging systems, placing the focus on the optical principles used to focus light onto a designated target (the photoresist-covered wafer).

BASIC PRINCIPLES OF LITHOGRAPHIC IMAGING SYSTEMS

An important component of any photolithographic imaging system used in semiconductor processing (and in various other fields where optics are used) is the optical elements such as lenses, mirrors, windows, prisms and filters (Willers 2013). Optical rays of flux can be converged or diverged using curved optical elements with refractive or reflection properties. The light is converged or diverged onto an active sensor such as a photodiode, or onto a passive layer of light-sensitive photoresist that changes its chemical properties upon illumination. The object and image relationships can be described by two primary systems: a thin lens and a thick lens, each with distinct characteristics, advantages and disadvantages. The imaging systems for thin and thick lenses are given in Figures 3.3 and 3.4, respectively.

Figure 3.3 represents an optical imaging system with its applicable parameters and characteristics for a system implementing a thin lens. A similar system, with the difference being that the lens is thick, is shown in Figure 3.4.

The optical imaging systems in Figures 3.3 and 3.4 show how an image of an object is formed in a plane, which is called the focal plane. The geometric parameters presented in Figures 3.3 and 3.4 are listed in Table 3.5.

The parameters in Figures 3.3 and 3.4, which are listed in Table 3.5, are related within a Cartesian plane and distances toward the right-hand side of each figure are represented as positive quantities, whereas distances toward the left-hand side are represented as negative quantities. The optical axis is the centerline of the optical imaging system and is normally defined by a zero field angle. For thick lenses, as shown in Figure 3.4, the system additionally has two principal planes, or nodal

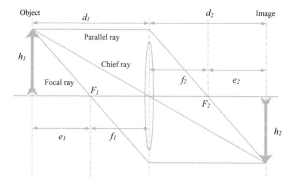

FIGURE 3.3 An optical imaging system showing the relationship between the object and the image, comprising a single thin lens.

points, as well as two focal planes, also referred to as focal points. If a unity index of refraction in both the object and image planes is assumed, the object and image properties in the imaging system are related such that

$$\frac{1}{d_2} = \frac{1}{d_1} + \frac{1}{f_2} \tag{3.2}$$

$$e_1 e_2 = -f_1^2 \tag{3.3}$$

and

$$m = \frac{h_2}{h_1} = \frac{d_2}{d_1} \tag{3.4}$$

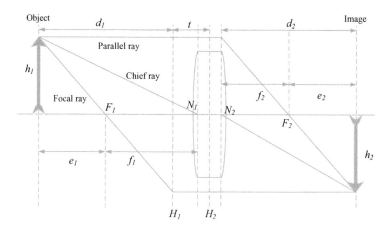

FIGURE 3.4 An optical imaging system showing the relationship between the object and the image, comprising a single thick lens.

TABLE 3.5

Optical Imaging System Parameters of Figures 3.3 and 3.4

Parameter	Description	Parameter	Description
d_1	Object distance	h_1, h_2	Object/image heights
d_2	Image distance	F_1, F_2	Focal points
f_1, f_2	Focal lengths	N_1, N_2	Nodal points
e_1, e_2	Extrafocal lengths	H_1, H_2	Principal points

where m is the magnification of the image with respect to the object, with a negative value for m representing an inverted image and a value for m smaller than 1 representing a scaled-down image. Conventional photolithography makes use of photons of light corresponding to wavelengths in either the UV or the deep-UV (DUV) regions of the electromagnetic spectrum (Cui 2008; Guy and Burwell 2013; Tummala 2001). Objective lenses are used to focus the light such that the aerial image that arrives at the wafer surface is a scaled-down depiction of the physical patterns that exist on the photomasks above it (Ciu 2008). A simplified representation of a semiconductor photolithographic optics configuration is shown in Figure 3.5.

From Figure 3.5, it is seen that illuminated light from a source is passed through a mechanical cover (mask) that comprises geometric structures. The light is intensified

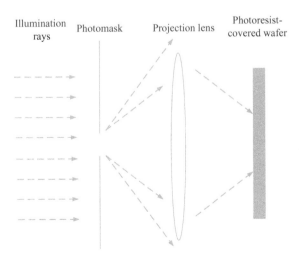

FIGURE 3.5 The basic configuration of a semiconductor photolithographic optics system where illumination at a specific wavelength is transferred through a lens system to an image plane, consisting of light-sensitive photoresist.

and focused using a lens system onto a wafer covered with light-sensitive photoresist. The fundamental principles to consider in lithographic optics include

- Diffraction.
- Partial coherence.
- DOF.
- Reflection and interference.
- Polarization dependence.

Quartz is highly transparent to both UV and DUV wavelengths and is commonly used to allow photons to pass through the light areas (spaces) of the photomasks, while a chrome layer, which completely reflects the photons, is used in the dark areas (lines) of the aerial image (Tummala 2001). As photons pass through the transparent regions of the photomask, the electromagnetic waves are spread out/diffracted, leading to interference with adjacent waves from other slits and blurring of the aerial image (Rudnitsky and Serdyuk 2012). Since the final aerial image arriving at the wafer is the superimposed product of the interfering waves, the optical resolution of the imaging system is ultimately limited by optical diffraction. As shown in Figure 3.6, the degree of interference is dependent on the lateral spacing (D) between openings in the mask; the further apart the slits are, the lower the degree of interference and, thus, the more resolved the aerial image is (Ciu 2008).

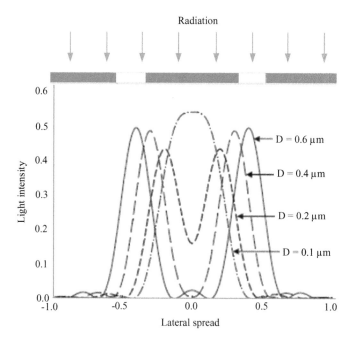

FIGURE 3.6 This image shows the resultant light intensity profile of the interfering beams of light that have been diffracted by two adjacent slits. The parameter D represents the lateral distance separating the slits in micrometers. (Adapted from Cui, Z. (2008). Nanofabrication by photons. *Nanofabrication: Principles, Capabilities and Limits*. Boston, MA: Springer US, 7–76.)

Short-exposure wavelengths in semiconductor photolithography can also lead to interfering standing waves. These waves are typically noticeable in a layer of the photoresist. Areas where positive/constructive interference occurs generate amplified exposure regions, leading to anomalies and inconsistent development of the photoresist. Standing waves generated in layers of the photoresist can be eliminated by using multiple wavelength sources or by post-baking the samples after exposure. Standing waves in the photoresist can also be enhanced by reflective water or aligner chuck reflections that lead to further miscalculations due to the increased presence of standing waves, especially along the sides of the applied photoresist. Standing waves are specifically prominent if the wafer is transparent at the exposure wavelength and can be reduced by using

- A flat-back alignment chuck manufactured by anodized aluminum.
- Anti-reflection coating.
- An optical absorber beneath the wafer.
- A transparent glass alignment chuck.

The effect of standing waves on light-sensitive photoresist is graphically depicted in Figure 3.7.

As shown in Figure 3.7, significant variations in photoresist development rate occur owing to variations in light intensity on layers of the photoresist, because of

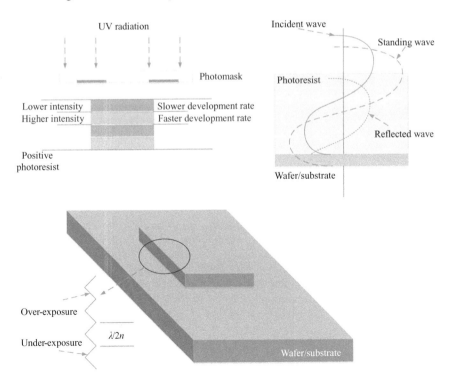

FIGURE 3.7 Graphical representation of the effect of standing waves on photoresist.

standing waves. The incident wave, a specific wavelength, reaches the wafer surface (covered in photoresist) and is reflected with a phase shift. The resulting exposure profile is also shown in Figure 3.7, where a saw-tooth anomaly exists within the sub-micron layers of the photoresist.

OPTICAL LITHOGRAPHY RESOLUTION

Using the laws of optics, including the law of refraction, also known as Snell's law, the minimum feature size due to optical limitations and the closest feature-to-feature lateral spacing (pitch) are determined by Ciu (2008) and Rudnitsky and Serdyuk (2012). The resolution of an optical imaging system is the smallest feature that can be printed and is of high quality and reproducible, given the process and equipment. In practical circumstances, process variations also limit the process resolution since decreased feature size has inherently less process latitude. The point spread function of a lens offers an acceptable figure of merit for printing single (very small) contact holes, whereas for dense lines and spaces, the smallest dimension of the pitch is limited by the number of diffraction orders that can pass through a lens. Dense lines and spaces are therefore essentially limited by the ratio of the wavelength and the NA of the lens, therefore λ/NA. The Rayleigh resolution criteria of an optical system can be derived with the imaging of equivalent lines and spaces for the pitch (p) to create a diffraction configuration with separate diffraction orders at spatial frequencies arranged in multiples of $1/p$. The largest spatial frequency able to pass through the lens is defined by NA/λ. Two diffraction orders are required in order to generate an image, assuming coherent illumination (a linear complex-amplitude imaging operation, usually illuminated by laser with complex valued point spread function). Therefore, the zero order and the two first orders should pass through the lens and the smallest pitch (p_{min}) that allows these criteria places the first diffraction order at the verge of the lens, such that

$$\frac{1}{p_{min}} = \frac{NA}{\lambda} \tag{3.5}$$

where λ is the wavelength of the incident light, and for equidistant lines and spaces. The resolution of the optical system is determined as half of the relationship, such that

$$R = \frac{p_{min}}{2} = \frac{\lambda}{2NA} \tag{3.6}$$

where R represents the resolution of the optical system. To account for variations in the assumptions that the lines and spaces are placed at equal distances apart and that the illumination is coherent, a proportionality factor is introduced in (3.6), such that

$$R = k_1 \left(\frac{\lambda}{NA} \right) \tag{3.7}$$

where k_1 represents this proportionality factor, whose value is dependent on the properties of the patterning process. These properties include parameters of the optics, photoresist and process latitude (linewidth in μm/nm versus exposure in mJ/cm^2, typically presented as Bossung curves) and $k_1 < 1$.

Photolithography employs standard lens practices to focus light from the source onto the intended target, typically a thin layer of photoresist on a wafer. The lens aperture can be derived from Figure 3.8, a simplified representation of a lens system.

The NA of the imaging tool is dependent on the focusing power of its objective lens, and is given by

$$NA = n \sin \theta \tag{3.8}$$

where n is the refractive index of the medium in which the lens is placed (typically 1, but this may differ when using immersion fluid lithography as discussed in this chapter). The amount of optical flux flowing through an optical system of a given focal length is another important characteristic in semiconductor lithography, defined as the f-number of the imaging system. The f-number is essentially a dimensionless geometric construct that comprises the diameter of the exit pupil and the focal length.

OPTICAL LITHOGRAPHY F-NUMBER AND NUMERICAL APERTURE

According to Figure 3.8, the measure of the light-gathering ability of a lens, and therefore its f-number or focal ratio (also referred to as the f-ratio, f-stop or relative aperture), is the focal length divided by the diameter of the lens, given as

$$f/\# = \frac{f}{D} \tag{3.9}$$

where,

$f/\#$	(or F#) is the f-number
f	is the focal length
D	is the diameter of the lens

High f-numbers result in lower irradiance, and vice versa, on the focal plane. Lower f-number lenses are therefore more complex, lead to higher cost and have

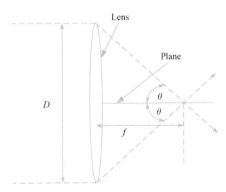

FIGURE 3.8 Simplified representation of an optical lens system used to determine the f-number of the lens.

superior performance to high f-number lenses. The f-number is only defined for systems at infinite conjugates, and for systems that have finite conjugates, the NA is used as a measure of the amount of flux flowing through the system at a given focal length (Willers 2013). The NA can be related to the f-number of a lens by

$$NA = \frac{D/2}{\sqrt{D^2/4 + f^2}}$$ (3.10)

which can be approximated to

$$NA \approx \frac{nD}{2f} = \frac{1}{2\left(\dfrac{f}{\#}\right)} = n\sin\theta$$ (3.11)

where,

    θ is the maximum angle of incidence of the light reaching the wafer

    n represents the refractive index of the medium amid the lens and the surface of the wafer

For dry lithography $n = 1$ (n of air) and for wet or liquid immersion lithography $n > 1$ (n of the immersion liquid – typically purified water). Liquid immersion lithography tools hence have higher NA values, which results in smaller CD and therefore smaller feature sizes. Historically, the semiconductor industry has improved the resolution capability of photolithography by improving the focusing power of the lenses as well as reducing the wavelengths used to expose photoresists (Ciu 2008; Guy and Burwell 2013; Tummala 2001). Another figure of merit of an optical lithography system is its DOF.

OPTICAL LITHOGRAPHY DEPTH OF FOCUS

The DOF is inversely proportional to the NA of the system; therefore, it decreases as the NA increases. The mathematical derivation of DOF is given by

$$DOF = k_2 \frac{\lambda}{(NA)^2}$$ (3.12)

where k_2 is a scaling coefficient that typically ranges between $0.5 < k_2 < 1$ and is process-dependent, and therefore does not have a well-defined value. Essentially, the DOF can be described by two scenarios: a small lens with small NA and a large lens with large NA, as presented in Figure 3.9.

    As feature size decreases, its sensitivity to focus errors increases dramatically (Mack 1995). This loss in focus is arguably one of the main contributors to the limitations of the use of optical photolithography to decrease the feature size of semiconductor technologies. The DOF is essentially the array of focus errors that a process can allow while still giving satisfactory lithographic results (Mack 1995). An error

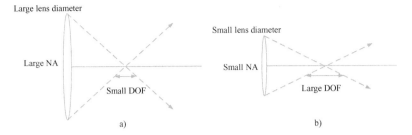

FIGURE 3.9 Representative scenarios to describe the DOF parameter. A lens with large diameter is presented in (a), which translates to a large NA but a small DOF. In (b), the small lens diameter translates to a larger DOF but a smaller NA.

or shift produced because of a shift in focus has two main effects on the geometries transferred to the wafer. These effects are

- A change in the profile of the photoresist, which includes variation in the linewidth (CD), variations in the sidewall angle and changes in the thickness of the photoresist at random locations (typically described as photoresist loss).
- Increased susceptibility to the sensitivity of the process to other processing errors such as exposure dose and development time.

An error in the curvature of the actual incident light wave compared with the desired incident light wave occurs if the light is defocused on the surface of the wafer (covered in photoresist). The optical path difference (OPD) is defined as the distance between the desired and the defocused light wave, represented by δ. A scenario that depicts the OPD in an optical imaging system is given in Figure 3.10.

From Figure 3.10, the OPD of an optical imaging system can be derived as

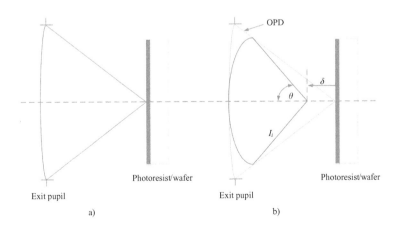

FIGURE 3.10 Scenario indicating the OPD of an optical imaging system, where (a) shows the incident light being in focus on the surface of the wafer (covered in photoresist) and (b) shows the light wave being out of focus by a distance δ.

$$OPD = n\delta(1 - \cos\theta) \qquad (3.13)$$

where θ is the angle of the light propagating from the exit pupil. For small lens NA (therefore small lens angles), the OPD of the imaging system is given by

$$OPD = \frac{\delta(\sin\theta)^2}{2} \qquad (3.14)$$

which is derived by ignoring higher-order Taylor series coefficients, as described in further detail by Mack (2007). The defocusing of the light disturbs the diffraction array and therefore reduces the resolution of the final image, described in Mack (2007) as a phase error of the radial location contained by the aperture. Near the resolution limit, small errors in mask CD causing significant errors are translated to the photoresist CD on the wafer, and the intensification of mask faults is defined by the mask error enhancement factor (MEEF), described by

$$MEEF = \frac{\Delta CD_{resist}}{\Delta CD_{mask}} \qquad (3.15)$$

where ΔCD_{resist} and ΔCD_{mask} are the variations in the CD of the photoresist and the mask, respectively (Schellenberg and Mack 1999). The MEEF ratio of modern semiconductors is being pushed higher and toward its limits of linearity as semiconductor optical lithography aspires to smaller k_1 factors in (3.7). If reticles are used in stepper equipment, the MEEF still relates to the configuration found on the reticle and the resultant pattern on the surface of the wafer, but it takes into account the imaging system reduction ratio from the optics, such that

$$MEEF = \frac{\Delta CD_{resist}}{\Delta \dfrac{CD_{reticle}}{M}} \qquad (3.16)$$

where,

$CD_{reticle}$ represents the critical dimensions of the reticle
M represents the reduction ratio, typically $M=4$ for DUV systems (Schellenberg and Mack 1999)

In a perfect linear imaging system, the MEEF equals 1.0, corresponding to perfect pattern transfer from the mask/reticle to the wafer.

OPTICAL LITHOGRAPHY PHOTOMASK PROTECTION (PELLICLES)

To reduce particle contamination on wafers in the semiconductor industry, photomasks are typically protected using a thin (typically in the 1 µm or thinner range) polymer film or membrane, termed a pellicle. The pellicle is stretched over a mask or reticle at a standoff height and glued onto the sides of the photomask. A typical pellicle placement is presented in Figure 3.11.

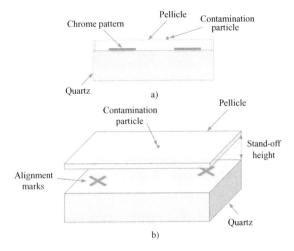

FIGURE 3.11   Pellicle placement on a photomask to limit the amount of particle contamination on the wafer. In (a), a side view of the pellicle is shown, and in (b) the three-dimensional structure of the same pellicle placement is shown.

Essentially, any small particle defects or dust that may settle on the pellicle will be out of focus and not transferred to the wafer because of the frame standoff height of the pellicle. Characteristic frame heights are 5–10 mm, with 6.35 mm being a typical value. The pellicle thickness is thin enough not to attenuate the incoming UV radiation, but its mechanical and optical behavior should be taken into account as part of the imaging system. Since typical DOF parameters are below 1 μm (see Table 3.4), standoff heights of a few millimeters can normally distort the image considerably. The distance between the pellicle and the mask places the particles in the far field (the Fraunhofer diffraction region) so that the pellicle can be treated as a pupil filter. Pellicle standoff heights must also be big enough to inhibit imperfections caused by significantly decreasing the intensity of the light. A generally acceptable height of the pellicle standoff can be determined by

$$t = \frac{4Md}{\text{NA}} \qquad (3.17)$$

where,

   M is the lens reduction
   d is the diameter of the contaminant (particle) (Levinson 2005)

Pellicles are often only attached to the chrome-side of the mask since the transparent glass 'bottom' serves a similar purpose to the reticle itself. Pellicle films are usually polymers, with nitrocellulose and forms of Teflon being the most common (Levinson 2005). It is important that the pellicle material is strong and resistant to stretching, completely transparent in the wavelength used by the imaging system and resistant to radiation damage. In terms of developments in 13.5 nm lithography, radiation damage of pellicles is a significant concern if the same material is used as for modern 193 nm lithography. Typical pellicles for the 193 nm wavelength are not designed to be transparent to 13.5 nm light and EUV would near-instantly obliterate these pellicles. ASML planned to build scanners that did not use pellicles, but at current feature sizes, a dust particle that cannot be seen with the naked eye would still be able to blot out hundreds of transistors, a risk that chipmakers are not comfortable taking (Courtland 2016). Therefore, pellicle design at 13.5 nm wavelengths poses an additional challenge for the shift toward EUV lithography.

Although the basic principles of lithographic imaging systems do not vary, there are multiple techniques to transfer an image from an object, typically a photomask, to a wafer. The most commonly used techniques are discussed in the following section.

LITHOGRAPHY TECHNIQUES

This section reviews commonly used techniques for transferring a pattern from an object (photomask) to a wafer, covered in photoresist, in order of their invention and industrialization, as presented in Table 3.1. These three techniques are contact lithography, proximity lithography and projection lithography. The advantages and disadvantages of each technique are also briefly reviewed, as well as the theoretical minimum pitch of the line-space pattern for each technique.

CONTACT LITHOGRAPHY

The simplest form of semiconductor photolithography is contact lithography. The mask is physically pushed against the wafer that is coated with photoresist during UV exposure, where the opening between the mask and the wafer should preferably be nil. A simplified contact lithography imaging system is presented in Figure 3.12.

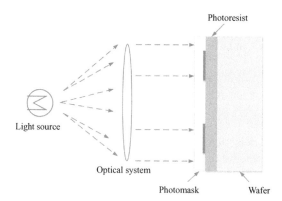

FIGURE 3.12   Contact photolithography where the photomask is placed directly on the photoresist during exposure. Note that this figure is rotated to better fit the content of the page. In reality, contact photolithography equipment is operated with the light source at the top and the wafer at the bottom, therefore a 90° clockwise rotation.

As shown in Figure 3.12, there is no gap between the photomask and the resist-coated wafer. The minimum pitch of the line-space pattern for contact photolithography is given by

$$2b = 3\sqrt{0.5\lambda d} \qquad (3.18)$$

where,

    $2b$ is the minimum pitch of the line-space pattern

    λ is the exposure wavelength

    d is the thickness of the photoresist

In contact photolithography, the mask plate can easily be damaged or can accumulate microscopic defects and contamination, which will transfer to the photoresist and compromise the final geometry of the images processed on the wafer, typically referred to as soft defects. Defects that cause a widening of the gap between the photomask and the photoresist, leading to improper focusing of the light onto the target, are known as hard defects. A primary disadvantage of contact lithography, experienced already during the 1970s, is low process yield due to the damage inflicted on the photomask and the wafer by this physical contact (Kapoor and Adner 2007). Contact lithography is generally the least expensive technique that also produces relatively high resolution, typically below 500 nm, and the resolution of contact lithography is limited by scattering effects that may occur in the photoresist because of its finite thickness. In addition, as modern wafers are becoming larger (refer to the discussion on 450 mm wafers in Chapter 2 of this book), the uniformity of mask-to-wafer contact is compromised and contact photolithography becomes less viable and prone to multiple defects on the wafer.

To alleviate some of the issues experienced with contact photolithography, proximity photolithography was developed, and is discussed in the following paragraph.

PROXIMITY LITHOGRAPHY

In proximity photolithography, the photomask and the resist-coated wafer do not come into direct contact during exposure. The photomask and the surface of the wafer are detached by a small opening to mitigate the defects that are created by direct contact (Kapoor and Adner 2007). The illumination system is optimized for collimated light (parallel light rays with minimum spreading during propagation – theoretically focused at infinity). Proximity lithography allows for a small gap between the mask and the wafer, typically in the range of 10–30 μm. The gap is small enough to ensure that the traveling light can be approximated as collimated light beams. The physical representation of proximity photolithography is given in Figure 3.13.

As shown in Figure 3.13, there is a gap between the photomask and the resist-coated wafer. The minimum pitch of the line-space pattern for proximity photolithography is given by

$$2b = 3\sqrt{\lambda(s + 0.5d)} \qquad (3.19)$$

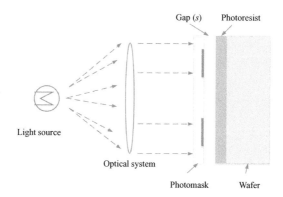

FIGURE 3.13 Proximity photolithography where the photomask is placed at a specified distance from the photoresist during exposure. Note that this figure is rotated to better fit the content of this page. In reality, proximity photolithography equipment is operated with the light source at the top and the wafer at the bottom, therefore a 90° clockwise rotation.

where s is the spacing between the mask and the photoresist. In proximity photolithography, a relatively small portion of the resolution is compromised because of the spacing between the mask and the wafer. To reduce (improve) the minimum line spacing of the lithography, the wavelength, the gap between the mask and the wafer or the photoresist thickness should be decreased. The advantages of proximity photolithography are that it reduces the number of defects and anomalies on the photomask due to contact photolithography. To eliminate hard and soft defects in photolithography, projection printing was developed. Projection lithography is discussed in the following paragraph.

PROJECTION LITHOGRAPHY

Projection printing provides high resolution and low defect densities and governs the photolithography practices used in modern semiconductor facilities. The physical representation of proximity photolithography is given in Figure 3.14.

As shown in Figure 3.14, there is a gap between the photomask and the resist-coated wafer. In addition, optical lenses are used to converge the light waves between the mask and the resist-coated wafer onto the wafer, and only small portions of the wafer (typically each die) are exposed at a time to increase the resolution of the system. An image of the patterned mask is projected onto the surface of the wafer, typically a few centimeters away from the mask. The optical lenses make it possible to achieve a reduction in the feature size on the photoresist. Commonly used projection systems use reduction optics of between 2× and 5× to reduce the image size projected onto the photoresist. An approximation of the attainable resolution of a projection lithography system takes a similar form as (3.7), since the projection lens is placed at a specific distance from the wafer, resulting in a field angle θ between the surface of the wafer and the projection lens. The resolution is therefore roughly approximated by

$$R = k_1 \left(\frac{\lambda}{\mathrm{NA}} \right) \tag{3.20}$$

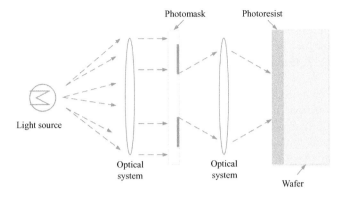

FIGURE 3.14 Projection photolithography where the photomask is placed at a specified distance from the photoresist during exposure and the light is focused onto the wafer using optical lenses. Note that this figure is rotated to better fit the content of this page. In reality, projection photolithography equipment is operated with the light source at the top and the wafer at the bottom, therefore a 90° clockwise rotation.

where k_1 represents the proportionality factor whose value is dependent on the properties of the patterning process. These properties include parameters of the optics, photoresist and process latitude (linewidth in μm/nm versus exposure in mJ/cm², typically presented as Bossung curves) and $k_1 < 1$. The Bossung curve analysis maps a control surface for CDs as a function of the variables of focus and exposure, and therefore the dose (Zavecz 2006). Most commonly, "the technique is used to calculate the optimum focus and dose process parameters that yield the largest DOF over a tolerable range of exposure latitude" (Zavecz 2006). Although higher resolution can be achieved using projection lithography, these systems are typically very complex, expensive and have a lower rate of wafer processing per hour. This is due to the longer time it takes to expose the entire wafer, since each die needs to be exposed separately as a result of the de-magnification, leading to complex and expensive steppers to achieve this task.

Projection printing was superseded by electron-beam (E-beam) patterning on photoresist, a technology that enabled submicron resolution in the semiconductor industry. E-beam lithography is discussed in the following paragraph.

E-BEAM LITHOGRAPHY

E-beam lithography essentially involves the direct patterning of photoresist on the wafer by using electron beams, thereby removing the requirement of a photomask (Kapoor and Adner 2007). The transition to E-beam lithography required significant changes in photolithography, especially in electron sources and compatible photoresists. The weakness of this technology is the low throughput (low volume) with current system implementations (Kapoor and Adner 2007). The minimum time (T) required to expose a specified area for a given dose is determined by

$$T = \frac{D \times A}{I} \tag{3.21}$$

where,

    D is the dose
    A is the area being exposed
    I is the E-beam current

Because of its limited throughput, E-beam lithography is not suited for high-volume production; furthermore, as feature sizes shrink, the number of incident electrons at a fixed dose also decreases. The effect is that shot noise increases and becomes a predominant limitation of the technology, leading to significant dose variations within a large feature population. E-beam lithography challenges in the industry directed the focus to the development of stepper technology (Kapoor and Adner 2007). This led to the g-line stepper operating at a high-intensity radiation wavelength at 436 nm (visible light) during 1978, as discussed in the following paragraph.

G-LINE LITHOGRAPHY

Two primary changes to photolithography were introduced with g-line lithography (436 nm) technology. These two changes were

- The use of a refractive lens to project light.
- To only project light onto a small area on the wafer, such that numerous exposures of the IC were required during the lithography process (Kapoor and Adner 2007).

The first commercially available wafer stepper operated at the Hg g-line, which is blue light in the visible part of the electromagnetic spectrum. Following g-line lithography, a shorter wavelength was introduced, and 405 nm steppers became commercially available, followed by 365 nm radiation wavelength (i-line), which was commercially available by 1985. I-line lithography is discussed in the following paragraph.

I-LINE LITHOGRAPHY

I-line lithography at 365 nm had achieved initial resolutions of approximately 800 nm by 1985, as listed in Table 3.1. I-line lithography steppers therefore enabled submicron semiconductor features while maintaining a DOF of 1 μm and above. I-line lithography dominated state-of-the-art lithography until DUV lithography was adapted as a commercial standard in the 1990s. I-line lithography was used for pattern definition at all levels with pitches of an active area of 800 nm, poly-silicon of 700 nm linewidth and metal layers of 800 nm (Montree et al. 1996). I-line steppers presented an NA of between 0.28 and 0.35 with a relatively high-volume throughput of about 45 wafers per hour, and alignment accuracy of ±250 nm (Horiuchi and Suzuki 1985).

DUV AND EUV

The use of excimer lasers at DUV (248 nm and 193 nm) further extended the CD to the sub-500 nm range. DUV is still used for modern semiconductor processing to produce features of 22 nm and below, with several advances and further improvements on radiation wavelength incorporated. EUV takes over the lithography responsibilities for features below 12 nm. The principles of DUV and EUV are reviewed in Chapter 4 of this book.

Photolithography is only as effective as the light-sensitive photoresist onto which the light wave is directed. The following section reviews the types of photoresist and their most crucial characteristics and specifications.

LIGHT-SENSITIVE PHOTORESIST

Photoresist is a light-sensitive material used in photolithography to transfer images from a photomask onto a wafer; exposure to wavelength-specific UV changes its chemical properties and its solubility. Photoresist is divided into two primary categories: positive photoresist and negative photoresist. Photoresist (both positive and negative) is not only distinguished by its solubility, it is characterized and typically qualified by features such as

- Its ability to distinguish between light and dark portions of the photomask; therefore, its *contrast*.
- How fine a line the photoresist can reproduce from an aerial image; therefore, its *resolution*.
- How much incident energy the photoresist requires to change its solubility; therefore, its *sensitivity*.
- Its absorbance of the wavelength of the light source; therefore, its *spectral response*.
- Its ability to protect the underlying material from etchants and its thermal stability; therefore, its *etch resistance*.

The contrast (γ), as described in the list of features that typically qualify photoresist, is given by

$$\gamma = \frac{1}{\beta + \alpha t_{resist}} \tag{3.22}$$

where,

β is a resist-specific constant
α is the absorption coefficient of the photoresist specified in cm^{-1}
t_{resist} is the thickness of the photoresist, specified in cm (or µm or nm)

Other important characteristics and requirements of photoresist are its

- Adhesion properties.
- Viscosity.
- Particulates and metal content.
- Process latitude.
- Consistency.
- Shelf life.

The adhesion properties of photoresist can be determined by using a goniometer test (derived from the two Greek words *gōnia*, meaning angle, and *metron*, meaning to measure), which estimates the surface energy of the wafer when the photoresist is applied. The energy on the surface can be found by applying a droplet of deionized water or the organic compound $(CH_2OH)_2$ (ethylene glycol) and determining the contact angle of the water drop. Incorporating Young's relation and a database of interfacial energy values, the surface energy can be estimated. The mathematical description of Young's relation used for estimating adhesion characteristics of photoresist is given by

$$\gamma_{SG} = \gamma_{SL} - \gamma_{LG} \cos\theta \qquad (3.23)$$

where,

γ_{SG} is the interfacial energy between the solid and gas (vapor)
γ_{SL} is the interfacial energy between the solid and the liquid
γ_{LG} is the interfacial energy between the liquid and the gas vapor
θ is the contact angle. Interfacial energy is described by J/m^2

The viscosity of photoresist and the spin speed of its application determine the thickness of the final layer of photoresist on the wafer. Photoresist viscosity is typically specified by manufacturers by its kinematic viscosity, also referred to as diffusivity of momentum. The stokes (St) is the unit of measurement for kinematic viscosity, often conveyed as centistokes (cSt), where $1 \text{ cSt} = 1 \text{ mm}^2.\text{s}^{-1} = 10^{-6} \text{ m}^2\text{s}^{-1}$. As a reference, the kinematic viscosity of water at 293 kelvin is 1 cSt. The thickness t of the photoresist coating on the wafer is primarily a function of the viscosity and the spin speed, such that its relation is given by

$$t = \sqrt{\frac{\mu}{\omega}} \qquad (3.24)$$

where,

ω is the angular velocity of the spinning chuck in rad/s
μ is the viscosity in Pa.s

Figure 3.15 is a representation of the angular velocity and wafer dimensions as typically implemented with the spin coating of the photoresist.

A typical representation of photoresist thickness as a function of angular spin speed (with its viscosity kept constant) as indicated in Figure 3.15 is presented in Figure 3.16.

As seen in Figure 3.16, the photoresist thickness decreases non-linearly with the increasing angular spin speed of the wafer. The angular spin speed in Figure 3.16 ranges between 1000 revolutions per minute (rpm) and 8000 rpm, which are typical angular spin speeds for photoresist, although most manufacturers provide information on the ideal angular spin speed for a required photoresist thickness. The spin speed also depends on the diameter of the wafer, where larger wafers should typically be spun at lower speeds to avoid excessive turbulence. Turbulence, not only from wafer-spin but also from airflow in cleanrooms and ambient humidity, affects

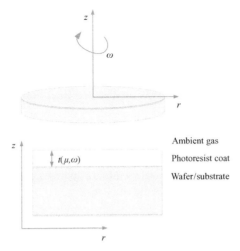

FIGURE 3.15 A typical representation of photoresist spinning implementations, indicating the angular velocity and substrate placement.

the evaporation rate and profile of the photoresist, which leads to variation in the photoresist thickness. Additional parameters typically supplied by manufacturers regarding photoresist are the

- Coating method (spin or spray, where high-viscosity photoresists are typically applied by spinning, whereas lower-viscosity photoresists can be applied by spraying methods).
- Solids content, given as a percentage.
- Kinematic viscosity in cSt (as described in this paragraph).

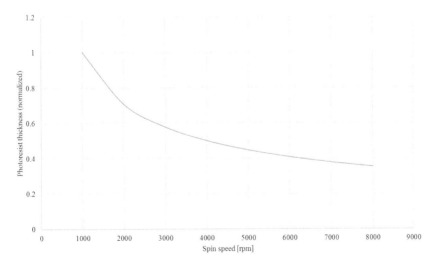

FIGURE 3.16 A typical representation of photoresist thickness as a function of angular spin speed versus constant photoresist viscosity. The photoresist thickness is normalized to one.

- Optimal exposure wavelength.
- Absorptivity specified in l/g*cm at the optimal exposure wavelength.
- Specific gravity at a specified temperature.
- Maximum water content.
- Particle count.
- Appearance (such that photoresists reaching their maximum shelf life can be distinguished).
- Solvent formula.
- Post-exposure bake requirements (temperature and time).
- Development and removal instructions.
- Filtration requirements, as discussed in this paragraph.
- Spin speed to achieve the desired thickness.

The contamination of wafers with insoluble particulates or metal content is a common problem due to sub-standard or contaminated photoresist. Two techniques are frequently used to remove particulates on a wafer resulting from contaminated photoresist, namely ultrasonic and mechanical scrubbing. Photoresist performance is therefore dependent on the formation of particles, or generally described as its ageing (beyond its specified expiry date or shelf life). Accelerated ageing occurs when photoresist is stored at above its recommended temperature and too high dilution and/or unsuitable solvents are used in conjunction with the photoresist. Storage time beyond its specified shelf life leads to thermal decomposition of the photo initiator concentration that causes a lower development rate and gradual darkening, which only marginally affects its UV absorption capabilities but is a sign that photoresist needs to be replaced to ensure consistency and avoid a possible decrease in adhesion. In addition, nitrogen forms in the solution that partly dissolves the resist and may cause bubbles to form during or after coating, leading to uneven photoresist coatings.

Photoresist is described by Rothschild et al. (1997) as the emulsion in the photolithography process. "Practically all photoresists developed for use at longer wavelengths are classified as single-layer resists and have uniform composition through their thickness" (Rothmans et al. 1997). Photoresist is applied by spin coating, where the rpm determine the thickness of the photoresist, followed by a baking procedure to remove remaining solvents. Photoresist developed for radiation exposure has several attributes that must be present. The attributes include that

- The photoresist employs photo-acid generators and modifies the dissolution characteristics of the photoresist.
- Positive resist (where the uncovered area is removed during development) has an aqueous base.
- Negative resist (where the covered area is removed during development) has an organic base.

The differences between positive photoresist and negative photoresist are outlined in the following paragraphs.

Positive Photoresist Patterning

Positive photoresist, typically having an aqueous developer base, undergoes an increase in its solubility in a developer. Regions that are directly below the transparent parts (spaces) of the photomasks are exposed to wavelength-specific UV light, as shown in Figure 3.17 (Bartczack and Galeski 2014; Bulgakova et al. 2014; Cui 2008; Guy and Burwell 2013; Indykiewicz et al. 2012; Ito 2005; Nouralishahi et al. 2008; Rudnitsky and Serdyuk 2012). The procedure of using positive photoresist to transfer images to a wafer is commonly referred to as positive-tone development (PTD).

The exposed regions as shown in Figure 3.17 are etched anisotropically (directionally dependent) using either wet chemical etching by immersion in a chemical bath, or dry etching such as plasma etching in the form of reactive ion etching (RIE) (Cui 2008; Guy and Burwell 2013; Tummala 2001). RIE uses strong electric fields of ions and activated neutrals to accelerate positively charged ions of chemically reactive species toward patterned wafers (Tummala 2001). The thin film regions that are exposed to the reactive ions are both physically (through ablation) and chemically (through chemical reactions) bombarded by the reactive ions to produce volatile products that are then pumped out of the RIE chamber (Tummala 2001). In the RIE process, cations are created from reactive gases that are accelerated toward the wafer and reacts chemically with the layer being etched. A higher vapor pressure results in higher volatility of the by-products that are produced. In contrast to exposed areas, the regions of the thin film that are directly underneath undissolved photoresist material are protected from the etching species and are not etched. Therefore, the etching process transfers the relief patterns from the photoresist layers onto the thin

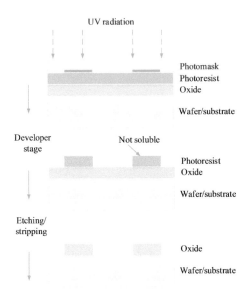

FIGURE 3.17 Optical and chemical properties of positive photoresist for a PTD step results in exposed photoresist increasing its water solubility and being removed during the development stage.

films directly below them (Bartczack and Galeski 2014; Bulgakova et al. 2014; Cui 2008; Guy and Burwell 2013; Indykiewicz et al. 2012; Ito 2005; Nouralishahi et al. 2008; Rudnitsky and Serdyuk 2012). The remaining areas of the photoresist films are then selectively removed (stripped), such that the photoresist material is completely removed from the wafer while leaving the patterned layers intact (Ciu 2008; Guy and Burwell 2013).

Positive photoresist used in PTD generally has very good adhesion to silicon, for example, although the adhesion of positive photoresist to various other materials, such as InGaAs, GaAs, InP, InSb or spin-on-glass (used as an alternative to ion implantation during the doping process) differs from that of silicon and its adhesion properties depend on the combination of materials. Positive photoresist tends to be more expensive to use when compared with negative photoresist and has a better step coverage and smaller (and therefore better) minimum feature size. Wet chemical etching of positive photoresist is not as good as when used with dry etching through plasma. The following paragraphs briefly review the properties and characteristics of negative photoresist.

NEGATIVE PHOTORESIST PATTERNING

In contrast to positive photoresist, the photochemistry of negative photoresist, typically having an organic developer base, results in a dramatic decrease in its solubility upon UV exposure. This results in the features of negative photoresist being inverted with respect to the transparent regions of the photomasks. The resist profile is therefore a negative image of the aerial profile, as shown in Figure 3.18 (Bartczack and

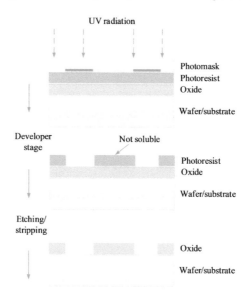

FIGURE 3.18 Optical and chemical properties of negative photoresist for an NTD step result in the exposed photoresist decreasing its water solubility and remaining on the sample during the developing stage.

Galeski 2014; Bulgakova et al. 2014; Cui 2008; Guy and Burwell 2013; Indykiewicz et al. 2012; Ito 2005; Nouralishahi et al. 2008; Rudnitsky and Serdyuk 2012). The procedure of using negative photoresist to transfer images to a wafer is commonly referred to as negative-tone development (NTD).

As shown in Figure 3.18, the portion of the negative photoresist that is exposed to UV light becomes practically insoluble in the photoresist developer. The portions of photoresist that are not exposed to UV light, and are therefore covered by the chrome patterns on the photomask, are dissolved by the developer.

Negative photoresist typically has good adhesion to silicon, whereas adhesion to other materials again depends on the combination of materials used, similar to the case of positive photoresist. Negative photoresist is typically less expensive than positive photoresist, but has lower step coverage and a larger (therefore poorer) minimum feature size.

COMPLETING THE PATTERNING PROCESS

By the end of each patterning sequence, circuit design features for a particular level have been permanently transferred onto a desired thin film, and the wafer can continue being processed to build other layers (Tummala 2001). By repeating the patterning process flow for subsequent levels (after performing the prerequisite processing steps between each patterning sequence and aligning each new patterning step to the wafer's current patterns), a fully integrated IC can be built from the bottom up (Tummala 2001). Modern-day ICs require several patterning steps, with each subsequent patterning level being adversely affected by any patterning defects that are encountered during prior patterning stages; hence, patterning limitations are considered one of the largest bottlenecks inhibiting the pace of device scaling at each technology node.

CONCLUSION

In 2004, the role that EUV would play in propelling the momentum of semiconductor manufacturing, specifically with respect to photolithography, was often discussed. There was still widespread uncertainty about EUV and the research and development required to refine this technology was still unknown. EUV promised to alleviate the challenges of printing decreasing sizes of lines, spaces and contacts needed for upcoming generations of chips and to ensure multi-billion-dollar industry growth for the near future. The predicament was immensely high-powered light sources producing too little output power, unreliable results and low wafer throughput. Conveniently, a new technology, known as immersion lithography, was being adopted by foundries globally. Immersion technology, back in 2004, was said to accomplish its life-extending sorcery by the addition of a small film of water flanked by the projection lenses of optical systems and the silicon wafers. Immersion technology also boasted a record-setting short development time, achieving in two years what most groundbreaking technologies had achieved in a few decades of development. Developers of the technology estimated that immersion technology would achieve linewidths of 65 nm by 2007 and 45 nm by 2009. These predictions were accurate and modern ICs are being produced with feature sizes of 12 nm using several lithography techniques, including immersion lithography, multiple patterning and phase-shifting masks. However, the limits of the 193 nm light source are beginning to present challenges that demand a new technology to replace the 193 nm wavelength. A primary concern is that light with a wavelength shorter than 193 nm is absorbed, rather than transmitted through, amorphous silicon dioxide lenses.

By 2016, foundries such as Fab 8 of GlobalFoundries cost in excess of US $12 billion and had not moved to EUV lithography using 13.5 nm light. EUV light is incapable of traveling in air and cannot be focused by lenses or conventional mirrors. Producing light at 13.5 nm still poses many challenges and requires large capital investments in research and development. Throughout the last decade, Moore's law was primarily driven by enhancements in 193 nm light generation and transfer. However, 13.5 nm EUV technology seems to be approaching high-volume commercial production status for leading-edge microprocessors and memory. During 2017, companies such as IBM and TSMC have produced 7 nm and 5 nm transistors using EUV in controlled environments. High-volume production facilities using EUV can potentially be realized within the next five years (Lapedus 2017). Although a decline in revenue is predicted for the first foundries to use EUV, its longevity and future success are still based on the premise of Moore's law.

This chapter presents the basic principles of semiconductor lithography and reviews characteristics such as the light sources used for photolithography since it first became a viable means to transfer geometries onto a semiconductor wafer. It also reviews the process flow for patterning a wafer, photolithography techniques and their associated advantages, disadvantages and limitations, as well as a discussion on photoresist, which must be compatible with the wavelength of photolithography. The chapter therefore serves as an introduction to new technologies and improvements that aim to build on the longevity of Moore's law.

REFERENCES

Bartczak, Z., Galeski, A. (2014). Mechanical properties of polymer blends. *Polymer Blends Handbook*, A. L. Utracki and A. C. Wilkie, Eds. Dordrecht: Springer Netherlands, 1203–1297.

Bourzac, K. (2016). Moore's Law's ultraviolet savior is finally ready. Retrieved 22 January from http://www.technologyreview.com

Bulgakova, S. A., Gurova, D. A., Zaitsev, S. D., Kulikov, E. E., Skorokhodov, E. V., Toropov, M. N., Pestov, A. E., Chkhalo, N. I., Salashchenko, N. N. (2014). Effect of polymer matrix and photoacid generator on the lithographic properties of chemically amplified photoresist. *Russian Microelectronics*, 43, 392–400.

Courtland, R. (2016). Leading chipmakers eye EUV lithography to save Moore's law. Retrieved 24 January from http://www.spectrum.ieee.org

Cui, Z. (2008). Nanofabrication by photons. *Nanofabrication: Principles, Capabilities and Limits*. Boston, MA: Springer US, 7–76.

Doolittle, A. (2008). Lithography and pattern transfer. *Georgia Tech – ECE 6450*. Retrieved 21 January 2017 from http://alan.ece.gatech.edu

Guy, O. J., Burwell, G., Castaing, A., Walker, K. D. (2013). Photochemistry in electronics. *Applied Photochemistry*, C. R. Evans, P. Douglas and D. H. Burrow, Eds. Dordrecht: Springer Netherlands, 435–465.

Horiuchi, T., Suzuki, M. (1985). An i-line wafer stepper with variable coherency and numerical aperture. *1985 Symposium on VLSI Technology*, 80–81, 1985.

Huff, H. R., Gilmer, D. C. (2005) *High Dielectric Constant Materials: VLSI MOSFET Applications.* Berlin Heidelberg: Springer.

Indykiewicz, K., Haj, M., Paszkiewicz, B. (2012). The new method of fabrication of submicron structures by optical lithography with mask shifting and mask rotation. *Central European Journal of Physics*, 11, 219–225.

Ito, H. (2005). Chemical amplification resists for microlithography. *Microlithography – Molecular Imprinting.* Berlin, Heidelberg: Springer Berlin Heidelberg, 37–245, 2005.

Kapoor, R., Adner, R. (2007). Technology interdependence and the evolution of semiconductor lithography. *Solid State Technology.* November 2007.

Lapedus, M. (2017). The race to 10/7 nm. Retrieved 20 November 2017 from http://semiengineering.com

Levinson, H. J. (2005). *Principles of Lithography*, 2nd ed., Bellingham, WA: SPIE Press.

Mack, C. A. (1995). Depth of focus. *Microlithography World*, Spring, 20–24.

Mack, C. A. (2007). *Fundamental Principles of Optical Lithography: The Science of Microfabrication.* Chichester: John Wiley & Sons, ISBN 1119965071.

Mack, C. A. (2011). Fifty years of Moore's Law. *IEEE Transactions on. Semiconductor Manufacturing,* 24(5), 202–207.

Mack, C. A. (2015). The multiple lives of Moore's Law. *IEEE Spectrum*, 52(4), 31–31.

Montree, A. H., Ansem, G., Huijten, L. H. M., Juffermans, C. A. H., de Laat, W. T. F. M., Lohmeier, M., Manders, B. S., Meijer, P. M., Paulzen, G. M., Roes, R. F. M., Webster, M. N., Zandbergen, P. (1996). High performance 0.3 μm CMOS technology using I-line lithography. *26th European Proceedings of Solid State Device Research Conference*, 597–600.

Moore, G. E. (1995). Lithography and the future of Moore's law. *Proceedings of SPIE*, 2437.

Nouralishahi, M., Wu, C., Vandenberghe, L. (2008). Model calibration for optical lithography via semidefinite programming. *Optimization and Engineering*, 9(19), 19–35.

Rothschild, M., Horn, M. W., Keast, C. L., Kunz, R. R., Liberman, V., Palmateer, S. C., Doran, S. P., Forte, A. R., Goodman, R. B., Sedlacek, J. H. C., Uttaro, R. S., Corliss, D., Grenville, A. (1997). Photolithography at 193 nm. *The Lincoln Laboratory Journal*, 10(1), 19–34.

Rudnitsky, A. S., Serdyuk, V. M. (2012). Integrated evaluation of diffraction image quality in optical lithography using the rigorous diffraction solution for a slot. *Technical Physics*, 57, 1387–1393.

Schellenberg, F. M., Mack, C. A. (1999). MEEF in theory and practice. *Mentor Graphics Deep Submicron Technical Publication.* Presented at the 19th Annual BACUS Symposium on Photomask Technology and Management, 16 September 1999.

Tummala, R. (2001). *Fundamentals of Microsystems Packaging.* New York: McGraw-Hill.

Willers, C. J. (2013). *Electro-Optical System Analysis and Design: A Radiometry Perspective.* Bellingham: SPIE Press, ISBN 9780819495693.

Zavecz, T. E. (2006). Bossung curves: An old technique with a new twist for sub-90-nm nodes. *SPIE 31st International Symposium on Advanced Lithography.* International Society for Optics and Photonics, 61522S–61522S.

4 Photolithography Enhancements

Wynand Lambrechts, Saurabh Sinha and Jassem Abdallah

INTRODUCTION

In the semiconductor manufacturing industry, transferring patterns from photomasks onto thin layers, or wafers, is done almost exclusively using photolithography (Chua et al. 2004). "Severe optical proximity effects and small DOF for isolated lines have posed challenges to sub-wavelength lithography for application to technology nodes of 100 nm and below using 248 nm and 193 nm scanners" (Chua et al. 2004). DOF criteria typically only apply to a feature at the resolution boundary or limitation of the imaging device, and Rayleigh equations are therefore insufficient to address the effect of numerical aperture and λ on the DOF (Chua et al. 2004). Source and mask optimizations have been suggested as an operational elucidation to prolong the life cycle of typical 248 nm and 193 nm lithography, given that these processes are computationally demanding (Peng et al. 2011).

The principles of photolithography were discussed and reviewed in Chapter 3 of this book. This chapter focuses on resolution enhancement techniques (RET) that push photolithography at constant wavelength to its limits, without necessarily varying or improving on the physical light source (therefore the associated wavelength). The figures of merit of a photolithographic process are its resolution, which translates to the minimum feature size that can be transferred to the wafer while maintaining high reliability; the accuracy by which sequential masks can be aligned; and the throughput, a measure of the efficiency through the amount of samples (wafers) that can be processed each hour. Photolithography is the most expensive manufacturing process step, accounting for up to 35% of the total processing cost. The remainder of the cost is subdivided into the categories of multi-level materials and etching (25%), cleaning and stripping (20%), furnaces and implants (15%) and metrology (10%). The procurement of photolithographic equipment is extremely expensive and there are various techniques to mitigate the cost of purchasing new-generation equipment. Increased wafer throughput and resolution enhancements are among the most commonly implemented techniques to recover some of the cost involved in photolithography. Among the concerns of photomasks are, again, the cost and time to fabricate them (typically this is outsourced by smaller foundries), contamination introduced by them, disposal of the masks, the difficulty in aligning several process-step masks and the possibility of defects introduced during handling.

As Moore's law progresses to smaller achievable nodes, new physical effects that arise and that have historically been ignored because of their insignificant impact on process yield are becoming more prevalent. Additional modifications to older-generation equipment are required to mitigate the physical defects introduced by smaller node technology. Essentially, these modifications are classified into two categories: distortion corrections and reticle enhancement. Distortion corrections compensate for alterations inherent in the manufacturing process steps such as lithography, etching or deposition. Reticle enhancements are alterations to improve the manufacturability or resolution of the process, with techniques such as phase-shifted masks and multiple patterning. These techniques, part of various RET, are reviewed in this chapter. In addition, optical proximity correction (OPC), which is a photolithography improvement procedure used to compensate for image mistakes resulting either from process defects or from diffraction, is also reviewed in this chapter. Essentially, a RET aims to optimize photolithography to reduce the CD of the system, ideally by keeping parameters such as the NA, λ and source parameters constant.

PHOTOLITHOGRAPHY RESOLUTION-ENHANCEMENT TECHNIQUES

This section reviews photolithography resolution-enhancement techniques such as immersion photolithography, extreme UV lithography (EUVL) and EBL. The first RET discussed in this chapter, immersion photolithography, replaces the traditional air opening between the lenses (optics) and the wafer with a liquid medium, essentially manipulating the refractive index that the light experiences on its way to the wafer. EUVL is a next-generation technology that departs from traditional deep-UV standards and uses UV light with a wavelength of 13.5 nm to expose patterns onto a substrate. This technology is not yet widely used but its implementation is imminent and this chapter reviews the advantages as well as the hurdles that have delayed the commercial use of EUVL. EBL scans a concentrated beam of electrons to draw customized contours and shapes on a substrate surface covered in electron-sensitive resist. EBL is an example of a maskless lithography technique and its characteristics are reviewed in this chapter.

IMMERSION PHOTOLITHOGRAPHY (SINGLE EXPOSURE)

Early projection tools emitted light with $\lambda = 436$ nm, and their single exposure resolutions were limited to length-scales in the order of a few micrometers (Tummala 2001; Mack 2015; Huff and Gilmer 2005; Guy et al. 2013). Because of technological advancements over the preceding decades, modern-day projection tools emit light with $\lambda = 193$ nm, make use of immersion lithography using ultrapure water, have NA values of 1.35 and have k_1 values approaching 0.25 (0.25 is considered the practical minimum limit for k_1) (Cui 2008). This allows 193 nm immersion lithography (193i) tools, with NA of 1.35 to resolve features with CD of 38 nm at a density of 80 nm pitch for single exposures (Cui 2008; Guy et al. 2013). Advanced photolithography patterning flows such as

- Sidewall image transfer (SIT).
- Self-aligned double patterning (SADP).
- Multiple exposures.
- Multiple patterning.
- Directed self-assembly (DSA).

have allowed the semiconductor industry to extend the life of the 193i lithography beyond the ~40 nm half-pitch node; this is discussed in a future section in this chapter (Cushen et al. 2015; Somervell et al. 2015; Tsai et al. 2012). Immersion lithography does, however, have several drawbacks and challenges of implementation. These issues include

- Ensuring high-quality water purification and techniques to avoid contamination. Contamination of purified water does not exclusively mean particulate or liquid contaminates, but includes nanoscale gaseous bubbles that create shadows in the light between the wafer and the light source.
- Special requirements for the interface between the water and lenses, which are typically made of inorganic calcium fluoride (CaF_2) glass.
- Special requirements for the interface between the water and the photoresist, since the water can be contaminated by dissociated nitrogen through the photochemical reactions between UV light and photoresist.
- Defects introduced from watermarks and stains from irregular drying of the wafer, which can create chip faults.
- Specialized equipment to contain the water during movement of the wafer stage.

It is shown in Chapter 3 that the resolution (R) of a photolithography process can be approximated by

$$R = k_1 \left(\frac{\lambda}{NA} \right) \tag{4.1}$$

where,

   k_1 represents the proportionality factor whose value is dependent on the properties of the patterning process
   λ represents the wavelength of the light
   NA   is the numerical aperture of the optics

Also presented in Chapter 3 is the equation to determine the NA of the imaging tool, which is given by

$$NA = n \sin \theta \tag{4.2}$$

where,

n is the refractive index of the medium in which the lens is placed

θ is the angle, as described in Figure 4.8 of Chapter 3

To increase the numerical aperture, the semiconductor industry adopted the use of immersion or wet lithography, which makes use of ultrapure water ($n = 1.47$ at 193 nm) rather than air ($n = 1$) between the objective lens and the wafer surface to increase the focusing power of the imaging system (Ciu 2008). Increasing n of the gap between the wafer and the lens to greater than unity results in a shortening of the exposure wavelength by a factor of n according to

$$\lambda_{\mathrm{eff}} = \frac{\lambda_0}{n} \qquad (4.3)$$

where λ_0 is the wavelength in air or in a vacuum, which is shortened to an effective wavelength λ_{eff} in a fluid with refractive index n greater than unity (Synowicki et al. 2004). Adapted from Synowicki et al. (2004), Table 4.1 presents the effect of the fluid refractive index on λ_{eff} for both exposure wavelengths at 193 nm and 157 nm.

As shown in Table 4.1, the effective wavelength (λ_{eff}) for 193 nm and 157 nm photolithography sources changes significantly based on the fluid refractive index introduced between the light source and the wafer. A 193 nm source wavelength exposed through ultrapure water with a refractive index of 1.4366 results in an effective wavelength that is significantly shorter than 193 nm, at 134.6 nm. This technique therefore enables the use of shorter wavelength photolithography without changing the wavelength at the source, leading to increased performance (improved resolution of the critical dimension of components) with minimal changes in system architecture, therefore at minimal cost.

TABLE 4.1

Effect of Fluid Refractive Index on the Effective Wavelength of an Optical Lithography System

Exposure Wavelength λ_0 (nm)	Fluid Refractive Index n	Effective Wavelength λ_{eff} (nm)
193	1.437	134.6
	1.500	128.9
	1.600	120.9
157	1.310	120.3
	1.326	118.9
	1.329	118.6
	1.362	115.7
	1.440	109.4

Source: Adapted from Synowicki, R. A., et al., *Semiconductor Fabtech*, 22, 55–58, 2004.

The refractive index of any material is defined by two parameters, the well-known refractive index n and the lesser-used parameter k, the extinction coefficient of the material, where both values are dependent on wavelength and temperature. The refractive index n is a direct indication of the minimum printable feature size of a lithography system, as determined through (4.3). The extinction coefficient k is a measure of the absorptive strength of a material (Synowicki et al. 2004). In terms of microelectronic component manufacturing, k is a direct indication of the wafer throughput, as the value of k determines the transparency and therefore the exposure time required for each wafer. The extinction coefficient is commonly also provided as the absorption coefficient α, where

$$\alpha = \frac{4\pi k}{\lambda} \qquad (4.4)$$

where λ is the wavelength. The absorption coefficient is typically defined by its units in cm^{-1}, therefore if λ is specified in nm, the absorption coefficient can be determined by first multiplying the wavelength by 10^7. In Dair et al. (2001), absorption coefficients α and penetration depths at 193 nm for balanced salt solutions are given. These adapted parameters are presented in Table 4.2.

As shown in Table 4.2, the penetration depth of light at a specific wavelength (in this case, at 193 nm) depends on the absorption coefficient; a higher absorption coefficient (in cm^{-1}) results in a shorter penetration depth (in μm). The species presented in Table 4.2 are adapted from the work presented by Dair et al. (2001) and specifically attributed to a 193 nm light source. In Smith et al. (2004), the effects of anions on the absorption peak of water are given for both 193 nm and 248 nm photolithography, and these results are adapted and presented in Table 4.3.

As shown in Table 4.3, the effects of anions in water on the absorption peak vary significantly based on the anion introduced in water. Bromine (Br^-), for example,

TABLE 4.2

Absorption Coefficients and Penetration Depths of Balanced Salt Solutions for a Source Wavelength of 193 nm

Compound	Species Name	Absorption Coefficient (cm^{-1})	Penetration Depth (μm)
NaCl	Sodium chloride	57.7	173
$Na_3C_6H_5O_7$	Sodium citrate	35.1	285
CH_3COONa	Sodium acetate	32.8	305
KCl	Potassium chloride	6.8	1477
$CaCl_2$	Calcium chloride	5.2	1926
$MgCl_2$	Magnesium chloride	2.0	5010

Source: Adapted from Dair et al., *Archives of Ophthalmology*, 119, 533–537, 2001 and arranged from lowest to highest penetration depth.

TABLE 4.3

Effect of Anion on Absorption of Water for 193 nm and 248 nm Photolithography

Anion in Water	Absorption Peak		Comments
	eV	nm	
I^-	5.48	227	Potential 248 nm alternatives (halogens)
Br^-	6.26	198	
Cl^-	6.78	183	
ClO_4^{-1}	6.88	180	Potential 193 nm and 248 nm
HPO_4^{2-1}	6.95	179	alternatives (phosphates and sulfates)
SO_4^{2-1}	7.09	175	
$H_2PO_4^-$	7.31	170	
HSO_4^-	7.44	167	

Source: Smith et al., *High Index Aqueous Immersion Fluids for 193 nm and 248 nm Lithography*, Rochester Institute of Technology, Center for Nanolithography Research, 2004.

introduced an absorption peak at 198 nm, whereas HSO_4 introduced an absorption peak at 167 nm. The measured UV absorbance spectra of sulfates and phosphates in water are also presented in detail in Smith et al. (2004). In addition, in Smith et al. (2004) the fluid absorbance α at 193 nm and 248 nm for various alternative species is presented. This list is adapted and presented in Table 4.4.

Similar to Table 4.1, an additional list of fluid refractive indices at 193 nm and 248 nm is presented in Table 4.5.

TABLE 4.4

Fluid Absorbance at 193 nm and 248 nm (Smith et al. 2004), Arranged from Smallest to Largest Fluid Absorbance

Fluid	λ_0	α (mm^{-1} at 193 nm)	α (mm^{-1} at 248 nm)
$Gd_2(SO_4)_3$ (1.5%)	195	0.0085	0
H_3PO_4 (20%)	192	0.0251	0.00213
H_2SO_4 (20%)	197	0.246	0.00183
NaH_2PO_4 (16%)	196	0.429	0.110
KH_2PO_4 (16%)	196	0.571	0.163
Cs_2SO_4 (40%)	199	0.706	0.0017
K_2SO_4 (8%)	199	1.03	6×10^{-4}
$MgSO_4$ (5%)	196	1.05	0
Na_2SO_4 (17%)	199	1.144	0.0014
HCl (20%)	210	2.91	0.0015
Na_2HPO_4 (16%)	208	4.72	0.0154

TABLE 4.5
Fluid Refractive Indices

Fluid	Refractive Index	
	at 193 nm	at 248 nm
Na_2SO_4 (30%)	1.479	1.423
Cs_2SO_4 (40%)	1.481	1.422
CsCl (60%)	1.561	1.466
Na_2SO_4 (30%)	1.479	1.423
NaHSO4 (44%)	1.473	1.418
H_3PO_4 (20%)	1.452	1.398
H_3PO_4 (40%)	1.475	1.420
H_3PO_4 (80%)	1.538	1.488
HCl (37%)	1.583	1.487

Source: Smith et al., *High Index Aqueous Immersion Fluids for 193 nm and 248 nm Lithography*, Rochester Institute of Technology, Center for Nanolithography Research, 2004.

In summary, photolithography for sub-45 nm 193i and sub-65 nm 248i technologies requires high index fluid development (Smith et al. 2004). Immersion fluid indices have been increased to 1.54 for 193 nm by introducing HCl (37%) and to 1.49 for 248 nm, also through HCl (37%). Immersion photolithography can therefore miniaturize technology nodes by lowering the effective wavelength between the light source and the wafer. Immersion lithography also has its limitations, depending on the immersion fluids used, and to further decrease the wavelength of the light exposed onto the wafer, techniques such as EUVL are aiming to facilitate Moore's law for future-generation technologies.

EXTREME UV LITHOGRAPHY

Extreme UV lithography operates on the same optical principles as other forms of optical lithography; however, EUV tools emit light with wavelengths of only 13.5 nm, which allows much smaller features to be patterned than, for example, 193i and 248i technologies, even with immersion lithography accounted for (Bakshi et al. 2007; Ciu 2008; Zhang et al. 2008). In Chapter 3 of this book, the primary characteristics and economic potential of EUV were reviewed, whereas in this section, the technology is reviewed from a technical perspective. Shown in Figure 4.1, laser-produced plasma (LPP) EUVL tools have complicated emission systems, which contain multiple elements that are controlled to ensure proper operation of the tool (Bakshi et al. 2007).

The LPP light source in Figure 4.1 comprises three key components, which are

- The laser driver.
- Plasma generation and exhaust.
- EUV light collection in a vacuum vessel.

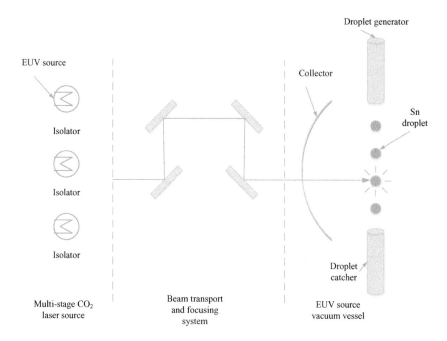

FIGURE 4.1 Basic configuration of an EUV source, showing the multiple EUV light sources, beam transport and focusing system as well as droplet generation (typically Sn – wavelength dependent) and the EUV source vacuum vessel.

As shown in Figure 4.1, a tiny (typically tin (Sn)) liquid droplet is produced from a device constructed for liquid jet technology (Endo 2014). A highly ionized Sn plasma produces a spherically uniform light (Endo 2014). The Sn liquid droplet is injected into a EUV emission point close to the nozzle tip with great placement accuracy. A short pulse laser disperses the droplet into equally distributed groups for improved coupling with the primary CO_2 laser irradiation. The rate of fall of the metal drops, the rate and power of the power supply, the temperature of the electrode and the timing of the laser beam must all be controlled to ensure proper operability of the EUV emission source (Bakshi et al. 2007). The LPP technology applies the CO_2 laser to the Sn droplets (typically 30 μm in diameter), which creates ionized gas plasma at electron temperatures of several tens of eV. The 13.5 nm radiation is then collected by a collector mirror coated with several layers of molybdenum (Mo) and silicon to selectively reflect the maximum amount of 13.5 EUV light and direct it to the intermediate focus position at the entrance of the scanner system. In addition to the complex emission systems, EUV tools must use reflective optics to direct and focus the photons, since EUV rays are highly absorbed by all materials; this precludes the use of lenses in EUV imaging systems (Bakshi et al. 2007; Ciu 2008; Zhang et al. 2008). The photons emitted by the EUV source must be collected via the use of a large concave EUV mirror, which redirects the photons toward subsequent reflective elements of the imaging system, as shown in Figure 4.2 (Ciu 2008; Wu 2011).

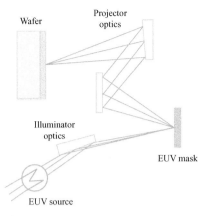

FIGURE 4.2 Rays reflected by the EUV masks are directed and focused by the reflective elements in the projector optics onto a wafer chip. (From Cui, Z., et al. *Nanofabrication by Photons. Nanofabrication: Principles, Capabilities and Limits*, Boston, MA, Springer US, 2008; Wu, B., *Science China Information Sciences*, 54, 959–979, 2011.)

In addition to the intricacies of the emission source, the control systems and the reflective optics, EUV systems must be kept under high vacuum and low temperature for long-term stability. During maintenance, EUV tools are vented to allow technicians to access the internal parts of the systems, but after the tool is fixed, it must be pumped again to high vacuum before wafers can be exposed. The turnaround time between maintenance and operation is typically long (Cui et al. 2008; Wu 2011) and reduces the wafer throughput when implementing these systems. Because of the low transparency of materials to soft X-rays, the EUV reticles or masks must be thin to allow as much light as possible to be transmitted through them to reach the wafer level (Wurm et al. 2007). This makes EUV reticles very fragile, and even with their decreased thickness, the intensity of the EUV light that reaches the wafer level remains low (Wurm et al. 2007) since the reflective elements in the EUV imaging system require multiple reflections to guide and focus the rays. At each reflection, the EUV photons are partially absorbed by the surfaces that are used to reflect the rays (Zhang et al. 2008; Wu 2011). The extremely low wafer throughput of EUV tools has been a key bottleneck in the implementation of EUV lithography in the high-volume manufacturing of ICs. Years of development by EUV equipment manufacturers were intended to increase the wafer throughput to practical levels (the current record is just over 1000 wafers per day), which was achieved by producing more powerful EUV emission sources to offset the absorption of the rays (Ciu 2008; Bakshi et al. 2007; Zhang et al. 2008; Wu 2011). The maximum repetition rate of EUV systems is restricted by interference of the Sn droplet from the intensifying micro-plasma cloud (Endo 2014). For higher average EUV power, the most cost-effective technique is to increase the maximum repetition rate, within its defined limitations.

In addition to the optical factors that limit wafer throughput, there are also material reasons for reduced wafer throughput, namely that EUV resists show very low sensitivity to incident photons such that relatively large doses of EUV energy are required to fully activate their photochemical reactions and allow pattern development. In contrast, standard photolithography resists and imaging technology allow wafer throughput to exceed 2500 fully exposed wafers per day (Bakshi et al. 2007; Ciu 2008; Zhang et al. 2008). Thus, EUV technology is limited by both optical and material factors and is therefore still under development, although it is considered the candidate technology to replace 193i and 248i lithography (Ciu 2008; Wu 2011).

Another limiting factor affecting EUV technology is the level of gaseous by-products (outgassing) released during the photochemical reactions that take place in EUV resists (Denbeaux et al. 2007; Hada et al. 2004; Santillan et al. 2007; Watanabe et al. 2003). These outgases are undesirable, since they condense as solid residues on cool surfaces. Metal surfaces used as reflective components of the EUV projection systems can be coated with films having higher levels of EUV absorption as well as causing diffuse photon scattering. Thus, outgases can cause costly damage to manufacturing equipment and lead to increased production downtime and higher operating costs (Bakshi et al. 2007; Bodermann et al. 2009).

Because of the high energy of EUV photons, heavy shielding in the form of thick metal walls must be used to protect process engineers from exposure to soft X-rays emitted by EUV scanners, but even with this shielding, the radiation exposure that each worker receives must be carefully monitored to avoid overexposure to harmful EUV rays (Ciu 2008).

A multitude of technical factors that have hindered the progress of EUV lithography have postponed its adoption at the high-volume manufacturing level as a substitute for 193i and 248i photolithography. This has had the knock-on effect of IC manufacturers having to come up with new processing methods to extend the life of current technologies long beyond what was expected by industry players. These advanced processing techniques will be discussed later in this chapter (Indykiewicz et al. 2012; Somervell et al. 2015; Tsai et al. 2012).

The following section discussed the maskless lithography technique, which uses electrons, as opposed to photons, to transfer patterns to a substrate, a process referred to as electron-beam lithography (EBL).

ELECTRON-BEAM LITHOGRAPHY

EBL is a maskless imaging system that uses electrons to induce solubility switches in resists as opposed to the traditional method of using photons (Cui 2008; Wu 2011; Zhou 2005). Usually, "EBL resists are high molecular-weight polymers dissolved in a liquid solvent" (Tseng et al. 2003). The structure of the polymer changes when open to radiation, which includes electron radiation. Subsequent to the exposure to radiation, similar to traditional photoresists, the resist is destabilized and the exposed resist becomes soluble in a developer (the mechanism describes positive resist; for negative resist, the unexposed resist is weaker and more soluble in the developer agent). The first resist discovered for use with electron radiation was polymethylmethacrylate, which has since been used for fabricating a variety of devices, including integrated circuits, photonic crystals and channels for nano-fluidics. EBL enables high resolution photolithography at a relatively low cost of implementation, lower than the total cost of EUVL, for example. The wavelength of the light in an optical system is the primary limitation of the attainable resolution. This limitation is noticeable for traditional photolithography using UV light, where EUV light beams at 13.5 nm aim to reduce feature size significantly, but at a high cost. EUV will inevitably create new opportunities in photolithography beyond the 13.5 nm wavelength limitation by using phase shifting and immersion lithography. The cost associated with these techniques remains high, and wafer throughput remains relatively low. EBL serves as a viable alternative up to the point where EUVL becomes readily available.

Electrons carry an electric charge; they are susceptible to deflection, confinement and focusing using electromagnetic fields. This property has been at the base of developing EBL systems. The potential of EBL to define nm-scale geometric patterns for microelectronic circuit manufacturing stems from the intrinsic short wavelength of electrons. The wavelength of an electron can be determined by the De Broglie wavelength, which is the wavelength associated with an object and is related to the momentum and mass of the object. The De Broglie wavelength equation is given by

$$\lambda = \frac{h}{p} \tag{4.5}$$

where,

h is Planck's constant (4.135×10^{-15} eV.s)
p is the momentum of the object

This momentum is typically determined by

$$p = mv \tag{4.6}$$

where,

m is the mass of the object
v is its velocity

The mass of an electron is 9.11×10^{-31} kg and it is assumed that the electron travels at a speed significantly lower than the fundamental limit of the speed of light. The velocity of the electron can be determined by its kinetic energy (E_k) (Pala and Karabiyik 2012) such that

$$E_k = \frac{1}{2} mv^2 \tag{4.7}$$

where the kinetic energy is specified in eV. This leads to the equation to determine the De Broglie wavelength of an electron through

$$\lambda = \frac{h}{\sqrt{2mE_k}} \tag{4.8}$$

which can be simplified to

$$\lambda = \frac{1.23}{\sqrt{E_k}} \qquad (4.9)$$

and is specified in nm. Figure 4.3 presents the wavelength of an electron if it is accelerated to energies of between 1 eV and 200 keV.

As shown in Figure 4.3, electron wavelength is approximately 1.23 nm for a kinetic energy of 1 eV. The electron wavelength decreases logarithmically at higher kinetic energy, to approximately 200 pm at 40 keV and 87 pm at 200 keV. From this deduction, the use of electrons in photolithography presents advantages over current photon-based lithography with much larger wavelength limitations. Limitations in electron-beam systems are therefore primarily determined by the beam column (discussed in this section) and the limits of the photoresist such as scattering, which enforces limitations on the obtainable resolution (Pala and Karabiyik 2012).

EBL systems can be defined by two primary classifications, depending on the beam shape. The two primary classifications are Gaussian and shaped beam systems. These classifications are depicted in Figure 4.4 shows the two primary classifications of EBL.

Gaussian and shaped beams. Gaussian beams consist of two scanning techniques, raster scan and vector scan; shaped beams are further categorized as fixed shapes or variable shapes (in terms of their geometry and size). Essentially, the difference between raster scan and vector scan is that raster scanning scans the entire sample from (for example) the top left to the bottom right corner, whereas vector scanning moves the electron beam between two defined points of the geometry primitives. The basic principles of raster canning and vector scanning are presented in Figure 4.5.

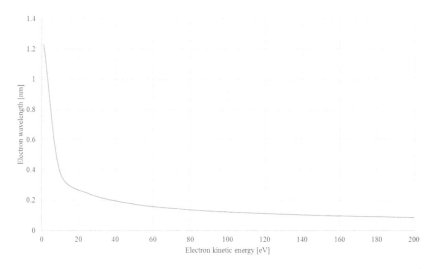

FIGURE 4.3   Electron wavelength specified in nm as a function of its kinetic energy (velocity) specified in eV. Electron kinetic energy is shown for the range of 1 eV to 200 keV.

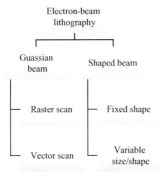

FIGURE 4.4 Primary classification of EBL systems according to beam shape. The primary classifications (Gaussian and shaped beams) are sub-classified by the scanning method.

The operation of an EBL resembles that of a scanning electron microscope (SEM), the primary dissimilarity being that the EBL transfers patterns onto a sample whereas the SEM reads the patterns from the sample. The most essential component in EBL systems is the electron source, where the electron is generated and accelerated toward its target. According to Pala and Karabiyik (2012), an ideal electron source should present

- High intensity.
- High uniformity.
- Low energy dispersion to reduce chromatic aberrations.
- Small spot size, which requires less optical correction and lowers the complexity of the system.
- High stability over typical writing times of several hours.
- Long extinction rate.

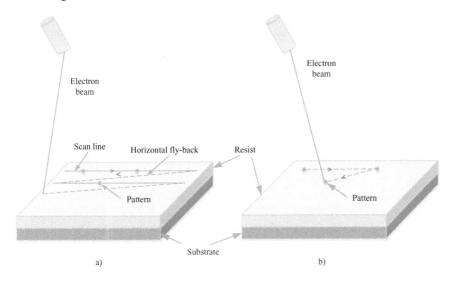

FIGURE 4.5 Basic principles of (a) raster scanning and (b) vector scanning.

Electrons can be emitted from a conducting material such as tungsten either by increasing the temperature or by applying a large electric field. The increased temperature should be high enough to overcome the work function and the electric field must be strong enough to force electrons to tunnel across the barrier. Electron beam sources such as tungsten thermionic, lanthanum hexaboride, thermal field aided emission and cold field emission sources, and the characteristics of each source, are given in detail in Pala and Karabiyik (2012).

Further terminology applicable to EBL, which defines the operation of such a device, is listed below (Altissimo 2015):

- Writing field – the largest area exposed without the electromechanical stage moving; it is a trade-off of the resolution of the system.
- Exposure element – the fixed number of exposure elements, which is determined by the resolution of the digital-to-analog converter used to generate stage movements.
- Exposure dosage – the amount of energy deposited on the substrate per unit area; it is expressed in terms of current deposited per unit area, $\mu C/cm^2$.
- System clock – also referred to as the writing speed, where a higher clock number relates to faster exposure (assuming the intensity is sufficient).
- Proximity effect – the finite number of electrons deposited next to the required spot size owing to elastic and inelastic scattering of electrons as they enter the substrate.
- Stitching – the process of adding exposure writing fields together through interferometric-controlled stage movements.
- Beam current – the quantity of incident electrons on the substrate per second, affecting the maximum obtainable resolution.

EBL tools use electromagnetic fields as opposed to mirrors and lenses to direct and focus electron beams toward wafers. The simplified operation of an EBL system is shown in Figure 4.6 (Zhou 2005). Unlike photolithography and EUV lithography.

EBL is a maskless imaging system; instead of a mask, EBL tools produce patterns by using electromagnetic fields to adjust the lateral position of the electron beam in mid-flight, as shown in Figure 4.6 (Cui 2008; Wu 2011; Zhou 2005). However, maskless imaging systems have unique constraints: since charges repel, the electrons within the confined beams are susceptible to electrostatic repulsion by other electrons as they travel toward the wafer (Schneider et al. 2012). This coulombic effect results in a broadening of the electron beam and, hence, a blurring of the aerial image (Cui 2008; Schneider et al. 2012; Wu 2011; Zhou 2005). In addition to beam broadening during the flight of the electrons, the high energy of the electrons provides them with sufficient momentum to penetrate deeply into the E-beam resist, even making it as far as the substrate's surface (Cui 2008; Schneider et al. 2012). This high-energy-induced penetration of electrons makes them vulnerable to both forward scattering (due to interactions with the top layers of the resist) and back-scattering (due to interactions with the substrate) (Schneider et al. 2012). Thus, the on-wafer resolution of EBL systems is limited by the combined blurring effects from electrostatic repulsion during the time-of-flight of the electrons, as well as scattering effects from the interaction

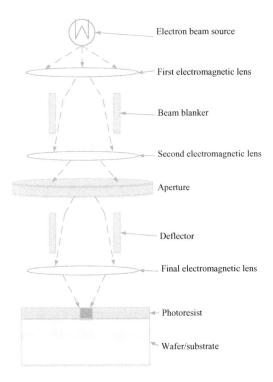

Electron beam source

First electromagnetic lens

Beam blanker

Second electromagnetic lens

Aperture

Deflector

Final electromagnetic lens

Photoresist

Wafer/substrate

FIGURE 4.6 This image shows a simplified view of the electron beam optics of a single-beam EBL tool (Zhou 2005). Electrons are emitted from the cathode (also called an electron gun) and then accelerated in the axial direction using an electric field.

between the high-energy electrons and the materials that they encounter as they reach the wafer (Cui 2008; Schneider et al. 2012; Wu 2011; Zhou 2005). By using thin resist thicknesses (to minimize electron scattering) and optimizing the operating parameters of EBL tools (such as total energy and beam current), resolutions of about 10 nm have been achieved (Cui 2008; Wu 2011; Zhou 2005). Although EBL imaging systems can theoretically achieve 1 nm beam confinement, material constraints still limit EBL resists to about 10 nm; hence, further research is needed to improve EBL technology down to the single-digit nanometer level (Ciu 2008; Zhou 2005).

Custom shapes and geometries are focused onto the electron-sensitive film (resist) on the wafer substrate through a scanning motion of the focused electron beams. The solubility of the resist changes at the points where the electron beam exposes the film. The electron beam optics within EBL imaging equipment are capable of confining the electron beams into very narrow widths, which allows them to achieve resolutions in the order of a few nanometers (Cui 2008; Wu 2011; Zhou 2005). Similar to electron microscopes, EBL tools should be kept under high vacuum during operation. Load locks are used to limit the loss of vacuum when wafers are transferred to and from EBL tools; the necessity to pump the systems back down to high vacuum after each wafer is patterned limits the wafer throughput of EBL tools (Wu 2011; Zhou 2005). Furthermore, EBL tools that only emit one electron beam at the electron source are only capable of raster scan exposure, leading to very long exposure times to pattern a wafer fully. This raster scan restriction limits wafer throughput, making single-beam EBL tools impractical for use in the high-volume manufacturing of ICs. The minimum time, T, to expose a given area for a given dose is determined by

$$T = \frac{D \times A}{I} \qquad (4.10)$$

where,

D is the dose
A is the area to be exposed
I is the electron-beam current

Fortunately, state-of-the-art EBL tools emit multiple electron beams, thus reducing the exposure times and leading to much higher wafer throughput than single-beam tools (Cui 2008; Wu 2011; Zhou 2005). However, state-of-the-art multi-beam EBL tools are still limited to wafer throughputs of 10–20 wafers per hour or 240–480 wafers per day, which is still low for use in the high-volume manufacturing of ICs (Wu 2011).

The following section reviews advanced photolithography processing techniques used to improve pattern density, optical resolution and wafer throughput by minimizing the cost typically involved in improving on photolithography techniques.

ADVANCED PHOTOLITHOGRAPHY PROCESSING TECHNIQUES

This section reviews three predominant advanced photolithography techniques that improve photomask resolution, feature density and throughput using practices that do not particularly modify the light source, the wavelength or the underlying technology used for photolithography. These techniques include pitch-split multiple exposure patterning, SIT and DSA patterning. Pitch-split multiple patterning involves the use of multiple photomasks to manufacture semiconductor ICs to split or decompose individual patterns into multiple levels. This technique can be used to improve the resolution of both DUV lithography and EUV lithography. SIT is based on spacer patterning, where a "spacer film layer is formed on the sidewall of a pre-patterned feature by deposition or a reaction of the film to a previous pattern" (Cui 2008; Indykiewicz et al. 2012; Nishimura et al. 2008). An etching process step follows to remove all the film material on the horizontal surfaces, leaving only the material on the sidewalls. DSA patterning refers to the incorporation of self-assembling constituents with customary manufacturing practices, enables current- and future-generation lithography capabilities to be enhanced and augmented and provides a strategy of nanoscale processing at reduced cost. There are two main types of DSA –/– one uses topgraphical features to physically confine the BCP film within pre-defined areas prior to patterning (grapho-epitaxy DSA or grapho-DSA); the other uses patterned areas on the wafer surface, which contain different chemical groups that induce preferential wetting of one or the other of the BCP domains at that site (chemical-epitaxy DSA or chemi-DSA). The characteristics of the BCP regulate the feature size as well as the uniformity of the resultant geometries.

The following paragraphs present the abovementioned advanced photolithography techniques.

PITCH-SPLIT MULTIPLE EXPOSURE PATTERNING

The resolution and feature density (pitch) of photolithography are both limited by diffraction as light passes through the photomask and travels toward the wafer. In order to achieve higher feature densities for optical lithography, the semiconductor industry has developed advanced processing techniques, such as pitch-split multiple patterning (double patterning, triple patterning, quadruple patterning and higher levels of patterning) (Cui 2008; Indykiewicz et al. 2012; Nishimura et al. 2008).

Pitch-split multiple patterning involves the splitting or decomposition of individual patterning levels into multiple masks, such that multiple exposures and etching steps are required to pattern an individual level fully. The decomposition of patterning levels is accomplished through the use of computational lithography software. Computation lithography therefore refers to the use of a computer algorithm to simulate the printing of lithography structures. Computation lithography uses various numerical simulations to improve the resolution and contrast of photomasks. These techniques include RET, OPC and source mask optimization (SMO). These techniques are briefly reviewed at the end of this chapter.

By dividing the design features of a specific level into multiple masks, the lateral spacing among design features on each mask is increased, which results in negligible blurring of the aerial image. Therefore, each exposure step is capable of patterning design features with high pattern fidelity, since the interference of diffraction fringes is minimized. In addition, if geometrically varying patterns such as L-shaped figures, straight bars or junctions exist, the patterns can be broken up such that each exposure can optimize a particular type of shape. An example of pattern decomposition is presented in Figure 4.7.

By using identical alignment marks on each mask, subsequent exposures can be aligned to ensure that, upon completion of all the exposures and pattern transfers, all the design features that are transferred into the patterning layer are in the correct relative positions and orientations for that level. Note that the decomposition shown in Figure 4.7 is idealized in that there is no OPC that takes diffraction limitations into account – it purely shows what a patterning level looks like when it is broken up into multiple masks. OPC principles are discussed at the end of this chapter.

There are two types of pitch-split multiple patterning. The first uses multiple exposures but only one etching step to transfer all the patterns ito a permanent layer. This type of multiple patterning is litho-freeze-litho-etch (LFLE) or litho-litho-etch (LLE), since two consecutive exposures are performed without an intermediate etch in between (Ando et al. 2008; Nakamura et al. 2010; Pieczulewski and Rosslee 2009; Robertson et al. 2011; Sugimachi et al. 2009). The second type of multiple patterning uses multiple exposures and multiple etches to pattern the complete level into permanent layers. This type of multiple patterning is designated as litho-etch-litho-etch (LELE). The basic sequence for LELE-type pitch-split double patterning is shown in Figure 4.8.

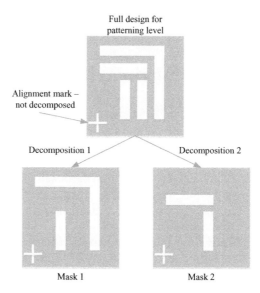

FIGURE 4.7   An ideal example of a patterning design level being decomposed into two separate masks. Note that the alignment marks are not decomposed, since the masks must line up with one another to reassemble the complete design for the patterning level properly.

Higher litho-etch pitch-split processes such as triple patterning (or LELELE) follow the same general processing mechanism as shown in Figure 4.8, but contain more patterning steps (Indykiewicz et al. 2012; Nishimura et al. 2008). The first three steps depicted in Figure 4.8 show a typical processing sequence for single-exposure patterning, except that the mask contains only half the design features for the level. Another difference is that the patterns are etched into a temporary layer (called the hard mask), which memorizes the first exposure patterns for future transfer into the patterning layer during the last step of the multiple patterning process. During the fourth processing step, the patterned hard mask is then overcoated with a thick, opaque film that planarizes (flattens) the wafer surface to avoid the topography of the hard mask interfering with the second exposure. A photoresist film stack is then deposited above the overcoat and a second mask (containing the remaining design features for the level) is aligned to the patterns in the hard mask to ensure that all features line up with each other once double patterning is completed.

During steps five through seven, the wafer is exposed and patterned with features from the second of the pitch-split masks; however, the features from the second exposure are selectively etched only into the photoresist stack and the overcoat film to avoid removal of the patterns in the hard mask. During the eighth and last step, the patterns from both exposures are etched down into the patterning layer and all of the temporary layers are removed to leave a densely patterned level with feature densities that are twice what could have been achieved had the entire level been patterned in a single exposure. In addition, LELE-type pitch-split patterning is compatible with EUV lithography so that it may be used to extend the resolution capability of EUV beyond what is currently achievable with a single-exposure EUV imaging system.

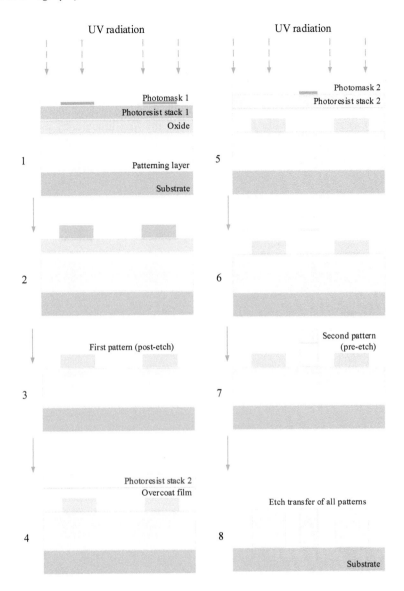

FIGURE 4.8 Process flow diagram depicting the processing sequence for the LELE-type pitch-split double patterning. Although the photolithography steps shown in the figure correspond to positive-tone photoresists, LELE is compatible with both positive-tone and negative-tone resists.

As stated earlier, LFLE/LLE pitch-split multiple patterning involves the use of two pitch-split exposures without an intervening etch step between them (Ando et al. 2008; Hori et al. 2008; Nakamura et al. 2010; Pieczulewski and Rosslee 2009; Robertson et al. 2011; Sugimachi et al. 2009). This is achieved by using a chemical agent to harden or *freeze* the patterns formed after the first photolithography step prior to coating and processing the second photoresist film stack. The general processing scheme for LFLE/LLE pitch-split double patterning is shown in Figure 4.9.

The first two steps in Figure 4.9 depict the usual photolithography process that is typical of other types of patterning. From step three to step four, a chemical agent is coated above the first exposure patterns, and then the wafer is heated to thermally induce chemical reactions that make the patterns insoluble to developer (commonly referred to as freezing). After the hardening of the first lithography patterns and

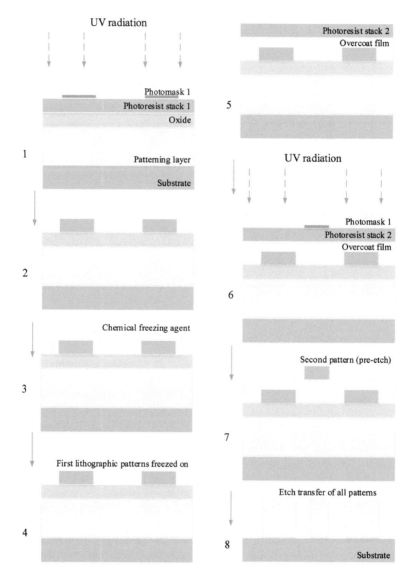

FIGURE 4.9   A process flow diagram depicting the processing sequence for the LFLE/LLE type of pitch-split double patterning.

rinsing off of residual freezing agent, the wafer can be safely coated with upper layers and the second photolithography step can proceed as normal without wiping out the first patterns. During the fourth step, the wafer is overcoated with a thick, UV-absorbent film to planarize the surface and prevent the topography of the first patterns interfering with the second photolithography process. Steps five and six represent the second photolithography patterning step. The last two steps show the selective etching of the two different patterns through all the layers of their respective film stacks and the transfer of the patterns into the permanent patterning layer. By comparing Figure 4.8 with Figure 4.9, it can be seen that, compared with LELE patterning, LFLE/LLE patterning has fewer film coats (no hard mask), fewer processing steps (a single etching step), a less complex patterning flow and, therefore, a cheaper processing flow. However, even though the freezing step protects the profile of the first patterns during subsequent processing, the second exposure does affect where shrinkage is observed. This changes the dimensions of the first patterns and introduces CD non-uniformity into the patterning level; the dimensions of the features patterned during the first and second pitch-split exposures do *not* match after processing has been completed (Nakamura et al. 2010; Sugimachi et al. 2009).

For this reason, LELE is still preferred over LFLE/LLE when high CD uniformity across a level is of paramount concern. In cases where processing simplicity and affordability are more important than CD uniformity across a level, LFLE/LLE is used. Regardless of whether LELE- or LFLE-type schemes are used, all pitch-split multiple patterning techniques result in higher pattern densities than can be reliably achieved by single-exposure patterning. The highest pitch that can be achieved by double-exposure patterning is twice what can be achieved by single-exposure patterning, with similar methodology for higher-level patterning exposures; hence, multiple patterning has allowed the semiconductor industry to extend the capabilities of photolithography into the sub-30 nm pitch nodes. However, multiple exposures require multiple masks, each of which is expensive to produce, and pitch-split patterning is more complex (more processing steps and the need for more mask-to-wafer alignments) than single-exposure patterning. Therefore, there are both economic and technical trade-offs when performing pitch-splitting.

The following paragraph reviews the advanced lithography technique of SIT to enhance resolution and feature density in microelectronic geometry patterning.

SIDEWALL IMAGE TRANSFER

The basic simplified processing sequence of SIT, which is also referred to as SADP, is shown in Figure 4.10 (Chen et al. 2014; Kanakasabapathy 2013). As shown in Figure 4.10, the first two steps of the SIT/SADP process follow the standard process flow of photolithography, where the resist used in the example of Figure 4.10 is a positive-tone resist. SIT/SADP can be used with both positive-tone and negative-tone resists. However, unlike multiple exposure pitch-split patterning, the post-lithography patterns are not etched into a memorization layer – instead, these patterns serve as templates for the spacer films, which are subsequently deposited on and conform to the templates through the process of chemical vapor deposition (CVD). This is performed during the third processing step shown in Figure 4.10. Since the spacers

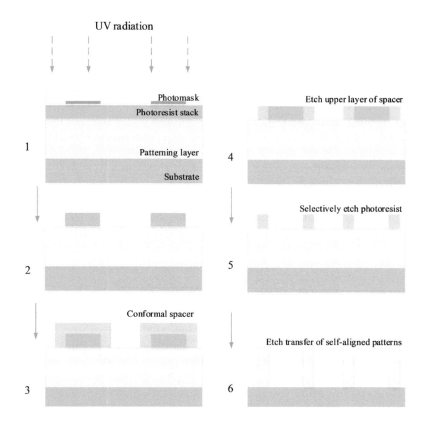

FIGURE 4.10   A process flow diagram depicting the basic processing sequence for the SIT process. It is also referred to as SADP.

are deposited by CVD onto the templates, there is near-perfect alignment of the spacer features around these templates. In contrast, mask-to-wafer alignment is a prerequisite for multiple exposure pitch-split patterning, which practically always has a finite error margin (Chen et al. 2014; Kanakasabapathy 2013; Nishimura et al. 2008; Xu et al. 2011). During the fourth and fifth steps, the top layer of the spacer film is etched to expose the resist patterns before they are selectively removed, leaving only vertically oriented spacer films adjacent to the former sidewalls of the templates (hence the name sidewall image transfer).

The spacer patterns are then transferred onto the patterning layer during the sixth and final processing step. Since the feature widths of the final patterns and the spacer layers are equivalent (the lateral dimensions of the final patterns are approximately equal to the thickness of the spacer layer), SIT/SADP patterning achieves resolutions that are beyond the diffraction limitations of the imaging system. SIT/SADP patterning is also compatible with pitch-split multiple patterning process flows such as LELE and LFLE and higher-order multiple patterning. For a combined multiple patterning/SIT flow, the multiple patterning will be performed first in order to produce the templates for the SIT spacers, but the templates will be patterned into a temporary hard mask, which will act as the pattern transfer layer. After the pitch-split templates have been formed, the final four processing steps of the SIT/SADP process, as shown in Figure 4.10, are performed. The final features are transferred from the pattern transfer layer into the patterning layer. State-of-the-art SIT/SADP processing is currently able to achieve feature resolutions below 20 nm using 193i photolithography tools (Chen et al. 2014; Kanakasabapathy 2013; Nishimura et al. 2008; Xu et al. 2011). Fortunately, the SIT/SADP processing technique is also compatible with other imaging systems such as EUV and EBL and can be used with next-generation imaging systems to extend feature-scaling capabilities of those imaging technologies as they eventually reach the resolution limits of the lithographic tools.

One major drawback of the SIT/SADP technique is that the lateral spacing among the features in the final patterns is not uniform, even if the templates (the first patterns) have a practically uniform pitch. This occurs because the self-aligned patterns are produced from the sidewalls of the templates, but the lateral spacing among self-aligned patterns is dependent on whether they are produced by the same template or not. To illustrate this drawback, consider the following example: if two self-aligned patterns are on either side of the same template, their lateral spacing will be equal to the width of the template. However, if the patterns correspond to adjacent templates, the lateral spacing among the patterns will be equal to the difference between the *pitch* and the *width* of the templates. Therefore, the pitch among self-aligned patterns from adjacent templates is the mandrel pitch, which translates to the template CD. This variation in the lateral spacing of the final patterns is referred to as pitch walking and is an undesirable characteristic of SIT/SADP processing in that it introduces non-uniformity in the performance of the device and adjacent devices, when processing has been completed.

DIRECTED SELF-ASSEMBLY PATTERNING

DSA is a hybrid patterning technique in which photolithography is used to outline guiding patterns that direct the morphology of BCP chains during thermally assisted self-assembly (Chen et al. 2014; Kanakasabapathy 2013; Nishimura et al. 2008; Xu et al. 2011). When BCP thin films are coated and annealed, the BCP chains phase-separate into nanoscale domains of specific geometries (the morphology of the phase-separated domains is determined by the relative composition of each polymer block of the BCP) (Bates and Frederickson 1990). For example, polystyrene-poly (methylmethacrylate) (PS-PMMA) BCPs with 50/50 mixtures will separate into lamellar domains (the domain geometry is planar) upon annealing, whereas PS-PMMA BCPs with compositions of 70/30 or 30/70 will phase-separate into hexagonal closed-packed (HCP) cylinder morphologies (Bates and Frederickson 1990). Both lamellar and HCP cylinder morphologies are relevant to complementary metal-oxide semiconductor (CMOS) architecture; vertical standing lamellar phases can be used to pattern densely packed fins for FinFET devices, whereas vertically oriented HCP cylinders can be used to pattern densely packed interlevel vias (Chan et al. 2014; Cheng et al. 2010; Cushen et al. 2015; Jeong et al. 2013; Ruiz et al. 2012; Son et al. 2013; Stoykovich et al. 2007; Tsai et al. 2012; Xiao et al. 2005). In order to ensure that the BCP domains form vertically oriented rather than horizontally oriented morphologies, a neutral orientation layer is used to prevent preferential surface wetting of either of the polymer blocks when the BCP films are coated onto the wafer

(Chan et al. 2014; Cheng et al. 2010; Cushen et al. 2015; Jeong et al. 2013; Ruiz et al. 2012; Son et al. 2013; Stoykovich et al. 2007; Tsai et al. 2012; Xiao et al. 2005). The neutral layer film consists of a random copolymer containing the same types of polymer chains as the BCP film. For example, if the BCP is PS-PMMA, then the neutral layer will be a random copolymer of PS and PMMA polymer chains so that both types of chains have equal wetting behavior on the neutral layer (Chan et al. 2014; Cheng et al. 2010; Cushen et al. 2015; Jeong et al. 2013; Ruiz et al. 2012; Son et al. 2013; Stoykovich et al. 2007; Tsai et al. 2012; Xiao et al. 2005).

The purpose of the guiding patterns is to direct the phase separation of the BCP chains upon annealing, such that the BCP domains form patterns of the desired dimensions, pitch and geometry to match the design level (Tsai et al. 2012; Jeong et al. 2013; Ruiz et al. 2012; Son et al. 2013). The DSA guiding patterns are patterned at a relaxed pitch, which allows optical lithography to be used without considering the diffraction limitations. The final feature sizes are driven by thermodynamic forces; imaging constraints do not limit the resolution of DSA patterning (Chan et al. 2014; Cheng et al. 2010; Cushen et al. 2015; Jeong et al. 2013; Ruiz et al. 2012; Son et al. 2013; Stoykovich et al. 2007; Tsai et al. 2012; Xiao et al. 2005).

Hence, DSA acts as a pitch-multiplication technique, in that the DSA patterns are at a higher pitch density than the guiding patterns (Somervell et al. 2015). Since the sizes of the domains are determined by the polymer chain sizes, the resolution of DSA is beyond the capabilities of the imaging system: using BCPs of appropriate polymer chain lengths, DSA can achieve domain sizes below 20 nm without the need for pitch-split double patterning or SIT/SADP.

There are two types of DSA patterning processing flows, namely

- Grapho-epitaxy-based DSA (commonly abbreviated as grapho-DSA)
- Chemical-epitaxy-based DSA (commonly abbreviated as chemi-DSA) (Azuma et al. 2015; Bates and Frederickson 1990; Cheng et al. 2010; Jeong et al. 2013; Liu and Chang 2009; Ruiz et al. 2012; Somervell et al. 2015; Son et al. 2013; Tsai et al. 2012).

Grapho-DSA uses physical relief features to confine the BCP films, as shown in Figure 4.11.

The first two processing steps in Figure 4.11 depict the typical steps of photolithography (although the imaging corresponds to positive-tone development, DSA is compatible with both positive- and negative-tone photolithography). The photoresist patterns act as the guiding patterns for the later phase separation of the BCP domains; however, in order for the DSA patterns to all be parallel to one another, the width of the guiding patterns must be an integer multiple of the natural periodicity of the BCP domains (the L_0) (Bates and Frederickson 1990; Cheng et al. 2010; Tsai et al. 2012). The photoresist stack is situated above a neutral layer such that after the development of the resist patterns, a BCP film can be coated above the neutral layer, as shown in the third processing step in Figure 4.11. In order to achieve vertical orientation of DSA patterns during the annealing, the thickness of the BCP film must match a half-integer multiple of the L_0 of the BCP ($0.5 \times L_0$, $1.5 \times L_0$, $2.5 \times L_0$ and so on) (Bates and Frederickson 1990). During the fourth processing step of Figure 4.11, the BCP film is annealed to thermally induce

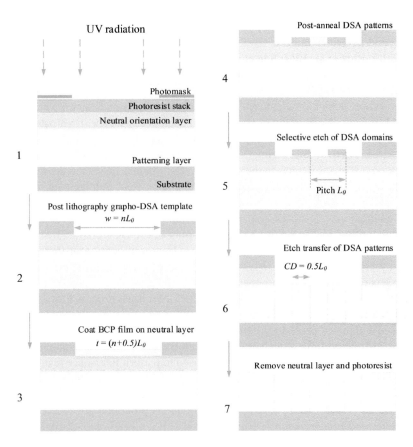

FIGURE 4.11 A process flow diagram depicting the basic processing sequence for grapho-DSA.

phase separation of the polymer chains into vertically oriented DSA patterns. The pitch of the DSA patterns matches the L_0 of the polymer, while the width of each polymer block's domain is equal to half of the L_0 (Bates and Frederickson 1990).

Since there are multiple DSA domains within a single guiding pattern, DSA essentially acts as a type of pitch-multiplication patterning. In addition, when the BCP composition matches the intended morphology of the DSA patterns, and the proportions of the guiding outline width and the thickness of the BCP are equal to the correct multipliers of L_0, as shown in Figure 4.11, the geometry of the DSA patterns will match the geometry of the guiding patterns (Jeong et al. 2013; Lui and Chang 2009; Peters et al. 2015; Ruiz et al. 2012; Son et al. 2013; Tsai et al. 2012). Therefore, grapho-DSA patterning is a type of sidewall image transfer process since the sidewalls of the guiding patterns direct the self-assembly of DSA patterns in the lateral direction (this occurs because of wetting of the resist sidewalls by one of the polymer blocks). The phase separation begins at the sidewalls of the guiding patterns. If the grapho-DSA guiding patterns have rectangular troughs with straight sidewalls, then the BCP (with 50/50 composition) coated and annealed within the guiding pattern will produce vertically oriented lamellar DSA patterns (commonly referred to as line/space patterns) with straight sidewalls (Jeong et al. 2013; Lui and Chang 2009; Peters et al. 2015; Ruiz et al. 2012; Son et al. 2013; Tsai et al. 2012). If the guiding patterns are cylindrical, then the final DSA patterns will be cylindrical; however, the DSA pattern sizes will be smaller since self-assembly of the BCP chains will form concentric circles within each guiding pattern (Jeong et al. 2013; Lui and Chang 2009; Peters et al. 2015; Ruiz et al. 2012; Son et al. 2013; Tsai et al. 2012). Hence, DSA patterning of holes leads to a thermodynamically driven reduction in the feature size of the guiding pattern, referred to as hole shrink (Somervell et al. 2015) A drawback of grapho-DSA patterning is that any defects in the guiding patterns induce defects in the self-assembly; for example, any sidewall roughness in the grapho-DSA guiding patterns will be replicated in the DSA patterns during annealing.

In contrast to grapho-DSA patterning, chemi-DSA patterning does not use physical confinement to guide the self-assembly of the BCP chains. Instead, photolithography is used in combination with surface modification of the wafer surface to produce very thin underlying guiding patterns that induce surface-assisted phase separation of BCP films that are coated and annealed on top of the guiding patterns (Chan et al. 2014; Somervell et al. 2015). Multiple process flows have been developed for chemi-DSA. These differ in the type of surface-assisted modification used and how it is produced, but all of them result in DSA patterns covering an entire substrate with the same pattern characteristics (feature size and pitch across the wafer). Therefore, when chemi-DSA is used to pattern line/space features using BCP films with 50/50 compositions, a *sea of fins* is produced after selective etching, and when chemi-DSA is used to pattern HCP cylinders using BCP films with either 60/40 or 40/60 compositions, a *sea of vias* is produced after selective etching (Bates and Frederickson 1990; Chan et al. 2014; Cheng et al. 2010; Somervell et al. 2015). Grapho-DSA patterning has the advantage of allowing different DSA patterns to be produced across a wafer during a single annealing step – this is because multiple types of guiding patterns can be patterned across the wafer prior to coating a BCP film.

An example of a chemi-DSA process flow is shown in Figure 4.12 (this type of chemi-DSA is sometimes referred to as lift-off type DSA or IBM/ARC flow chemi-DSA). The first two processing steps of the lift-off-type chemi-DSA flow consist of regular photolithography patterning, as shown in Figure 4.12. Although Figure 4.12 depicts a positive-tone photoresist, the chemi-DSA is compatible with both positive- and negative-tone photolithography. During the third processing step in Figure 4.12 reactive neutral orientation layer (referred to as a brush) is coated onto the wafer surface and then reacts with the underlying patterning layer (Chan et al. 2014; Cheng et al. 2010). Note that the brush does not form a continuous, conformal coat across the wafer and the sidewalls of the resist are not covered by the brush. In addition, the reactive chemical groups in the brush do not react with the resist. During the fourth process step, the resist is lifted off (removed) along with the brush that was lying above the top surfaces of the resist patterns (Cheng et al. 2010). The excess layers of brush, the unreacted upper layers, are rinsed off, leaving behind a wafer with very thin patterns of brush material; these guiding patterns will later direct the chemical-assisted self-assembly of the BCP domains during the annealing step

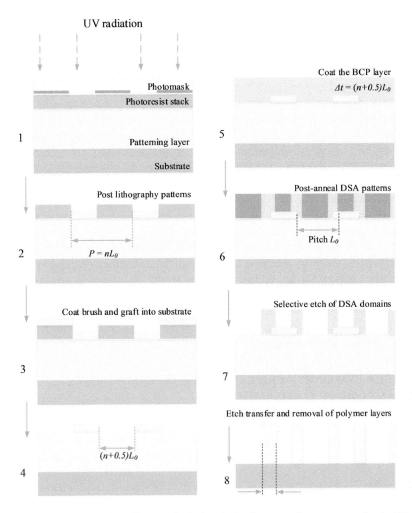

FIGURE 4.12 Process flow diagram depicting the basic processing sequence for the lift-off type of ARC flow chemi-DSA.

(Chan et al. 2014; Jeong et al. 2013). The pitch of these guiding patterns matches an integer multiple of the L_0 of the BCP and the width of the guiding patterns and the space between them match a half-integer multiple of L_0 ($0.5 \times L_0$, $1.5 \times L_0$, $2.5 \times L_0$ and so on) (Bates and Frederickson 1990). During the fifth processing step, a BCP is coated to a thickness that is a half-integer multiple of its L_0 to ensure that, during phase separation, the polymer block domains are vertically oriented. During the sixth and seventh steps, the BCP is annealed to thermally induced phase separation, and one of the BCP domains is then selectively etched to leave behind dense DSA patterns with widths equal to half the L_0 at a pitch of L_0 across the entire wafer surface (Chan et al. 2014; Cheng et al. 2010; Jeong et al. 2013).

Thermodynamic principles allow both grapho- and chemi-DSA patterning to produce dense features due to pitch multiplication of the guiding patterns. Since DSA patterning is not limited by diffraction, it is capable of feature sizes and feature densities that are beyond the capabilities of the photolithography tools. Researchers have successfully patterned sub-10 nm features at sub-20 nm pitch using DSA patterning; furthermore, DSA is compatible with next-generation imaging technologies (such as EUV and EBL) (Jeong et al. 2013; Son et al. 2013). In addition, DSA is compatible with pitch-split double patterning, since LELE and LFLE can be used to form the DSA guiding patterns. By performing DSA after multiple patterning of guiding patterns, the final DSA patterns can be produced at high feature densities (tighter pitches) than with 193i lithography-based multiple patterning processing. Since DSA is compatible with EUV and EBL technologies, it can be used in future nodes even when optical lithography technology becomes obsolete. However, as mentioned earlier, DSA is susceptible to the transfer of defects from the guiding patterns (Kim 2014).

In addition, DSA is increasingly sensitive to other defects, such as foreign contaminants, merging/micro-bridging, pattern collapse, swelling and/or delamination of guiding patterns (Somervell et al. 2015). Metrology of chemi-DSA guiding patterns and final DSA patterns is also difficult because polymers tend to build up charge when exposed to an electron beam, as occurs in a scanning electron microscope. Because of the potential for DSA in the scaling of feature sizes for future-generation semiconductor manufacturing, the semiconductor industry is devoting abundant resources to solve the problems related to a high defect count of DSA patterns, as well as difficulties related to metrology.

OPTICAL PROXIMITY CORRECTION AND RESOLUTION ENHANCEMENT TECHNIQUES

As feature dimensions and feature density continue to scale in line with Moore's law, imaging limitations due to diffractive optics become more apparent (Cui 2008; Rudnitsky and Serdyuk 2012). The reduction in patterning fidelity is due to the increased interference of propagating waves that form aerial images as patterning requirements approach the resolution capabilities of imaging equipment. A formal problem statement derived in Cobb (1998) follows: given a sought-after geometry on the sample/wafer, the problem is to get a mask design where the final geometric pattern left behind after the entire lithography procedure is as close as possible to the anticipated shape, including the specification of its feature density and contrast. In order to account for imaging limitations, computational lithography electronic design automation (EDA) software is used to model the flight of electromagnetic waves and their interference as they are diffracted by a virtual mask containing digital patterns that are intended to be patterned on the wafer (Cai et al. 2008; Kawahira et al. 1999; Lin et al. 1999; Lin et al. 2011; Nouralishahi et al. 2008; Rudnitsy and Serdyuk 2012; Schneider et al. 2012; Shibuya 1997; Yi and Chang 2007; Yoshikawa et al. 2015). With empirical optical models, wafer-level aerial image defects are predicted for different shapes of dimensions, as exemplified by the diagrams in Figure 4.13. OPC results are then fed into algorithms that back-calculate the requisite mask-level design features (serifs) that would mostly closely produce the desired aerial image at the wafer level (Cai et al. 2008). Counteracting these limitations is referred to as OPC RET. OPC software is furthermore able to account for the imaging limitations of EBL (beam

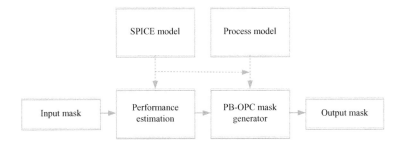

FIGURE 4.13   A process flow diagram depicting the simplified methodology of the PB-OPC method presented in Teh et al. (2010).

broadening, forward and backward scattering), hence OPC RET is an essential tool for circuit designers and patterning engineers in the quest to scale devices ever smaller (Cui 2008; Schneider et al. 2012; Yi and Chang 2007). Because of the significant computing requirements that rigorous OPC models utilize, semiconductor manufacturers must provide high-end mainframe clusters to their engineers in order to support OPC RET. Since purchasing or renting time on high-end computing clusters is expensive, OPC RET escalates the cost of device scaling (Hutcheson 2009; Sullivan 2007).

The benefits of OPC include more accurate CD and improved edge assignment. Furthermore, OPC presents the capability to alter process windows for different kinds of structures so that the intersection of the process window is increased. This permits more dependable pattern transfer at lesser k_1 values. The technique, however, does not warrant electrical performances against distortion as a result of the added or subtracted shapes on the optimized photomask. Additional post-layout simulation tools and inputs from process models are required to ensure optimal and expected electrical behavior. Such a mask generation algorithm is presented in Teh et al. (2010), where the process flow to generate an output mask is achieved through a performance-based OPC methodology (PB-OPC). The simplified representation of this methodology is given in Figure 4.13.

As shown in Figure 4.13, the performance-based OPC methodology presented in Teh et al. (2010) receives a desired input mask design for a specific circuit. The performance of this design is estimated (simulated) using reference design libraries, typically provided by a foundry for a specific process. The system then enters a recursive loop, referred to as PB-OPC mask generation, which iterates the modified mask design until the electrical circuit performance congregates with the desired performance. Once convergence is achieved, the output mask, which is used as an alternative to the input mask, is generated. The performance deviation error (PDE) of the generated output mask is determined by

$$\text{PDE} = \frac{P_{\text{post-lithography}} - P_{\text{desired}}}{P_{\text{desired}}} \times 100\% \qquad (4.11)$$

where,

P_{desired} refers to the desired circuit performance

$P_{\text{post-lithography}}$ refers to the real circuit performance subsequent to each iterative loop (Teh et al. 2010)

In the IC microelectronic component manufacturing industry, the benefits offered by convergence to an OPC output mask can be measured by

- Increased yield for a particular minimum feature size from improved process windows.
- Improved circuit performance from linewidth consistency.
- The implementation of design rules of smaller features (Cobb 1998).

A simplified functional diagram of an OPC system with steps that include fragmentation to generate variables, cost function, edge updates and simulation updates is presented in Figure 4.14. An OPC system, as shown in Figure 4.14, first fragments photomask polygons into edge and corner segments, which will be stepped through by fine control systems. Fragmentation disrupts edges into reduced edge subdivisions to permit increased degrees of freedom for the duration of OPC movement. The number of vertices produced by OPC affects the algorithm speed and resources required to perform this operation. Several variations of fragmentation algorithms exist, including

- 'Constant distance fragmentation' leading to reduced edge subdivisions of fixed length.
- Adaptive fragmentation, which only fragments polygons if a higher degree of freedom is required.
- Intra-feature fragmentation, a subcategory of adaptive fragmentation, which causes fragmentation to be added alongside the two edges that link a junction.
- Inter-feature fragmentation, also a subcategory of adaptive fragmentation, which causes edges to be created beside an edge where junctions are sufficiently close to interact.

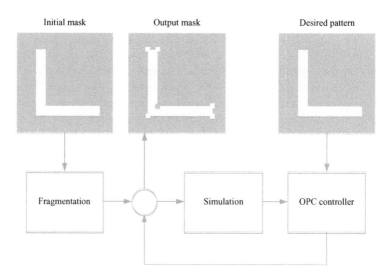

FIGURE 4.14   Simplified functional diagram of an OPC system.

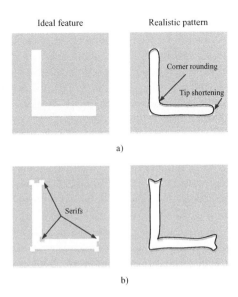

FIGURE 4.15   This flow diagram depicts the basic steps by which OPC models are used to predict and correct for the diffraction-induced limitations of lithography tools.

The following step involves sporadically chosen controller sites fitted at specific positions on the sample/wafer. The fragmented patterns are the optimization parameters and can be manipulated and offset from their distinctive locations. As the OPC advances, tiny perturbation polygons are added or subtracted from the mask, which results in a comparable photomask with improved edge patterns at specified locations on the wafer.

Even with the incorporation of advanced patterning techniques, such as pitch-split multiple patterning and SIT/SADP processing, poorly resolved features become more prevalent. Frequently encountered patterning defects include corner rounding, micro-bridging and merging (Cai et al. 2008; Somervell et al. 2015). Examples of corner rounding and line end (or tip) shortening are shown in Figure 4.15.

In Figure 4.15a, the solid white shapes depict the mask design features, while the black lines symbolize the outlines of the OPC-predicted patterns. The rectangular shapes in Figure 4.15b are examples of serifs, which are used to correct the design features so that the actual wafer-level patterns are as close as possible to the intended patterns.

Although OPC and RET enable current-generation (as well as future-generation) lithography equipment to predict and correct diffraction-induced limitations, certain material and processing-based patterning limitations, which lower the resolution and pattern density of photomask pattern transfers, still exist. These material and processing-based patterning limitations are reviewed in the following section.

MATERIAL AND PROCESSING-BASED PATTERNING LIMITATIONS

Although OPC models are effective at simulating the effects of diffraction on pattern fidelity, they are unable to account for the patterning limitations that arise from material properties (for example, the specific properties of the photoresist used) or those arising from various processing parameters such as the plasma conditions used during the etch transfer of patterns. Photoresist formulations contain a variety of components (photosensitive polymer, photo-acid generator and surfactants), the exact composition of which are proprietary in nature (Bartczak and Galeski 2014; Bulgakova et al. 2014; Guy et al. 2013; Ito 2005). In addition, each photoresist formulation possesses different patterning characteristics and is optimized for specific types of patterns (for example, line/space, vias, two-dimensional patterns or unidirectional patterns), different feature densities (dense or isolated) and different process flows (for example, LELE, LFLE, grapho-DSA or chemi-DSA). Patterning technicians carry out a large number of patterning trials in order to evaluate photoresist samples that are provided by chemical suppliers (Mesawich et al. 2009). Thus, OPC RET does not absolve semiconductor manufacturers of the need to carry out costly and time-consuming rounds of trial and error runs to determine the optimal materials and processing conditions to achieve the highest pattern fidelity for each patterning level (Nouralishahi et al. 2008; Yoshikawa et al. 2015).

Although the specific material properties of every photoresist are confidential and held by the chemical supplier, the same types of photoresist for a particular wavelength (positive-tone or negative-tone) have some generic properties (Bartczak and Galeski 2014; Bulgakova et al. 2014; Guy et al. 2013; Ito 2005). For any wavelength used by an exposure tool, the photosensitive polymer in the photoresist will have the same intrinsic qualities. This is because only specific polymeric chains have the requisite solubility, mechanical rigidity and optical transparency (at a particular wavelength) to allow it to be used in a photoresist formulation (Bulgakova et al. 2014; Guy et al. 2013; Ito 2005). The photosensitive polymers used in 193 nm photolithography resists (also called ArF resists) are based on methacrylate-based main chains, but their molecular weights vary with their formulation (Bulgakova et al. 2014; Guy et al. 2013; Ito 2005; Morita 2016). Upon exposure to 193 nm wavelength light, the photo-acid generator chemical groups undergo photochemical reactions that release strong acids into the resist. These acids attack protective functional side groups in the methacrylate-based polymers, resulting in a solubility switch (insoluble to soluble for positive-tone resists and vice versa for negative-tone resists) on a developer. Reaction kinetics for the photochemical activation of the acids, the chemical amplification of the acids and the dissolution rate of the soluble regions of the polymer chains are all dependent on both physical (for example, temperature) and chemical (for example, acid concentration) conditions. Over time, empirical models have been used to generalize the mechanical, thermodynamic and kinetic characteristics of these different types of chemical reactions (Bulgakova et al. 2014; Guy et al. 2013; Ito 2005; Lin et al. 1999; Morita 2016). These empirical models require experimental data from patterning trials in order to determine the values of the chemical and physical parameters used in these models (Yoshikawa et al. 2015; Mesawich et al. 2009). Once metrology data are fed into the empirical models, the patterning characteristics of the corresponding resist can be estimated to predict important patterning characteristics, such as

- The illumination doses required to fully activate the photo-acid generators in the photoresist.
- The contrast curves in the resist after exposure.

- The acid profile and acid diffusion within the resist.
- The dissolution rates of the soluble regions of the resist in developers.
- The line edge roughness of patterns after development.
- The standing waves in the resist after the post-exposure bake (Guy et al. 2013; Ciu 2008; Ito 2005; Bulgakova et al. 2014; Bartczak and Galeski 2014; Morita 2016).

Trial and error patterning runs are still necessary to verify the predictions of the chemical models, as well as to accurately determine the patterning defects due to material properties such as pattern collapse, swelling and delamination of resist material during the development of patterns, all of which affect the process window of each resist (Mesawich et al. 2009; Somervell et al. 2015; Yoshikawa et al. 2015).

The following section reviews metrology-based OPC verification and design rule checks (DRC) implemented in photolithography techniques to characterize, predict and mitigate processing errors.

METROLOGY-BASED OPC VERIFICATION AND DESIGN RULE CHECKS

Because of the well-established background of the electromagnetic equations used in OPC models and the computing power available from modern computing clusters, OPC RET is able to predict the wafer-level aerial images with fairly reasonable accuracy, depending on the computer resources assigned to each task. To verify the accuracy of optical models, as well as to account for the patterning defects arising from the extraneous factors mentioned in the previous section, patterning experiments must be performed and characterized using SEM and transmission electron microscopy (TEM) (Schneider et al. 2012; Yoshikawa et al. 2015). With the goal of facilitating OPC verification, circuit designers create mask features specifically designed to test the accuracy of OPC predictions for different shapes, dimensions, illumination characteristics and photoresist types (Lin et al. 1999; Shibuya 1997; Yoshikawa et al. 2015). Among the errors arising from practical patterning experiments is the mask enhancement error factor (MEEF), the ratio that measures the difference in dimensions between the mask and the wafer. Once wafers are patterned and measured using electron microscopy, MEEF data are used to scale design features accurately from the mask to a wafer and vice versa. Other design test structures, such as wafer characterization index features, are used to optimize the illumination conditions for dense and isolated features across a wafer. The optimal conditions for dense and isolated features are very different, meaning that a single illumination condition cannot be used to pattern both types of features with high fidelity. Using metrology data from multiple patterning experiments along with improvements in OPC model predictions, design engineers are able to work hand in hand with patterning engineers to define design rules for the design features of present and future technology nodes. Design rules affect every level of circuit design, such as

- The minimum acceptable lateral spacing between line edges (the tip-to-tip distance).
- The minimum pitch between dense features.

- The minimum CD for lines and spaces.
- The minimum hole diameter of vias.

These rules are then specified in algorithms, which automatically check all the design elements of each patterning level to avoid overly aggressive patterning requirements that would inadvertently lead to poor production yields. All violations found by DRC are highlighted and rejected as acceptable design elements when designing masks; in this way, impractical design elements are curtailed at the design phase. Only when all the design elements of a patterning level have passed all DRCs is a mask approved for manufacturing. Although DRC leads to improved yields, there are no guarantees that DRC-approved features will be easy to pattern with high fidelity. This is because DRC typically relies on historical experimental data from previous nodes. Therefore, feature scaling always pushes the boundaries of patterning capabilities; hence, semiconductor manufacturers, equipment providers and chemical suppliers must continue to develop new materials, innovative processing flows and advances in equipment to support the design and manufacture of next-generation devices for future technology nodes (Banerjee 2007; Cusumano and Yoffie 2016; Czerniak 2015; Davis 2015; Golio 2015; Holt 2016; Mack 2015; Shalf and Leland 2015; Vardi 2014).

CONCLUSION

This chapter evaluates various photolithography enhancement techniques on a process level as well as future-generation technologies, which will enable feature sizes and resolution enhancements to ensure the longevity of Moore's law. The first part of the chapter reviews photolithography resolution-enhancement techniques such as immersion photolithography, EUVL and EBL. These techniques offer the largest degree of lithography improvements that can be obtained by current- and future-generation technologies. Each of these technologies, its underlying drivers, advantages, disadvantages and limitations are evaluated in this chapter. Following these enhancement techniques, the chapter also reviews advanced photolithographic techniques that improve upon current-generation lithography equipment and offer improvements at minimal cost. These processing techniques include pitch-split multiple exposure patterning, SIT and DSA patterning. The section provides details of process flow diagrams for all processing techniques discussed.

To summarize the contribution of presenting various lithography enhancement techniques, this chapter also assesses OPC and RET, since a reduction in patterning fidelity is due to increased interference in propagating waves that form the aerial images as patterning requirements approach the resolution capabilities of imaging equipment. Material and processing-based patterning limitations further restrict the resolution and pattern density with high fidelity of decreasing node sizes. Limitations and important considerations are discussed in this chapter. The final section reviews metrology-based OPC verification and DRC implementation to reduce the number of defective devices and to improve yield as well as wafer throughput in high-volume IC production environments.

REFERENCES

Altissimo, M. (2015). E-beam lithography for micro-/nanofabrication. Retrieved 17 February 2017 from http://www.ncbi.nln.nih.gov

Ando, T., Takeshita, M., Takasu, R., Yoshii, Y., Iwashita, J., Matsumaru, S., Abe, S., Iwai, T. (2008). Pattern freezing process free litho-litho-etch double patterning. *Proceedings SPIE 7140, Lithography Asia 2008*, 71402H-71402H-8.

Azuma, T., Seino, Y., Sato, H., Kasahara, Y., Kobayashi, K., Kubota, H., Kanai, H., Kodera, K., Kihara, N., Kawamonzen, Y., Nomura, S., Miyagi, K., Minegishi, S., Tobana, T., Shiraishi, M. (2015). Electrical yield verification of half-pitch 15 nm patterns using directed self-assembly of polystyrene-block-poly (methyl methacrylate). *Journal of Vacuum Science & Technology B*, 33, 06F302.

Bakshi, V., Lebert, R., Jaegle, B., Wies, C., Stamm, U., Kleinschmidt, J., Schriever, G., Ziener, C., Corthout, M., Pankert, J., Bergmann, K., Neff, W., Egbert, A., Gustafson, D. (2007). Status report on EUV source development and EUV source applications in EUVL. *23rd European Mask and Lithography Conference (EMLC)*, 1–11.

Banerjee, S. (2007). New materials and structures for transistors based on spin, charge and wavefunction phase control. *AIP Conference Proceedings*, 931, 445–448.

Bartczak, Z. Galeski, A. (2014). Mechanical properties of polymer blends. *Polymer Blends Handbook*, A. L. Utracki and A. C. Wilkie, Eds. Dordrecht: Springer Netherlands, 1203–1297.

Bates, F. S., Fredrickson, G. H. (1990). Block copolymer thermodynamics: Theory and experiment. *Annual Review of Physical Chemistry*, 41, 525–557.

Bodermann, B., Wurm, M., Diener, A., Scholze, F., Gross, H. (2009). EUV and DUV scatterometry for CD and edge profile metrology on EUV masks. *25th European Mask and Lithography Conference (EMLC)*, 1–12.

Bulgakova, S. A., Gurova, D. A., Zaitsev, S. D., Kulikov, E. E., Skorokhodov, E. V., Toropov, M. N., Pestov, A. E., Chkhalo, N. I., Salashchenko, N. N. (2014). Effect of polymer matrix and photoacid generator on the lithographic properties of chemically amplified photoresist. *Russian Microelectronics*, 43, 392–400.

Cai, Y., Zhou, Q., Hong, X., Shi, R., Wang, Y. (2008). Application of optical proximity correction technology. *Science in China Series F: Information Sciences*, 51, 213–224, 2008.

Chan, B. T., Tahara, S., Parnell, D., Delgadillo, P. A. R., Gronheid, R., de Marneffe, J., Xu, K., Nishimura, E., Boullart, W. (2014). 28 nm pitch of line/space pattern transfer into silicon substrates with chemo-epitaxy directed self-assembly (DSA) process flow. *Microelectronic Engineering*, 123, 180–186.

Chen, F. T., Chen, W., Tsai, M., Ku, T. (2014). Sidewall profile engineering for the reduction of cut exposures in self-aligned pitch division patterning. *Journal of Micro/Nanolithography, MEMS, and MOEMS*, 13, 011008.

Cheng, J. Y., Sanders, D. P., Truong, H. D., Harrer, S., Friz, A., Holmes, S., Colburn, M., Hinsberg, W. D. (2010). Simple and versatile methods to integrate directed self-assembly with optical lithography using a polarity-switched photoresist. *ACS Nano*, 4, 4815–4823.

Chua, G. S., Tay, C. J., Quan, C., Lin, Q. (2004). Performance improvement in gate level lithography using resolution enhancement techniques. *Microelectronic engineering*, 75, 155–164.

Cobb, N. B. (1998). *Fast optical and process proximity correction algorithms for integrated circuit manufacturing*. Doctoral Thesis. Graduate Division of the University of California at Berkeley.

Cui, Z. (2008). *Nanofabrication by Photons. Nanofabrication: Principles, Capabilities and Limits*. Boston, MA: Springer US, 7–76.

Cushen, J., Wan, L., Blachut, G., Maher, M. J., Albrecht, T. R., Ellison, C. J., Willson, C. G., Ruiz, R. (2015). Double-patterned sidewall directed self-assembly and pattern transfer of sub-10 nm PTMSS-b-PMOST. *ACS Applied Materials & Interfaces*, 7, 13476–13483.

Cusumano, M. A., Yoffie, D. B. (2016). Extrapolating from Moore's Law. *Communications ACM*, 59, 33–35, January 2016.

Czerniak, M. (2015). What lies beneath? 50 years of enabling Moore's Law. *Solid State Technology*, 58, 25–28.

Dair, G. T., Ashman, R. A., Eikelboom, R. H., Reinholz, F., van Saarloos, P. P. (2001). Absorption of 193- and 213-nm laser wavelengths in sodium chloride solution and balanced salt solution. *Archives of Ophthalmology*, 119, 533–537.

Davis, S. (2015). The IoT and the next 50 years of Moore's Law. *Solid State Technology*, 58, 13.

Denbeaux, G., Garg, R., Waterman, J., Mbanaso, C., Netten, J., Brainard, R., Fan, Y. J., Yankulin, L., Antohe, A., DeMarco, K., Jaffe, M., Waldron, M., Dean, K. (2007). Quantitative measurement of EUV resist outgassing. *23rd European Mask and Lithography Conference (EMLC)*, 1–5.

Endo, A. (2014). Extendibility evaluation of industrial EUV source technologies for kW average power and 6.x nm wavelength operation. *Journal of Modern Physics*, 5, 285–295.

Golio, M. (2015). Fifty years of Moore's Law. *Proceedings of IEEE*, 103, 1932–1937.

Guy, O. J., Burwell, G., Castaing, A., Walker, K. D. (2013). Photochemistry in electronics. *Applied Photochemistry*, C. R. Evans, P. Douglas and D. H. Burrow, Eds. Dordrecht: Springer Netherlands, 435–465.

Hada, H., Hirayama, T., Shiono, D., Onodera, J., Watanabe, T., Lee, S., Kinoshita, H. (2004). Outgassing characteristics of low molecular weight resist for EUVL. *2004 International Microprocesses and Nanotechnology Conference, 2004. Digest of Papers,* 234–235.

Holt, W. M. (2016). 1.1 Moore's law: A path going forward. *IEEE International Solid-State Circuits Conference (ISSCC)*, 2016, 8–13.

Hori, M., Nagai, T., Nakamura, A., Abe, T., Wakamatsu, G., Kakizawa, T., Anno, Y., Sugiura, M., Kusumoto, S., Yamaguchi, Y., Shimokawa, T. (2008). Sub-40-nm half-pitch double patterning with resist freezing process. *Proceedings SPIE 6923, Advances in Resist Materials* and *Processing Technology XXV*, 69230H-69230H-8.

Huff, H. R., Gilmer, D. C. (2005) *High Dielectric Constant Materials: VLSI MOSFET Applications.* Berlin, Heidelberg: Springer.

Hutcheson, G. (2009). The economic implications of Moore's law. *Into the Nano Era*, 1st ed., H. Huff, Ed. Berlin, Heidelberg: Springer, 11.

Indykiewicz, K., Haj, M., Paszkiewicz, B. (2012). The new method of fabrication of submicron structures by optical lithography with mask shifting and mask rotation. *Central European Journal of Physics*, 11, 219–225.

Ito, H. (2005). *Chemical Amplification Resists for Microlithography. Micro-Lithography – Molecular Imprinting.* Berlin, Heidelberg: Springer, 37-245.

Jeong, S., Kim, J. Y., Kim, B. H., Moon, H., Kim, S. O. (2013). Directed self-assembly of block copolymers for next generation nanolithography. *Materials Today*, 16, 468–476.

Kanakasabapathy, S. K. (2013). Sidewall image transfer using the lithographic stack as the mandrel. US8455364 B2, 4 June 2013.

Kawahira, H., Hayashi, N., Hamada, H. (1999). PMJ' 99 panel discussion review: OPC mask technology for KrF lithography. *Proceedings SPIE 3873, 19th Annual Symposium on Photomask Technology*, 318.

Kim, S. (2014). Stochastic simulation studies of line-edge roughness in block copolymer lithography. *Journal of Nanoscience and Nanotechnology*, 14, 6143–6145.

Lin, B., Yan, X., Shi, Z., Yang, Y. (2011). Erratum to: A sparse matrix model-based optical proximity correction algorithm with model-based mapping between segments and control sites. *Journal of Zhejiang University Science C*, 12, 614–614.

Lin, H., Lin, J. C., Chiu, C. S., Wang, Y., Yen, A. (1999). Sub-0.18-µm line/space lithography using 248-nm scanners and assisting feature OPC masks. *Proceedings of SPIE*. 3873, *19th Annual Symposium on Photomask Technology*, 307.

Liu, T., Chang, L. (2009). Transistor scaling to the limit. *Into the Nano Wra*, 1st ed., H. Huff, Ed. Berlin, Heidelberg: Springer .

Mack, C. A. (2015). The multiple lives of Moore's Law. *IEEE Spectrum*, 52(4), 31.

Mesawich, M., Sevegney, M., Gotlinsky, B., Reyes, S., Abbott, P., Marzani, J., Rivera, M. (2009). Microbridge and e-test opens defectivity reduction via improved filtration of photolithography fluids. *Proceedings of SPIE, Advances in Resist Materials and Processing Technology XXVI*, 72730O.

Morita, H. (2016). Lithography process simulation studies using coarse-grained polymer models. *Polymers Journal*, 48, 45–50.

Nakamura, T., Takeshita, M., Yokoya, J., Yoshii, Y., Saito, H., Takasu, R., Ohmori, K. (2010). Process feasibility investigation of freezing free litho-litho-etch process for below 32 nm hp. *Proceedings of SPIE 7639, Advances in Resist Materials and Processing Technology XXVI*, 76392-76392-8.

Nishimura, E., Kushibiki, M., Yatsuda, K. (2008). Precise CD control techniques for double patterning and sidewall transfer. *Proceedings of SPIE 6924, Optical Microlithography XXI*, 692425-692425-8.

Nouralishahi, M., Wu, C., Vandenberghe, L. (2008). Model calibration for optical lithography via semidefinite programming. *Optimization and Engineering*, 9(19), 19–35.

Pala, N., Karabiyik, M. (2012). Electron beam lithography. *Encyclopedia of Nanotechnology*, 718–740.

Pease, R. F., Chou, S. Y. (2008). Lithography and other patterning techniques for future electronics. *Proceedings of the IEEE*, 96(2), 248–270.

Peng, Y., Zhang, J., Wang, Y., Yu, Z. (2011). Gradient-based source and mask optimization in optical lithography. *IEEE Transactions on Image Processing*, 20(10).

Peters, A. J., Lawson, R. A., Nation, B. D., Ludovice, P. J., Henderson, C. L. (2015). Simulation study of the effect of molar mass dispersity on domain interfacial roughness in lamellae forming block copolymers for directed self-assembly. *Nanotechnology*, 26, 385301, September 2015.

Pieczulewski, C. N., Rosslee, C. A. (2009). Litho-freeze-litho-etch (LFLE) enabling dual wafer flow coat/develop process and freeze CD tuning bake for >200wph immersion ArF photolithography double patterning. *Proceedings of SPIE 7520, Lithography Asia* 75201H-75201H-12.

Robertson, S. A., Wong, P., Biafore, J. J., Smith, M. D., Vandenbroeck, N., Wiaux, V. (2011). Characterization of a 'thermal freeze' litho-litho-etch (LLE) process for predictive simulation. *Proceedings of SPIE 7973, Optical Microlithography XXIV*, 79730-79730-11.

Rudnitsky, A. S., Serdyuk, V. M. (2012). Integrated evaluation of diffraction image quality in optical lithography using the rigorous diffraction solution for a slot. *Technical Physics*, 57, 1387–1393.

Ruiz, R., Wan, L., Lille, J., Patel, K. C., Dobisz, E., Johnston, D. E., Kisslinger, K., Black, C. T. (2012). Image quality and pattern transfer in directed self-assembly with block-selective atomic layer deposition. *Journal of Vacuum Science & Technology B*, 30, 06F202.

Santillan, J. J., Kobayashi, S., Itani, T. (2007). Outgas quantification analysis of EUV resists *Microprocesses and Nanotechnology, 2007. Digest of Papers*, 436–437.

Schneider, M., Belic, N., Sambale, C., Hofmann, U., Fey, D. (2012). Optimization of a short-range proximity effect correction algorithm in E-beam lithography using GPGPUs. *Algorithms and Architectures for Parallel Processing: 12th International Conference,*

ICA3PP 2012, Fukuoka, Japan, 4–7 September 2012, *Proceedings, Part I*, Y. Xiang, I. Stojmenovic, B. O. Apduhan, G. Wang, K. Nakano, A. Zomaya Eds. Berlin, Heidelberg: Springer, 41–55.

Shalf, J. M., Leland, R. (2015). Computing beyond Moore's Law. *Computer*, 48, 14–23.

Shibuya, M. (1997). Resolution enhancement techniques for optical lithography and optical imaging theory. *Optical Review*, 4, 151–160.

Smith, B. W., Fan, Y., Zhou, J., Bourov, A., Zavyalova, L., Piscani, E., Park, J., Summers, D., Cropanese, F. (2004). High index aqueous immersion fluids for 193 nm and 248 nm lithography. Rochester Institute of Technology: Center for Nanolithography Research.

Somervell, M., Yamauchi, T., Okada, S., Tomita, T., Nishi, T., Kawakami, S., Muramatsu, M., Iijima, E., Rastogi, V., Nakano, T., Iwao, F., Nagahara, S., Iwaki, H., Dojun, M., Yatsuda, K., Tobana, T., Negreira, S. R., Parnell, D, Rathsack, B., Nafus, K., Peyre, J., Kitano, T. (2015). Driving DSA into volume manufacturing. *Proceedings of SPIE, 9425, Advances in Patterning Materials and Processes XXXI*, 94250Q.

Son, J. G., Son, M., Moon, K., Lee, B. H., Myoung, J., Strano, M. S., Ham, M., Ross, C. A. (2013). Sub-10 nm graphene nanoribbon array field-effect transistors fabricated by block copolymer lithography. *Advanced Materials*, 25, 4723–4728.

Stoykovich, M. P., Kang, H., Daoulas, K. C., Liu, G., Liu, C., de Pablo, J. J., Muller, M., Nealey, P. F. (2007). Directed self-assembly of block copolymers for nanolithography: Fabrication of isolated features and essential integrated circuit geometries. *ACS Nano*, 1, 168–175.

Sugimachi, H., Kosugi, H., Shibata, T., Kitano, J., Fujiwara, K., Itou, K., Mita, M., Soyano, A., Kusumoto, S., Shima, M., Yamaguchi, Y. (2009). CD uniformity improvement for double-patterning lithography (litho-litho-etch) using freezing process. *Proceedings SPIE 7273, Advances in Resist Materials* and *Processing Technology XXVI*, 72731D-72731D-7.

Sullivan, R. F. (2007). The impact of Moore's law on the total cost of computing and how inefficiencies in the data center increase these costs. *ASHRAE Transactions*, 113(5), 457–461.

Synowicki, R. A. Pribil,, Cooney, G. Herzinger, C. M. Green, S. E. French, R. H. Yang, M. K. Lemon, M. F Burnett, J. H. Kaplan, S. (2004). Immersion fluids for lithography: Refractive index measurements using prism minimum deviation techniques. *Semiconductor Fabtech*, 22, 55–58.

Teh, S. H., Heng, C. H., Tay, A. (2010). Performance-based optical proximity correction methodology. *IEEE Transactions on Computer-Aided Design of Integrated Circuits and Systems*, 29(1), 51–64.

Tsai, H., Miyazoe, H., Engelmann, S., To, B., Sikorski, E., Bucchignano, J., Klaus, D., Liu, C., Cheng, J., Sanders, D., Fuller, N., Guillorn, M. (2012). Sub-30 nm pitch line-space patterning of semiconductor and dielectric materials using directed self-assembly. *Journal of Vacuum Science & Technology B*, 30, 06F205.

Tseng, A. A., Chen, K., Chen, C. D., Ma, K. J. (2003). Electron beam lithography in nanoscale fabrication: Recent development. *IEEE Transactions on Electronics Packaging Manufacturing*, 26(2), April 2003.

Tummala, R. (2001). *Fundamentals of Microsystems Ackaging*. New York: McGraw-Hill.

Vardi, M. Y. (2014). Moore's Law and the sand-heap paradox. *Communication ACM*, 57, 5, May 2014.

Watanabe, T., Hamamoto, K., Kinoshita, H., Hada, H., Komano, H. (2003). Resist outgassing characteristics in EUVL. *2003 International Microprocesses and Nanotechnology Conference, Digest of Papers*, 288–289.

Wu, B. (2011). Next-generation lithography for 22 and 16 nm technology nodes and beyond. *Science China Information Sciences*, 54, 959–979.

Wurm, S., Seidel, P., Peski, C. V., He, L., Han, H., Kearney, P., Cho, W. (2007). EUV mask infrastructure challenges. *23rd European Mask and Lithography Conference (EMLC)*, 1–12.

Xiao, S., Yang, X., Edwards, E. W., La, Y., Nealey, P. F. (2005). Graphoepitaxy of cylinder-forming block copolymers for use as templates to pattern magnetic metal dot arrays. *Nanotechnology*, 16, S324.

Xu, P., Chen, Y., Chen, Y., Miao, L., Sun, S., Kim, S., Berger, A., Mao, D., Bencher, C., Hung, R., Ngai, C. (2011). Sidewall spacer quadruple patterning for 15nm half-pitch. *Proceedings of SPIE 7973, Optical Microlithography XXIV*, 79731Q-79731Q-12.

Yi, H., Chang, J. (2007). Proximity-effect correction in electron-beam lithography on metal multi-layers. *Journal of Material Science,* 42, 5159–5164.

Yoshikawa, S., Fujii, N., Kanno, K., Imai, H., Hayano, H., Miyashita, H., Shida, S., Murakawa, T., Kuribara, M., Matsumoto, J., Nakamura, T., Matsushita, S., Hara, D., Pang, L. (2015). Study of defect verification based on lithography simulation with a SEM system. *Proceedings of SPIE* 9658, *Photomask Japan 2015: Photomask and Next-Generation Lithography Mask Technology XXII*, 96580V.

Zhang, C., Katsuki, S., Horta, H., Imamura, H., Akiyama, H. (2008). High-power EUV source for lithography using tin target. *2008 IEEE Industry Applications Society Annual Meeting*, 1–4.

Zhou, Z. J. (2005). Electron beam lithography. *Handbook of Microscopy for Nanotechnology*, N. Yao, Z. L. Wang, Eds. Boston, MA: Springer US, 287–321.

5 Future Semiconductor Devices

Exotic Materials, Alternative Architectures and Prospects

*Wynand Lambrechts, Saurabh Sinha,
Jassem Abdallah and Jaco Prinsloo*

INTRODUCTION

The traditional scaling of complementary metal-oxide semiconductor (CMOS) technology is becoming increasingly complex and exponentially expensive in such a way that many foundries are unable to keep up with the market-leading companies such as Intel, Samsung and GlobalFoundries. There is considerable interest in increasing device-level performance by altering the materials used in the conductive channel of active devices and introducing dramatic process enhancements that deviate from the historical approaches of smaller wavelength photolithography. The Internet-of-things (IoT) is deemed one of the primary driving factors of future-generation scaled electronics, with the mobile smartphone and random access memory possibly the largest contributors within this category.

This chapter reviews the five most promising next-generation technologies that aim to drive Moore's law in the foreseeable future, combined with traditional enhancements in photolithography and multi-patterning. These five technologies are summarized in Figure 5.1.

As shown in Figure 5.1, the five next-generation technologies that are listed and discussed in this chapter are graphene-based electronic circuits, optoelectronic waveguides and photonic crystals, molecular electronics, spintronics and quantum computing.

Graphene is a transparent (allowing 97% of visible light to pass through it) monolayer allotrope of carbon (one of the most abundant elements in the world), and similar to diamond and charcoal, an atomic-scale hexagonal lattice of carbon atoms. Graphene exhibits distinctive and adaptable physical properties, which make it a favored carbon-based material for future-generation electronic circuits. Graphene structures in semiconductor metal-oxide semiconductor field-effect transistor (MOSFET) have evolved in recent years. The simplified structure of a MOSFET device with epitaxial graphene grown on a silicon-carbide (SiC) substrate is presented in Figure 5.2.

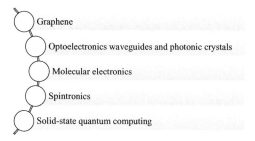

FIGURE 5.1   The five next-generation semiconductor scaling technologies that promise to drive the foreseeable future of Moore's law.

As shown in Figure 5.2, graphene is typically epitaxially grown on SiC substrates among the source, gate dielectric and drain terminals of the MOSFET. Epitaxial graphene is created by thermal decomposition subsequent to hydrogen etching of hexagonal SiC at high temperatures of 1350°C in a vacuum (De Heer et al. 2007). Graphene is grown on both the silicon- and carbon-terminated faces as a multilayered structure. Being flexible and strong (as is its carbon allotrope, diamond), graphene is used outside of the semiconductor industry as an element within a resin to manufacture solid structures such as carbon-fiber sports equipment, and is even used in motor vehicle manufacturing. In the semiconductor industry, graphene-based active devices present several advantages such as

- High mobility of up to 200,000 cm^2/Vs at room temperature in a vacuum.
- High saturation velocity of 4×10^5 m/s.
- High transconductance.
- Ambipolarity in its unmodified state.
- Enabling the implementation of active devices operating in the high-GHz and THz regime, if combined with photodetectors and optical interconnects, thus allowing extremely high bandwidth electronic circuits.

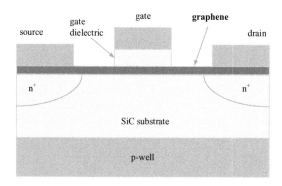

FIGURE 5.2   A simplified representation of a CMOS MOSFET with graphene grown on the semi-insulating SiC substrate.

As with most emerging technologies, graphene also presents challenges related to its successful integration and engineering into existing technologies. The challenges of graphene semiconductor engineering include

- The necessity to contrive constant and undeviating non-zero band gap material (since graphene has no band gap in its natural state).
- Complexity and difficulty in its manufacturability.
- Its suitability for integration with existing silicon-based CMOS technologies.
- Difficulty with electrical contact to metals.

Experimental systems are not adequately developed and reproducible to permit the recognition of high-performing and dependable devices using graphene (Fiori and Iannaccone 2013).

OPTOELECTRONIC WAVEGUIDES AND PHOTONIC CRYSTALS

Optoelectronic devices transport, detect or manipulate light waves and form an integral part of electrical-to-optical or optical-to-electrical transfer of information signals. Optoelectronic theory is based on the quantum mechanical effect of light waves on semiconductor materials. Optoelectronic integrated systems use enabling technologies to achieve various goals, which allow its successful integration. These enabling technologies include

- Light-emitting diodes for electroluminescence.
- Avalanche photodiodes converting light signals to electrical ones by a mechanism called the photoelectric effect.
- Single chips with combined optic circuits.
- Laser diodes.
- Optic-electronic circuits combining photonic and electronic integrated circuits to realize full system-on-chip integrations.
- *pn*-photodiodes with an additional intrinsic region, which causes impact ionization under large reverse-bias conditions.

If specific or external signal attributes are not considered (such as modulation type or signal strength), the quality of the electromagnetic (EM) waves propagating through a medium are predominantly subject to the permittivity (ε) and the permeability (μ) of the broadcast medium. The resistance that the medium offers to the movement of EM waves is defined by the intrinsic impedance η and given by

$$\eta = \sqrt{\frac{\mu}{\varepsilon}} \tag{5.1}$$

The Friis free-space equation also considers the specific characteristics of the signal propagating through the medium, based on a similar principle when compared with any sound-of-light transponder. This equation is typically implemented to equate the required distance and power ratios within a system, and is given by

$$P_r(d) = \frac{P_t G_t G_r \lambda^2}{(4\pi)^2 d^2 L}$$ (5.2)

where,

$P_r(d)$	is the power in the receiver antenna (in watts) at a defined distance d (in m, and with an inverted square proportionality to the received power)
P_t	is the transmitted power (in watts)
G_t and G_r	are the dimensionless gains of the transmitter and receiver, respectively
λ	represents the transmitted signal wavelength in m
L	is the system loss factor ($L \geq 1$)

The wavelength and the frequency (f in Hz) of the optical signal are correlated by

$$f = \frac{c}{\lambda}$$ (5.3)

where c is the speed of light in a vacuum in m/s. EM radiation interacts with materials and all matter by various means through the frequency bands. An optical waveguide restricts the spatial region in which light signals can propagate. Typically, the waveguide material comprises an area with a higher refractive index (n) than its surrounding material. There are two primary categories of waveguides: non-planar or channel waveguides and planar waveguides. Channel waveguides permit light waves to travel only in a single direction, therefore restricting light wave travel in two directions. A planar waveguide only restricts light in one direction and therefore allows light to travel in two different directions. In addition, optical fibers are also optical waveguides, used most commonly in the commercial sector, and waveguides can also be written into transparent media such as glass or crystals. Planar (or slab) waveguides for integrated photonics such as laser chips are used to integrate electrical circuits with optical interconnects or light sources. The basic construction of an optical waveguide is presented in Figure 5.3.

The transverse field pattern of a waveguide is known as the waveguide mode and its amplitude and polarization profiles remain constant in the longitudinal direction. For a waveguide of two-dimensional transverse optical confinement, there are two degrees of freedom in the transverse xy-plane. For the planar waveguide, the mode fields do not depend on the y-coordinate.

Photonic crystals, also known as photonic band gap materials, are periodic optical nanostructures that affect the motion of photons in solids and can be described by the macroscopic Maxwells' equations

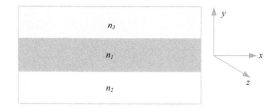

FIGURE 5.3 A simplified representation of a planar waveguide for integrated photonics.

$$\nabla.\mathbf{B} = 0 \qquad \nabla \times \mathbf{E} + \frac{\delta \mathbf{B}}{\delta t} = 0 \qquad (5.4)$$

and

$$\nabla.\mathbf{D} = \rho \qquad \nabla \times \mathbf{H} - \frac{\delta \mathbf{D}}{\delta t} = \mathbf{J} \qquad (5.5)$$

where,
  E and *H*   are the macroscopic electric and magnetic fields, respectively
  D and **B**   are the displacement and magnetic induction fields
  ρ and **J**   are the free charge and current densities

Joannopoulos et al. (2011) further derive what they define as the master equation, which together with the equation for divergence completely defines the harmonic modes as spatial patterns **H(r)**, such that

$$\nabla \times \left(\frac{1}{\varepsilon(r)} \nabla \times \mathbf{H}(\mathbf{r}) \right) = \left(\frac{\omega}{c} \right)^2 \mathbf{H}(\mathbf{r}) \qquad (5.6)$$

where,
  ε   is the permittivity of the material
  ω   is the periodic frequency of the light waves
  c   is the speed of light in a vacuum

The crystal lattice of the photonic crystal enables and defines the periodic arrangement of waves within the crystal. A photonic crystal has macroscopic media with varying dielectric constants and a periodic dielectric function, also referred to as a periodic index of refraction. If the dielectric constants of the materials in the crystal are sufficiently different and if the absorption of light by the materials is minimal, then the refractions and reflections of light from all the various interfaces can produce many of the same phenomena for photons that atomic potential produces for electrons in electronic crystals (Joannopoulos et al. 2011). The complexity of material composition in the *xyz*-Cartesian plane determines the periodicity of a photonic crystal in each direction. A simplified representation of photonic crystals, as described in Joannopoulos et al. (2011), is given in Figure 5.4.

The periodic electromagnetic media presented in Figure 5.4 show the planes in which light can be trapped and transmitted through various dielectric materials. The difference in color among adjacent optical components in Figure 5.4 shows the difference in dielectric properties of these materials. Light can therefore be trapped or transmitted either in cavities within these structures or as waveguides of multiple cavities. The intrinsic operation of photonic crystals falls outside the scope of this book, and a detailed and in-depth description of all aspects concerning the phenomena of photonic crystals is provided in Joannopoulos et al. (2011).

MOLECULAR ELECTRONICS

To scale microelectronic components down further, a bottom-up approach using molecular electronics has been undertaken since unavoidable silicon-based technology

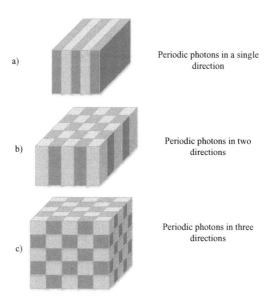

a)   Periodic photons in a single direction

b)   Periodic photons in two directions

c)   Periodic photons in three directions

FIGURE 5.4 A simplified representation of photonic crystal periodic electromagnetic media complexity showing periodicity in (a) a single direction, (b) two directions and (c) three directions. (Adapted from Joannopoulos et al., *Photonic Crystals: Molding the Flow of Light*, 2nd ed., Princeton University Press, 2011.)

limitations have been identified. As a result, the charge-transport of molecules is being studied. Molecular electronic devices can be categorized into three primary classes of devices, as a function of the physical size of the device. These categories are

• Single-molecule devices.
• Self-assembled molecular electronics.
• Thin-film molecular electronics.

In electronics applications, molecular structures have four major advantages that make them a sought-after solution in the electronics industry. The first of these advantages is their sheer size, with molecule sizes between 1 nm and 100 nm, functional nanostructures that present the ability to create small structures with increased efficiency, lower cost and improved power dissipation (Heath and Ratner 2003). Second, specific intermolecular interactions among molecules enable nanoscale self-assembly. This means that the electronic behavior of these components can be modified on a molecular scale. The third advantage is the fact that many molecules have multiple distinct stable geometries where each (isomer) can present distinct optical and electrical properties. Finally, by varying the composition and geometry of these structures, it is possible to tailor the transport mechanism of the molecules extensively, as well as its binding, optical and structural properties.

This chapter briefly introduces molecular electronics with respect to the demands of Moore's law, which requires that the minimum feature sizes of IC components continue to shrink while feature density increases with each technology node.

SPINTRONICS

Spintronics uses the spin of an electron to realize electronic devices of non-volatile, low dynamic power consumption, zero standby power and high device density. Spintronics aims to achieve node sizes of sub-22 nm without the compromises in short-channel effects that traditional CMOS semiconductor components exhibit below this feature size. The discovery of the giant magnetoresistance (GMR) effect paved the way for spintronics in its current capacity. GMR was studied in a structure containing a non-magnetic layer between two ferromagnetic layers, also referred to as a spin valve (Verma et al. 2016). In this experiment, it was shown that the ferromagnetic layer structure with the anti-parallel magnetization state contributes a higher value of resistance compared with the parallel magnetization state (Verma et al. 2016). Following the discovery of GMR, a similar resistance change effect was observed in magnetic tunnel junctions (MTJs) and termed the tunnel magneto-resistive effect (TMR) (Verma et al. 2016). This discovery led to new classes of memory technology known as magneto-resistive random access memory (MRAM). MRAM relies on two magnetic fields generated by two orthogonal current-carrying metal lines, which are referred to as the write word line and the bit line.

In a broad sense, the relevant developments in spintronic materials are subdivided into two categories: those devices that enable the conversion of electrical into magnetic information, the write function, and those that convert magnetic information into electrical information, the read function. These write and read functions are the crux of memory devices, used traditionally in logic circuits. Spin switches use the MTJ mechanism to read and perform write functions using the giant spin-hall effect (GSHE). Ganguly et al. (2017) describe these functions in further detail, and this chapter aims to identify the technologies of spintronics, which can potentially enable the continuation of Moore's law, albeit using a speculative approach and dependent on certain technological breakthroughs that are still being researched. Current challenges such as long and asymmetric write latency, high write current and poor sense margins still exclude the use of spintronics as a traditional CMOS replacement, and spintronics is currently (at the time of writing) still considered an emerging technology.

SOLID-STATE QUANTUM COMPUTING

The final section presented in this chapter reviews solid-state quantum computing, referred to only as quantum computing for convenience. The power of quantum computing is derived from the use of quantum bits, which can exist in both a zero and a one state, simultaneously. This is called the quantum superposition state and leads to computing power that essentially doubles with the addition of a quantum bit. A register of N quantum bits can therefore contain 2^N computational basis states at the same time, and operation is permitted simultaneously on all corresponding values. Two of the most advanced and promising quantum computing technologies rely on trapped ions or superconducting circuits (Skuse 2016). The realization of quantum computing would bring about a significant advancement in computing power that could support stronger encryption for more secure communications, as well as more complex simulations for scientific purposes. For quantum computing to be realized as a viable replacement to traditional on-off state computing, several key challenges must be overcome. These challenges include

- The ability to strongly epitomize the information of quantum bits.
- The complete isolation of quantum computing devices from external magnetic and electric fields.
- Efficient and accurate techniques to evaluate the result of a quantum operation.

In the final section of this chapter, a review of the basic operation of quantum computing is presented with regard to the states of quantum bits and the potential applications of quantum computing, as well as algorithms that have been applied successfully to quantum computing.

The first future technology discussed and reviewed in this chapter, with reference to Figure 5.1, is graphene-based electronic circuits.

GRAPHENE-BASED ELECTRONIC CIRCUITS

Because of the obstacles preventing standard CMOS devices from easily extending Moore's law, the semiconductor industry has recently begun to explore alternative types of devices. Some of these new devices involve the use of exotic materials such as graphene (Mohanram and Yang 2010; Xiying et al. 2012; Service 2015; Singh et al. 2015). Graphene is a two-dimensional form of carbon with the thickness of a single monolayer of carbon atoms covalently linked in a hexagonal comb structure, similar to that of graphite. Graphene's monolayer thickness makes it both flexible and transparent, thus providing the potential for future use in making flexible, transparent electronics or devices with ultrathin form factors (Xiying et al. 2012; Service 2015; Singh et al. 2015; Zheng and Kim 2015). Graphene has semiconducting properties with zero energy band gap and charge carrier mobility values one order of magnitude larger than doped silicon substrates (Mohanram and Yang 2010, Czerniak 2015).

In addition, graphene has electron mobility values higher than metallic conductors; thus, graphene is also being investigated as an alternative conducting material for use in electronic devices such as organic solar cells, liquid crystal displays (LCDs), organic light-emitting diodes (OLEDs), supercapacitors and organic field-effect transistors (OFETs) (Zheng and Kim 2015, Zhao et al. 2015). However, its lack of a band gap prevents intrinsic graphene from being switched off, thus current semiconducting materials used in CMOS transistors cannot be replaced by unmodified graphene since there would be no method to turn off the logic gate (Xiying et al. 2012). Fortunately, the chemical structure of graphene can be functionalized to produce modified graphene layers (such as graphene oxide) with the appropriate electronic properties to be used in field-effect transistors (Zhao et al. 2015; Zheng and Kim 2015). In addition to chemical functionalization, a band gap can be introduced into graphene as a result of strong electronic interactions between graphene and adjacent materials with compatible work functions (Xiying et al. 2012; Singh 2015; Zheng and Kim 2015).

Although the modification methods discussed above have been successful in introducing band gaps into graphene layers, the modifications result in much lower carrier mobility values than those of intrinsic graphene functions (Xiying et al. 2012; Singh 2015; Zhao et al. 2015; Zheng and Kim 2015). An alternative method exists for introducing a band gap into graphene while maintaining relatively high carrier mobility values; this method relies on the quantum confinement of the graphene structure via the physical restriction of its chemical structure into narrow (less than 10 nm wide), one-dimensional strips called graphene nanoribbons.

There are two types of graphene nanoribbon chemical structures (Dröscher et al. 2014; Ruffieux et al. 2016). The first type has edges that resemble a row of arm-chairs while the second type has zigzag-like edges. Graphene nanoribbons can be manufactured in two ways. The first method uses a top-down *subtractive* approach in which lithography and reactive etching are used to pattern and then selectively remove regions of a graphene sheet to produce nanoribbons (Ruffieux et al. 2016). The second processing method is based on a bottom-up *additive* approach, which relies purely on synthetic chemistry; reactive chemical precursors of graphene are polymerized and cyclized to form nanoribbon end-products.

Although the pattern and etch approach to producing nanoribbons is attractive from the standpoint of processing simplicity, it is susceptible to the same factors that limit the resolution of photolithography, namely that the wavelength used must be small enough to avoid Rayleigh scattering and allow the widths of the graphene nanoribbons to be restricted to less than 10 nm in size (Mack 2011, 2015). Thus, the 193 nm wavelength of standard immersion lithographic tools is too large to be used for this approach, but the 13 nm wavelength of EUV imaging tools is sufficiently small to allow 10 nm-wide nanoribbons to be patterned reliably (Bakshi et al. 2007; Zhang et al. 2008; Bodermann et al. 2009). Unfortunately, EUV lithography suffers from low output intensities, poor photoresist sensitivity and high line edge roughness, hence state-of-the-art EUV technology is still not ready to support the volume manu-facturing of graphene nanoribbons (Bakshi et al. 2007; Zhang et al. 2008; Bodermann et al. 2009). E-beam lithography has a sufficiently high resolution to pattern graphene nanoribbons, but its raster scan requirement results in an unacceptably low through-put to support viable manufacturing of graphene-based devices (Tummala 2001).

The bottom-up (synthetic chemistry) approach to producing graphene nanorib-bons is elegant in that the chemical structure of the nanoribbon precursor can be selected to produce the relevant type of nanoribbon strip (zigzag or armchair) with the desired width (the number of aromatic rings from edge to edge in the nanoribbon is dependent on the initial precursor) (Ruffieux et al. 2016). However, the bottom-up synthesis of graphene nanoribbons lacks the processing simplicity of the pattern and etch approach in that it requires specialized knowledge of organic chemistry to design the chemical scheme. In addition, controlling the physical and chemical conditions of the reactor vessel in order to produce repeatable results (nanoribbons of the same dimensions in every reaction batch) would be non-trivial. Unlike the patterning approach, synthesizing nanoribbons does not offer an easy means of directed production of nanoribbons at specific locations (there is no pattern place-ment). Uncontrolled nanoribbon geometries are incompatible with the circuit design principles used to build electronic devices (Ruffieux et al. 2016).

Thus, both the bottom-up and top-down methods of producing graphene nanorib-bons need further development to support the large-scale manufacturing of inte-grated graphene circuits. Although graphene transistor research is still at the proof-of-concept stage, the electronic properties of intrinsic graphene layers, gra-phene nanoribbons and functionalized graphene sheets (for example, graphene oxide and other graphene-based materials) are so impressive that there is great interest from both academia and industry in bringing graphene technology to maturity (Mohanram and Yang 2010; Xiying et al. 2012; Singh et al. 2015).

OPTOELECTRONICS: WAVEGUIDES AND PHOTONIC CRYSTALS

Optoelectronic circuits use both optical and electrical components in an integrated system (Tummala 2001; Zimmermann 2010; Miller 2009; Orcutt et al. 2013). Optical interconnects, such as optical fibers, have the advantage of using the fastest signals. These signals are immune to electromagnetic interference and less susceptible to heat dissipation than CMOS ICs (Tummala 20011 Tong 2014b). In addition, individual opti-cal interconnects can support multiple optical signals of different wavelengths (to avoid electromagnetic interference). This property is referred to as wavelength-division multi-plexing and it is a key motivating factor for pursuing optical signaling (Tummala 2001; Chung et al. 2012; Hristova et al. 2016). In contrast to optical interconnects, individual electrical interconnects can only support one electrical signal at a time. Hence, for an equal number of interconnects, optical systems can carry much more information than electrical systems, which is the reason why optical fibers serve as the cabling for the internet's backbone (Tummala 2001; Chung et al. 2012; Orcutt et al. 2013). Although the signals in optical interconnects travel faster than those in electrical wiring, the speed of optical switching is much lower than that of electrical switching. This switching delay hinders the performance of purely optical circuits such that over short distances (as are found in microchips), their overall end-to-end speed is lower than that of CMOS circuits. To overcome switching delays and other limitations of purely optical systems, optoelectronic ICs use a hybrid of optical and electrical components to take advantage of each technology's strengths where they are better suited (Miller 2009; Zimmermann 2010; Orcutt et al. 2013; Filipenko et al. 2015). Optoelectronic technology has matured over several decades and fully functional optoelectronic circuits have been demon-strated (Tummala 2001; Miller 2009; Zimmermann 2010; Orcutt et al. 2013).

There are multiple types of planar optical interconnects that may be built on a substrate, such as embedded/buried optical waveguides, optical rib waveguides, slab waveguides and others, as shown in Figure 5.5 (Miller 2009; Zimmermann 2010; Orcutt et al. 2013; Filipenko et al. 2015). Optical waveguides can be made from a variety of materials, including organic polymers, semiconductors and dielec-tric materials (Zimmermann 2010; Miller 2009; Orcutt et al. 2013; Tong 2014a). Optical waveguides work via the confinement and propagation of discrete modes of electromagnetic waves inside patterned waveguide cores (depicted as material 1 in Figure 5.6 and as the darker regions of Figure 5.5) (Heebner et al. 2008; Wang et al. 2016). The waveguide cores have higher refractive indices than the surrounding (cladding) materials, and confinement of the light within the cores is accomplished via total internal reflection (Heebner et al. 2008; Wang et al. 2016).

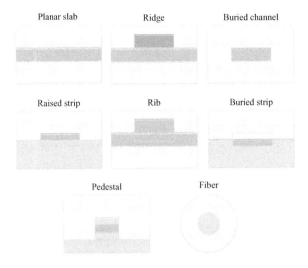

FIGURE 5.5 The different geometries of optical waveguides viewed end-on. All the types of waveguides shown (except for the optical fiber) are planar optical waveguides. The darker shaded regions are the waveguide cores, to which the propagating waves are largely confined. (From Wang et al., Basic analysis on optical waveguides. *Progress in Planar Optical Waveguides*, X. Wang, C. Yin, Z. Cao, Eds. Berlin, Springer, 1–16, 2016; Heebner et al., Optical dielectric waveguides. *Optical Microresonators: Theory, Fabrication, and Applications*, J. Heebner, R. Grover, T. Ibrahim, Eds. New York, NY, Springer, 9–70, 2008.)

As shown in Figure 5.5, the buried channel waveguide is formed with a high-index waveguide core buried in a low-index surrounding medium. The waveguide core can have any cross-sectional geometry but it is often shaped as a rectangle. A strip waveguide is formed by loading a planar waveguide, which already provides optical confinement in the x-direction, with a dielectric strip of index $n3 < n1$ (see Figure 5.3) or a metal strip to facilitate optical confinement in the y-direction. A ridge waveguide looks similar to a strip waveguide but has an additional ridge on top of its planar structure with a high refractive index. The ridge waveguide exhibits strong optical confinement, since it is surrounded on three sides by low refractive

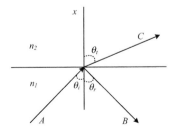

FIGURE 5.6 This figure is a classic depiction of the refraction of light at the interface of two materials as it is transmitted from the first material (1) to the second material (2). (From Wang et al., Basic analysis on optical waveguides. *Progress in Planar Optical Waveguides*, X. Wang, C. Yin, Z. Cao, Eds., Berlin, Springer, 1–16, 2016.)

index cladding material or air. The rib waveguide has a structure that is also similar to the strip and ridge waveguide, but the strip material has the same high refractive index as the core layer beneath it.

If we depict the refraction of light at the interface of two materials as light being transmitted from a higher refractive index material to a lower refractive index material, as shown in Figure 5.6, Snell's law then applies (Heebner et al. 2008; Wang et al. 2016), Where n_1 is the refractive index of material 1, n_2 is the refractive index of material 2, θ_i is the angle of incidence, θ_r is the angle of reflection and θ_t is the angle of transmission. The relationship between these parameters is described by

$$n_1\sin\theta_i = n_2\sin\theta_t \tag{5.7}$$

where, for total internal reflection to occur at the core/cladding interface, the value of θ_i must be larger than the critical angle (θ_c) in the waveguide core (Heebner et al. 2008; Wang et al. 2016). When total internal reflection occurs, none of the light is transmitted between the two materials. Instead, all of the incident light is reflected at the interface, as shown in Figures 5.6 and 5.7, and $\theta_i=\theta_r$. At the critical angle (θ_c), light will refract along the interface instead of passing into the second material; therefore, when $\theta_i=\theta_c$, then $\theta_t=\pi/2$ (therefore 90°). After substituting this value into (5.7), the critical angle can be determined, as shown in Wang et al. (2016), by

$$\theta_c = \sin^{-1}\left(\frac{n_2}{n_1}\right) \tag{5.8}$$

and, therefore, for light to propagate in a waveguide core,

$$\theta_c > \sin^{-1}\left(\frac{n_2}{n_1}\right). \tag{5.9}$$

where,

n_1 is the refractive index of the waveguide core
n_2 is the refractive index of the waveguide cladding

The side view of the total internal reflection of light propagating in an optical waveguide core is presented in Figure 5.7.

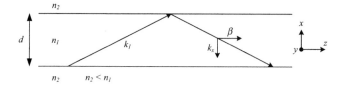

FIGURE 5.7   The side view image of the total internal reflection of light propagating in an optical waveguide core. (From Heebner et al., Optical dielectric waveguides. *Optical Microresonators: Theory, Fabrication, and Applications*, J. Heebner, R. Grover, T. Ibrahim, Eds. New York, NY, Springer, 9–70, 2008.)

Another requirement for light propagation in an optical waveguide is that the phase of the electromagnetic wave as it traverses from one edge of the waveguide to the other and back must be an integer multiple of 2π ($360°$) (Heebner et al. 2008). This requirement is summarized by the TE mode of the wave and the TM mode (Heebner et al. 2008). For a given set of values for the optical waveguide height (d) and wave frequency (ω), there is only a discrete set of electromagnetic fields (called modes) that satisfies the transverse wave requirements for light propagation (Heebner et al. 2008). Thus, not all angles of incidence larger than the critical angle can support light propagation in a waveguide core:

$$2d\left(n_1^2 k_0^2 - \beta_m^2\right)^{1/2} - 4\tan^{-1}\left[\frac{\left(\beta_m^2 - n_2^2 k_0^2\right)^{1/2}}{\left(n_1^2 k_0^2 - \beta_m^2\right)^{1/2}}\right] = m2\pi \quad \left(\text{for TE mode}\right)$$

(5.10)

and

$$2d\left(n_1^2 k_0^2 - \beta_m^2\right)^{1/2} - 4\tan^{-1}\left[\left(\frac{n_1}{n_2}\right)^2 \frac{\left(\beta_m^2 - n_2^2 k_0^2\right)^{1/2}}{\left(n_1^2 k_0^2 - \beta_m^2\right)^{1/2}}\right] = m2\pi \quad \left(\text{for TM mode}\right)$$

(5.11)

where,
  m is the integer multiple of the round-trip phase shift of the propagating ray
  β_m is the longitudinal component of the wave vector (Figure 5.7)
  k_0 is the wave number of the vacuum $=\omega/c$, where ω is the frequency of the electromagnetic wave and c is the speed of light in a vacuum

Once the transverse characteristics of the propagating modes are found (in TE and TM modes), these properties are substituted into the wave equations to fully simulate the propagating modes in all three dimensions (Heebner et al. 2008; Wang et al. 2016).

Well-established numerical methods for efficiently solving the wave equations for different optical waveguide geometries have been developed over several years, and commercially available software from multiple vendors currently exists for modeling the optical properties of optical waveguides (Heebner et al. 2008; Wang et al. 2016). As stated earlier, optical waveguide technology is mature, and numerous types of waveguides have been fabricated using a variety of materials, which are selected depending on the application being targeted (for example, polymer waveguides are compatible with roll-to-roll processing and can be used to produce flexible optoelectronic circuits) (Zimmermann 2010; Tong 2014a).

$$\nabla^2 E + k^2 E = 0 \tag{5.12}$$

$$\nabla^2 H + k^2 H = 0 \tag{5.13}$$

$$k = \frac{n\omega}{c} \qquad\qquad (5.14)$$

where,

    E   refers to the vector of the electric field
    H   refers to the vector of the magnetic field
    k   is the wave number of the material
    n   is the refractive index of the material

Another type of planar waveguide is called a photonic crystal waveguide. Unlike optical waveguides, photonic crystal waveguides do not rely on total internal reflection to confine and propagate electromagnetic waves (Stoffer et al. 2000; Filipenko et al. 2015). Instead, photonic crystals waveguides use nanostructured materials to tune the band gap of the material in order to allow specific frequencies of light to pass only along pre-defined paths (Stoffer et al. 2000; Filipenko et al. 2015). The nanostructures of photonic crystals consist of periodic arrays of porous layers intermixed with non-porous channels that define the paths where light within the band gap is able to travel. Mathematical models are used to calculate the required photonic crystal nanostructure geometry (including the required dimensions of the pores and channels) in order to tune the band gap of the waveguide material as desired (Stoffer et al. 2000; Filipenko et al. 2015). The electromagnetic waves propagating within photonic crystal waveguides are able to make 90° turns while remaining tightly confined within the channels (there is negligible leakage of the wave into the material surrounding the waveguide channels) (Miller 2009; Zimmermann 2010; Orcutt et al. 2013; Filipenko et al. 2015). In contrast to photonic crystal waveguides, optical waveguides are not capable of the confinement of light around an orthogonal bend; this gives photonic crystal waveguide designs the advantage of supporting higher interconnect density and more complex designs than optical waveguides for a fixed amount of real estate on a substrate (Tummala 2001; Tong 2014a). In addition to directional control, the ability of photonic crystal nanostructures to determine which frequencies of light may pass along their channels means that they simultaneously act as optical interconnect materials and optical filters (Filipenko et al. 2015).

Both planar optical waveguides and planar photonic crystal waveguides are compatible with standard CMOS fabrication techniques (photolithography, reactive ion etching, chemical vapor deposition, diffusion). This allows these optical interconnects to be integrated with CMOS components to form optoelectronic ICs on inexpensive substrates such as silicon, making them economically viable (Miller 2009; Yanjun et al. 2009; Tong 2014b). In addition, the lateral dimensions of planar optical interconnects are in the same order of magnitude as the wavelength of the electromagnetic waves that will propagate within them (for example, optical waveguides are usually at least a micrometer wide) (Wang et al. 2016). These dimensions are much larger than the minimum feature sizes of CMOS circuits, indicating that the patterning optical interconnects do not pose any technical challenges, unlike those being experienced by standard CMOS manufacturers (Mack 2011,2015). Unfortunately, some of the components of optical systems must be individually placed during packaging, leading to a relatively high manufacturing cost and low production throughput in comparison to purely electrical-based circuitry (Tummala 2001; Yanjun et al. 2009; Orcutt et al. 2013). These pick-and-place components are relatively bulky, making them incompatible for use on the limited real estate of a microchip die; hence, complete optoelectronic ICs are currently limited to off-chip (backplane motherboard level) applications (Tummala 2001; Yanjun et al. 2009; Orcutt et al. 2013).

In order to reduce the manufacturing costs of optoelectronic ICs as well as to reduce their footprint and make them compatible with on-chip (at the wafer level) applications, researchers have been actively investigating methods for building entire optoelectronic ICs on single substrates (monolithic integration) (Yanjun et al. 2009). Although III–V semiconducting substrates such as gallium arsenide (GaAs) are optically active and are thus compatible with the monolithic integration of optoelectronic circuits, the cost of III–V wafers is prohibitively expensive, making them impractical for use in the high-volume manufacturing of microchips. Until such a time that optoelectronic IC manufacturing can be done either monolithically on an inexpensive substrate such as silicon, or the cost of III–V semiconductor substrates drops to a reasonable level, optoelectronic circuitry will remain an option for off-chip applications only (Tong 2014a). One factor that may bring forward the adoption of optoelectronic ICs as a substitute for CMOS ICs is that the profitability of manufacturing CMOS is beginning to decrease with each technology node (Mack 2011). This drop in profit margin is due to the difficulty in producing devices with ever-smaller feature sizes at ever-increasing feature densities while maintaining high production yields, as demanded by Moore's law (Hutcheson 2009; Mack 2011,2015). Should CMOS manufacturing become either prohibitively expensive or unsustainable (for example, owing to low yields), the switch from CMOS circuitry to optoelectronic circuitry may become imperative rather than just an option.

MOLECULAR ELECTRONICS

Moore's law requires that the minimum feature sizes of IC components continue to shrink while feature density increases with each technology node (Moore 1995). However, as discussed earlier, the ability of semiconductor manufacturers to produce nanoscale features with high pattern fidelity (little to no defects) becomes more difficult as feature sizes continue to decrease (Kawahira et al. 1999; Lin et al. 1999; DeHon 2009; Mesawich et al. 2009; Bret et al. 2014; Yoshikawa et al. 2015). The increasing complexity of circuit designs and increases in feature density with each technology node compound this problem. As a result, defect density increases and manufacturing yield decreases, leading to a reduction in manufacturing throughput and thus reductions in the rate of revenue (Sullivan 2007; Hutcheson 2009; Vardi 2014). Below a certain feature size, it becomes both impractical and uneconomical to continue producing ICs using conventional manufacturing methods (Vardi 2014). As a result of the practical problems related to the conventional top-down approach to producing ICs, semiconductor researchers have begun to investigate bottom-up approaches to building circuitry through the use of logic components made up of single molecules (Choi and Mody 2009; Stan et al. 2009; Cui et al. 2015). Since individual molecules represent the minimum physical size that IC features can eventually reach, ICs built using molecular components would have the highest feature density possible (theoretically, the feature density could be as high as 10^{12} devices per cm^2) (Choi and Mody 2009; Stan et al. 2009; Cui et al. 2015).

However, molecular components are too small to be patterned by even the most advanced lithographic techniques; therefore, molecular sizes are beyond the resolution capability of all known lithographic technologies, including EUV lithography (Bakshi et al. 2007; Zhang et al. 2008; Bodermann et al. 2009). To overcome the resolution limitations of top-down manufacturing (typically with regard to patterning), researchers use either self-assembly or chemical synthesis to build molecular semiconducting components from the bottom up. After synthesis, the molecular components are coupled to larger components such as interconnects in order to complete the integrated circuit (Choi and Mody 2009; Stan et al. 2009; Cui et al. 2015). Figure 5.8 summarizes the strengths and weaknesses of top-down and bottom-up approaches to building IC components and contrasts the capabilities of each approach against the requirements for producing molecular electronic devices (Choi and Mody 2009; Stan et al. 2009; Cui et al. 2015).

One of the leading candidate materials for use in building molecular electronic devices is the carbon nanotube (CNT) (Appenzeller 2008; Franklin et al. 2012). A CNT is a cylindrically shaped macromolecule composed only of carbon atoms, which are covalently linked to one another via a hexagonal lattice. Depending on the arrangement of the hexagonal lattice structure, CNTs may have either semiconducting or metallic properties. Both types of CNTs are of use in ICs since semiconducting CNTs may serve as the material by which a field-effect transistor could be made, while metallic CNTs might serve as replacement materials for interconnects (Appenzeller 2008; Franklin et al. 2012). Both types of CNTs have superior properties to the conventional materials currently used in CMOS devices; semiconducting CNTs have higher carrier mobility than silicon, while metallic CNTs have better conductivity than

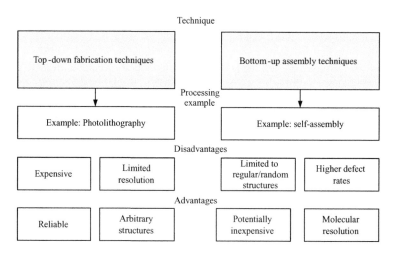

FIGURE 5.8 This figure compares the key characteristics of top-down and bottom-up approaches to manufacturing semiconductor devices against the requirements for producing molecular electronics. (From Stan, M., Rose, G., Ziegler, M., Hybrid CMOS/molecular integrated circuits. *Into the Nano Era*, 1st ed., Berlin Heidelberg, Springer, 257, 2009.)

copper (Appenzeller 2008; Franklin et al. 2012). Although CNTs are a very attractive material, researchers have found it challenging to develop viable process integration schemes that allow CNTs to be manufactured reliably and placed where they are
desired in large-scale ICs. Indeed, the entire field of molecular-based electronics is
still in the nascent stage in that, currently, all types of molecular logic components are
painstakingly built step by step using specialized tools, such as scanning probe microscope tips (Choi and Mody 2009; Stan et al. 2009; Cui et al. 2015). Hence, this field is
technologically limited and the realization of the mass-scale production of molecular
components with large-scale integration is many years, if not decades, away (Choi and
Mody 2009; Stan et al. 2009; Cui et al. 2015) However, it is hoped that by studying
the basic properties of molecular circuits now, the semiconductor industry may learn
enough about their capabilities and the challenges of producing them to prepare itself
for when CMOS feature sizes eventually become too small to pattern reliably in the
future (Choi and Mody 2009; Stan et al. 2009; Cui et al. 2015).

SPINTRONIC DEVICES

Since the inception of CMOS-based ICs, the charge of an electron has been the means
by which binary information was stored or converted from one binary state to another
(Tummala 2001). However, apart from the charge that an electron bears, it has an
intrinsic spin state, which also has two modes (up and down) (Banerjee 2007). The
field of spintronics was conceived with the notion that by exploiting the intrinsic spin
states of an electron along with the charge, the electronics industry could experience a
dramatic improvement in the capabilities of ICs, especially with regard to the storage
capacity of memory components (Banerjee 2007; Zahn 2007; Kulkarni 2015).

One particular field that has dominated the interest of researchers is metallic systems consisting of layered structures comprising alternating magnetic and
non-magnetic materials (Banerjee 2007). This alignment of magnetic layers that are
separated by non-magnetic layers leads to a phenomenon, discovered in 1986, that
is known as interlayer exchange coupling (IEC). Furthermore, this particular layout
of material types brings to light interesting transport properties and phenomena of
the materials. One such phenomenon is called GMR (Zahn 2007). GMR involves a
change in electrical resistivity that occurs under the influence of an external magnetic field. This change in electrical resistivity of the material is due to a change
in the relative orientation of the internal magnetic layers, which is caused by an
external magnetic field. By altering the orientation of these magnetic layers within
the material, the GMR effect makes it possible to convert two-state magnetic information systems into electrical equivalent systems. This is done by relating parallel
and anti-parallel magnetic orientations to binary representations, as used in digital
electronics with ones and zeroes (Zahn 2007).

Following experiments which led to the discovery of GMR, further experiments
with alternating ferromagnetic and non-magnetic layers were performed. These experiments led to the discovery of an effect called TMR (Zahn 2007). TMR is the phenomenon in which spin-polarized electrons tunnel between ferromagnetic layers through
a non-magnetic barrier that acts as an insulating layer between the ferromagnetic layers. A device consisting of such layers is termed an MTJ (Zahn 2007). While GMR

systems exhibit a large change in electrical resistivity under the influence of an external magnetic field, TMR systems exhibit a large change in voltage across the MTJ while operating with electrical currents of small magnitudes. Since these effects are to be exploited in the development of new technologies, particularly nanotechnologies, the dimensions of these materials become critical parameters in the design of such systems. The sensitivity to the dimensionality of materials used in these applications is revealed by considering the number of metallic surface atoms and their nearest neighbors. Upon decreasing the dimensions of a system, the number of nearest neighbors decreases while keeping the number of electronic states constant, resulting in reduced bandwidth between electronic states. Consequently, the local density of states also increases and the tendency toward magnetism is strengthened (Zahn 2007; Kulkarni 2015).

Metallic multilayered systems, consisting of alternating magnetic and non-magnetic metal layers, typically exhibit the phenomenon of magnetic IEC (Zahn 2007). This exchange coupling is caused by the influence of the electron cloud in the non-magnetic layer on the magnetic layers, which causes alignment of the magnetic moments in the magnetic layers of the system. A particularly interesting fact is that the moment alignment (parallel or anti-parallel) depends on the non-magnetic inter-layer's thickness. Furthermore, the strength of the interlayer coupling oscillates with the change in interlayer thickness. This fact was demonstrated by Grünberg in 1986 and is illustrated in Figure 5.9.

The black-and-white stripes shown in Figure 5.9b mark the parallel and anti-parallel alignment regions of the top and bottom layers in an Fe/Cr/Fe tri-layer system (Zahn 2007). It can be noted that the regions containing parallel and anti-parallel alignments alternate with increasing thickness of the Cr-interlayer. These alternations can be explained by using Friedel oscillations as an analogy. The magnetic layer in the system induces a magnetic disturbance onto the non-magnetic layer that acts as a spacer between the magnetic layers. This magnetic disturbance is screened by the electron gas cloud in the non-magnetic layer, resulting in a damped magnetization

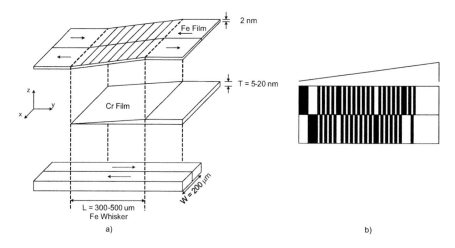

a)   b)

FIGURE 5.9 (a) An experimental setup used to determine the IEC strength and period of oscillation of an Fe/Cr/Fe system, and (b) the experiment result.

density wave. This induced magnetization wave causes the other magnetic layer, which is located directly adjacent to the non-magnetic layer, to interact with this magnetization wave. This effect, for the case of a magnetic point defect and magnet layer, is illustrated in Figure 5.10.

It can be noted from Figure 5.10 that the intensity of the induced magnetization density decreases with increasing distance from the cause of the magnetic perturbation (Zahn 2007). In the case of a point defect (left-hand illustration in Figure 5.10), the decay is inversely proportional to the cube of the distance from the defect. Conversely, for a planar perturbation (right-hand illustration in Figure 5.10), the decay is inversely proportional to the square of the distance from the magnetic plane. These relationships are denoted by (Zahn 2007)

$$\left|\bar{k}\right| \propto \frac{1}{r^3} \qquad (5.15)$$

and

$$\left|\bar{k}\right| \propto \frac{1}{z^2} \qquad (5.16)$$

where $\bar{k}$ represents the decay oscillation vector.

The GMR effect also occurs in metallic multilayer systems, as described above. As discussed previously, the changes in magnetization orientation directions result in a large change in electrical resistivity. Figure 5.11 illustrates this change in electrical resistivity (Kulkarni 2015).

The occurrence of the GMR phenomenon is mostly due to the influence of electronic states in the different types of layers of a metallic multilayer system. The first of these influences is encountered in the magnetic layers, where spin split electronic states undergo matching. The second of these influences is due to the spin-degenerated states encountered in the non-magnetic interlayer. This occurrence is particularly prominent in Co/Cu and Fe/Cr systems (Zahn 2007, Kulkarni 2015).

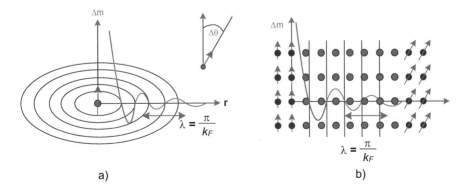

FIGURE 5.10 Illustration of the oscillations due to the induced magnetization density wave within a non-magnetic metal interlayer as a result of the presence of a magnetic point defect (a) and a magnetic layer (b).

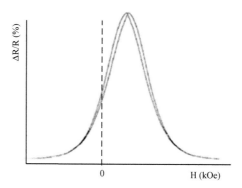

FIGURE 5.11   Electrical resistivity change due to an applied magnetic field. (From Kulkarni, S. K., *Nanotechnology: Principles and Practices*, Springer International Publishing, 2015.)

Since GMR systems are very sensitive to changes in magnetic fields, these systems have found prominent applications in hard disk technologies, with particular use in reading sensors (Kulkarni 2015). Such systems have an advantage over older hard disk reading heads in that the sensor devices can be miniaturized to a great extent, while making an increased data transfer rate and storage capacity on hard disk drives possible. Furthermore, sensor stability is increased by the fact that the electron spin cannot be easily influenced by material impurities and external disturbances such as magnetic interference. One particular type of technology that is already in wide use in current hard disk technology is the spin valve (Kulkarni 2015). Other promising technologies, such as the spin field-effect transistor (SFET), are also on the horizon because of advancements in spintronics. In the case of the SFET, first proposed in 1990, the source and the drain (or collector) consists of ferromagnetic or half-metal materials. The magnetic orientations within the SFET materials make it possible to control the flow of current through the device. Although the SFET has been realized since its proposition, technologies that use it are still being developed and it is expected that utilization of the SFET will lead to faster and more efficient data processing technologies (Kulkarni 2015).

SOLID-STATE QUANTUM COMPUTING

THE PRINCIPLES OF QUANTUM COMPUTING

Quantum computing is a new approach to mathematical logic, which is analogous to Boolean logic but uses the principles of quantum physics (Banerjee 2007; DiVincenzo 2009; Bergou and Hillery 2013; Akama 2015). The fundamental difference between a digital computer as we know it today and a quantum computer lies at the very basic level of transferring information. Classical Boolean logic uses entities called bits, which can have only two possible states: 1 or 0. A two-state quantum system uses entities called qubits, which is short notation for quantum bits (DiVincenzo 2009; Bergou and Hillery 2013; Akama 2015). Unlike a bit, which has two distinct states, a qubit can have a continuum of possible states, which is any superposition of 1 and 0. Although a qubit will have a superposition state between 1 and 0, the superposition is destroyed upon the observation of the qubit's state and the observed state will be either 1 or 0. Since the qubit can have a continuum of possible states when not observed, the qubit's state can be represented by the following wave function (Bergou and Hillery 2013):

$$|\Psi\rangle = a|0\rangle + b|1\rangle. \tag{5.17}$$

In (5.17), the coefficients a and b are arbitrary complex numbers, since complex coefficients have both a magnitude and a phase. The modulus squared of these coefficients gives the probability of the qubit being found in a particular state upon observation (Bergou and Hillery 2013). The phase of the coefficients is also of importance, but will be explained later. The wave function thus represents a superposition between the two states $|0\rangle$ and $|1\rangle$, and $|\Psi\rangle$ therefore satisfies both $|0\rangle$ and $|1\rangle$. The two quantum states can be rewritten as (Bergou and Hillery 2013)

$$|0\rangle = \begin{bmatrix} 1 \\ 0 \end{bmatrix} \tag{5.18}$$

and

$$|1\rangle = \begin{bmatrix} 0 \\ 1 \end{bmatrix}. \tag{5.19}$$

The qubit's state can be visually represented by the Bloch sphere in Figure 5.12. In the Bloch sphere, two angles are considered in the state of the qubit: the polar

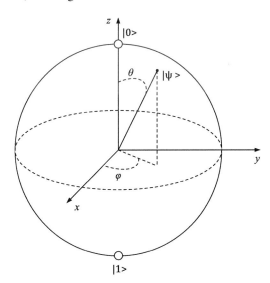

FIGURE 5.12 The Bloch sphere representing the qubit and its state. (From Bergou, J. A., Hillery, M., *Introduction to the Theory of Quantum Information Processing*, New York, Springer, 2013.)

angle with $0 \le \theta \le \pi$ and the azimuthal angle with $0 \le \varphi \le 2\pi$. Thus, (5.17) can be rewritten as follows (Bergou and Hillery 2013):

$$|\Psi\rangle = \cos\left(\frac{\theta}{2}\right)|0\rangle + e^{i\varphi}\sin\left(\frac{\theta}{2}\right)|1\rangle. \tag{5.20}$$

It can thus be concluded that the two qubit states $|0\rangle$ and $|1\rangle$ are located on opposite poles of the sphere, for angles $\theta = 0°$ (north pole) and $\theta = 180°$ (south pole), respectively. Furthermore, the state of n qubits can be represented by (Bergou and Hillery 2013)

$$|0\rangle \otimes \ldots |0\rangle \otimes |0\rangle = |0\ldots 00\rangle$$

$$|0\rangle \otimes \ldots |0\rangle \otimes |1\rangle = |0\ldots 01\rangle$$

$$\vdots \tag{5.21}$$

$$|1\rangle \otimes \ldots |1\rangle \otimes |1\rangle = |1\ldots 11\rangle.$$

Considering the Hilbert space H, each qubit has a value that is observable and, with Hamilton operator N (Akama 2015),

$$N = \begin{bmatrix} 0 & 0 \\ 0 & 1 \end{bmatrix}. \tag{5.22}$$

As depicted in (5.22), the use of the Hamilton operator enables the calculation of the probability that a qubit is found in state $|\Psi\rangle$ (Akama 2015):

$$\langle N \rangle = \langle \Psi|N|\Psi \rangle = \begin{bmatrix} a* & b* \end{bmatrix}\begin{bmatrix} 0 & 0 \\ 0 & 1 \end{bmatrix}\begin{bmatrix} a \\ b \end{bmatrix} = |b|^2 \tag{5.23}$$

where $a*$ and $b*$ are the unitary conjugates of a and b. In the case of multiple qubits, represented by wave functions in a composite system, the state of such a composite system is calculated as the tensor product of the individual qubits $|\Psi\rangle \otimes |\varphi\rangle$ (Akama 2015). With $|x\rangle$ denoting the state of the qubit, where x is a binary number of n digits, the state of the general n-qubit can be expressed by

$$|\Psi\rangle = \sum_{x=0}^{2^N - 1} c_x |x\rangle. \tag{5.24}$$

Thus, for a composite system with n qubits, it can be shown that n qubits can simultaneously represent 2^n states. Intuitively, it can therefore be concluded that a qubit carries far-more information compared with a bit (Akama 2015). The state of a qubit can therefore be denoted by a two-dimensional vector within a space of complex numbers, which evolves by rotation. This means that the evolutions of a qubit can be calculated by applying a 2×2 rotation matrix to the vector described by the wave function in (5.17) (DiVincenzo 2009).

QUANTUM GATES

The basis of any quantum algorithm is quantum logic gates. These quantum gates form the quantum circuit used to perform computation. Classical computers use circuits consisting of logic gates that are designed by means of Boolean algebra. From Boolean logic, it is known that only two basic logic operations can be applied to a single bit: identity and inversion. An identity operation does not alter the state of a bit in any way, while the inversion operation (denoted by the operator NOT) inverts the state of the bit. These two basic logic operations can be extended to quantum logic through the use of matrices, as given by the identity operator (DiVincenzo 2009)

$$\begin{bmatrix} 1 & 0 \\ 0 & 1 \end{bmatrix} \tag{5.25}$$

and the NOT operator

$$\begin{bmatrix} 0 & 1 \\ 1 & 0 \end{bmatrix}. \tag{5.26}$$

These matrices represent two fundamental quantum gates used to perform computation on qubits. Such computations are therefore specified by unitary matrices, which transform a set of qubit states at a certain point in time to another set of qubit states at another (later) point in time. Because of this unitary nature of computation, the computation performed by quantum gates on any qubit must be reversible. This implies that the input state of any quantum gate can be inferred directly from the output state of that particular gate. However, this fact has some significant consequences in the design of quantum computation circuits. Because of this unitary and reversible nature of computation in quantum circuits, certain classical logic gates have no quantum versions. It can be thus concluded that, for a unitary (reversible) transformation, a quantum gate must have equal amounts of inputs and outputs (Akama 2015). For example, consider the logic AND gate that requires at least two binary inputs. The CMOS AND gate is represented in Figure 5.13. The truth table of the CMOS AND gate is given in Table 5.1.

Since the AND gate's output is determined by the product of two binary inputs, there are three possible combinations of inputs (00, 01 or 10) that result in the output state 0. Although a logic high (1) output of the AND gate is only possible by the input combination 11, it is impossible to determine which of the three abovementioned input combinations to the AND gate resulted in a 0 output state at a certain point in time. The AND gate thus results in a non-unitary transformation and is therefore not reversible, implying that it does not have a quantum version (Bergou and Hillery 2013). Conversely, the NOT gate does have a quantum version. The NOT gate and its truth table are presented in Figure 5.14. The CMOS NOT gate is equivalent to the CMOS inverter circuit presented in Chapter 1.

This is because the input state to the NOT gate can easily be derived from its output state, which is simply an inverse operation. The operation of the NOT gate is formally denoted by (Bergou and Hillery 2013)

$$\alpha|0\rangle + \beta|1\rangle \rightarrow \alpha|1\rangle + \beta|0\rangle. \tag{5.27}$$

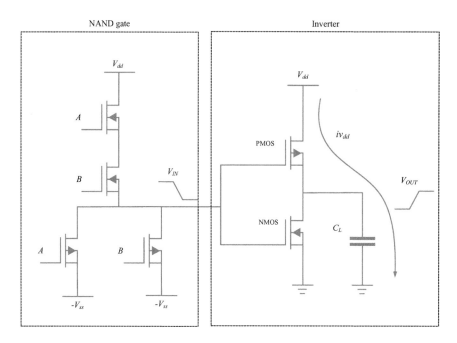

FIGURE 5.13 The CMOS NAND gate combined with a CMOS inverter, resulting in the CMOS AND gate.

By using a two-component column vector to represent the qubit state, we arrive at

$$\alpha|0\rangle + \beta|1\rangle = \begin{bmatrix} \alpha \\ \beta \end{bmatrix} \tag{5.28}$$

and the Pauli matrix σ_x can be used to represent the quantum NOT gate (Bergou and Hillery 2013)

$$\sigma_x = \begin{bmatrix} 0 & 1 \\ 1 & 0 \end{bmatrix} \begin{bmatrix} \alpha \\ \beta \end{bmatrix} = \begin{bmatrix} \beta \\ \alpha \end{bmatrix}. \tag{5.29}$$

The quantum NOT gate is one of a series of three Pauli gates, which will be considered later in this chapter. Two derivations of the quantum NOT transformation matrices that are of particular interest in quantum logic are the square root and fourth root of the NOT transformation matrix. These are given by

$$\sqrt{\text{NOT}} = \begin{bmatrix} \cos\left(\dfrac{\pi}{4}\right) & \sin\left(\dfrac{\pi}{4}\right) \\ -\sin\left(\dfrac{\pi}{4}\right) & \cos\left(\dfrac{\pi}{4}\right) \end{bmatrix} \tag{5.30}$$

TABLE 5.1

The Truth Table of the CMOS AND Gate

Input A	Input B	Output NAND	Output AND
0	0	1	0
0	1	1	0
1	0	1	0
1	1	0	1

and

$$\sqrt[4]{\text{NOT}} = \begin{bmatrix} \cos\left(\dfrac{\pi}{8}\right) & \sin\left(\dfrac{\pi}{8}\right) \\ -\sin\left(\dfrac{\pi}{8}\right) & \cos\left(\dfrac{\pi}{8}\right) \end{bmatrix}. \tag{5.31}$$

An additional example is the Hadamard gate, which consists of a single qubit (similar to the NOT gate). The Hadamard gate performs the transformations denoted by (Bergou and Hillery 2013)

$$H|0\rangle = \frac{1}{\sqrt{2}}\left(|0\rangle + |1\rangle\right) \tag{5.32}$$

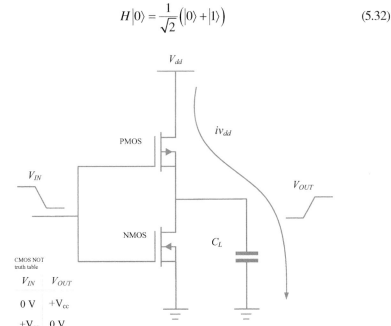

CMOS NOT truth table	
V_{IN}	V_{OUT}
0 V	$+V_{cc}$
$+V_{cc}$	0 V

FIGURE 5.14 The CMOS NOT gate driving a load capacitor C_L at the output. The CMOS NOT gate inverts the input signal, therefore a logical 0 input is measured as a logical 1 at the output terminal and a logical 1 at the input is measured as a logical 0 at the output terminal.

and

$$H|1\rangle = \frac{1}{\sqrt{2}}\left(|0\rangle - |1\rangle\right). \tag{5.33}$$

The Hadamard gate is an example of a quantum gate that takes a computational basis state and outputs a superposition state. The transformation between these two states is essentially a one-qubit rotation, which maps the two-qubit basis states to two superposition states with equal weight. Because of the nature of the transformation performed by the Hadamard gate, it is often used as an initial step in quantum algorithms (Akama 2015). Another representation of the Hadamard equation can be obtained by writing it in matrix form. This results in a matrix known as the Hadamard matrix, which is given by (Akama 2015):

$$H = \frac{1}{\sqrt{2}}\begin{bmatrix} 1 & 1 \\ 1 & -1 \end{bmatrix}. \tag{5.34}$$

Furthermore, another important property of the Hadamard gate is revealed when applying it twice in succession. This particular series of two successive operations results in the final state of the transformation operation being exactly the same as the initial state. Therefore, applying the Hadamard gate twice in succession yields an identity operation, as shown by

$$H^2 = I = \begin{bmatrix} 1 & 0 \\ 0 & 1 \end{bmatrix}. \tag{5.35}$$

The controlled-NOT (C-NOT) gate consists of two inputs and two outputs. The upper qubit is referred to as the control qubit and the lower qubit is referred to as the target qubit.

The control qubit's state is never changed by the gate, while the target qubit's state is dependent on the control qubit's state. In particular, when the state of the control qubit is $|0\rangle$, the state of the target qubit remains unchanged. Conversely, when the control qubit's state is $|1\rangle$, the target qubit's state is inverted. Equation (5.35) formally denotes the C-NOT gate's properties, where the first and the second qubits represent the control and target qubits, respectively (Bergou and Hillery 2013):

$$\begin{array}{cc} |0\rangle|0\rangle \rightarrow |0\rangle|0\rangle & |0\rangle|1\rangle \rightarrow |0\rangle|1\rangle \\ |1\rangle|0\rangle \rightarrow |1\rangle|1\rangle & |1\rangle|1\rangle \rightarrow |1\rangle|0\rangle. \end{array} \tag{5.36}$$

If the control and target qubits are represented by a and b, respectively, it is worth noting that the resulting state of the target qubit can be given by

$$b' = a \oplus b \tag{5.37}$$

which is the definition of the exclusive-OR (XOR) gate. It is also worth noting that the C-NOT gate is reversible, which means that the C-NOT gate has a classical logic equivalent. By adding another control-qubit line to the C-NOT gate, a new quantum gate called the controlled-controlled-NOT (CC-NOT) gate emerges, represented by Figure 5.15. This gate is also known as the Toffoli gate.

When the two control-qubit lines are both in a high state, the third qubit's state is inverted. It can also be noted that the CC-NOT gate is reversible and is thus unitary. Another particularly prominent property of the CC-NOT gate is that it is a universal reversible logic gate. This implies that any Boolean logic circuit can be realized by using CC-NOT gates. Consider the AND gate (which is not reversible). This gate can be realized by using a CC-NOT gate where the third qubit line's state = 0, since both control-qubit lines need to be asserted in order to invert the state of the third qubit line (Akama 2015).

Some quantum gates have more unique and extended properties. The fan-out gate is such an example. The fan-out gate branches/replicates a single input line to two identical output lines. While the functionality of this gate seems trivial, it has useful applications in quantum circuit design situations where the number of lines in a circuit needs to be increased or expanded (Akama 2015). Another example of a quantum gate consisting of two qubits is the swap gate. As its name implies, this gate swaps its two input lines as outputs, exhibiting the property given by (Akama 2015)

$$|\Psi\rangle_a |\varphi\rangle_b \rightarrow |\varphi\rangle_a |\Psi\rangle_b. \tag{5.38}$$

The Fredkin gate, which is also called a controlled-swap (C-SWAP) gate, is another important quantum gate. It can be observed in Figure 5.16 that, for the Fredkin gate, the number of inputs is equal to the number of outputs. Furthermore, it has the property of conserving the number of 0 and 1 inputs. This property is also known as invariance.

The Fredkin gate is a particularly important gate in quantum circuitry since it is also a universal quantum gate that is able to realize other gates. Since it is a C-SWAP gate, its properties are similar to the C-NOT gate; when the first qubit line (control line) is asserted, the output states of the second and third qubit lines are swapped. For any input other than 1 on the control-qubit line, the output states remain the same as the input states (Akama 2015).

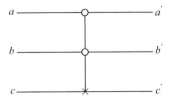

FIGURE 5.15 The CC-NOT gate symbol. (From Akama, S., *Elements of Quantum Computing: History, Theories and Engineering Applications*, Springer International Publishing, 2015.)

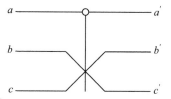

FIGURE 5.16 The Fredkin gate symbol. (From Akama, S., *Elements of Quantum Computing: History, Theories and Engineering Applications*, Springer International Publishing, 2015.)

Recall that the quantum NOT gate is represented by one of a series of three Pauli matrices, as introduced in (6.23). These Pauli matrices are known as the Pauli-X, Pauli-Y and Pauli-Z matrices, representing quantum gates with the same names as these matrices. Each Pauli matrix (gate) also denotes one-qubit rotation operations along each of the three axes in the Bloch sphere in Figure 5.12 (Akama 2015). These quantum gates will now be briefly described.

The Pauli-X gate represents the quantum NOT gate. Recall that the quantum NOT operation is simply an inverse operation on a single qubit. If the position of qubit state $|0\rangle$ in the Bloch sphere in Figure 5.12 is considered, an inverse operation on the qubit results in the qubit state $|1\rangle$. Note that this operation is essentially a rotation of π radians around the x-axis of the Bloch sphere. This operation is denoted by (Akama 2015)

$$\sigma_x = \begin{bmatrix} 0 & 1 \\ 1 & 0 \end{bmatrix} \begin{bmatrix} \alpha \\ \beta \end{bmatrix} = \begin{bmatrix} \beta \\ \alpha \end{bmatrix}. \tag{5.39}$$

The quantum NOT gate is therefore also known as the Pauli-X gate. The Pauli-Y matrix, representing the Pauli-Y gate, represents a π radian rotation around the y-axis in the Bloch sphere. This gate maps the qubit state $|0\rangle$ to $i|1\rangle$ and the qubit state $|1\rangle$ to $-i|0\rangle$. The Pauli-Y gate operation is denoted by (Akama 2015)

$$\sigma_y = \begin{bmatrix} 1 & 0 \\ 0 & -1 \end{bmatrix} \begin{bmatrix} \alpha \\ \beta \end{bmatrix} = \begin{bmatrix} \beta \\ \alpha \end{bmatrix}. \tag{5.40}$$

Lastly, the Pauli-Z gate performs a π radian rotation around the Bloch sphere's z-axis. Application of this gate to the basis state $|0\rangle$ leaves the state unchanged, while application to the basis state $|1\rangle$ maps it to $-|1\rangle$. Apart from this operation being a rotation around the z-axis of the Bloch sphere, it also causes a phase shift in the qubit state. Therefore, the Pauli-Z gate can be considered as a type of phase shift gate. The Pauli-Z gate operation can be represented by (Akama 2015)

$$\sigma_z = \begin{bmatrix} 1 & 0 \\ 0 & -1 \end{bmatrix} \begin{bmatrix} \alpha \\ \beta \end{bmatrix} = \begin{bmatrix} \alpha & 0 \\ 0 & -\beta \end{bmatrix}. \tag{5.41}$$

However, the Pauli-Z gate can be considered as a special case of a phase shift gate, which is an application of the rotation matrix in (Akama 2015):

$$R_{\varphi} = \begin{bmatrix} 1 & 0 \\ 0 & e^{i\varphi} \end{bmatrix}. \tag{5.42}$$

The angle φ in (5.42) corresponds to the azimuthal angle in the Bloch sphere illustration of Figure 5.12. From the rotation matrix, it can be noticed that the phase shift gate leaves the basis state $|0\rangle$ unchanged, while changing the phase of the basis state $|1\rangle$. Recall that the modulus squared of the coefficients in the wave function denotes the probability of a qubit being found in a certain state upon observation. This implies that although the phase shift gate alters the phase of a qubit state, it does not affect the probability of the qubit being observed in a particular state. Common applications of the phase shift gate are with azimuthal angles $\varphi = \pi/4$, $\varphi = \pi/2$ and $\varphi = \pi$. Note that the last of these phase shifts describes the Pauli-Z gate.

APPLICATIONS OF QUANTUM COMPUTING

Several methods for realizing the basic element of a quantum computer, the qubit, have been proposed since it was first conceived. These methods include utilizing nuclear spins in solute organic molecules and on semiconductor surfaces, atomic energy level individual ions and neutral atoms, photons in optical cavities, electron spins and orbital states. Although these methods are all ingenious and plausible in principle, their range and scope further imply the infancy of quantum technology. However, important research and theoretical advancements are already under way, which have made it possible to lay the foundation for using these methods to realize quantum computing technology (DiVincenzo 2009). Quantum computers will use quantum algorithms for solving any quantum problem. These quantum algorithms have the following attributes and features (DiVincenzo 2009):

- First, a given problem statement is translated into a series of quantum logic gates.
- A set of qubits associated with the quantum logic gates is initialized to a state of all-zeroes.
- Applying the quantum logic gates to the all-zero initialized qubits, the qubits' quantum states are manipulated according to the layout of the quantum logic gates.
- Measurement of the qubits causes the superposition states of the qubits to be destroyed and results in a discrete, two-state outcome – 1 or 0.

One of the most prominent computational problems is prime factorization. Other areas for which quantum algorithms will be useful in finding solutions are optimization problems, iterated function evaluation and complex quantum physics problems. Another interesting application is quantum cryptography, which involves the transmission and processing of information privately and securely via processes that are fundamentally impossible on a non-quantum computational platform (DiVincenzo 2009; Akama 2015). Several quantum algorithms have been proposed to solve these types of computational problems. We will now briefly explore some of these quantum algorithms.

SHOR'S ALGORITHM

Shor's quantum algorithm is perhaps the single most prominent advance in the field of quantum computing. His algorithm, proposed in 1994, posed a quantum method for prime factorization. At present, essentially no algorithm exists for performing high-speed prime factorization using classical computers. This fact is the very foundation on which the effectiveness of cryptographic technology rests. Shor's quantum algorithm is so prominent because of its massive implications for digital security (Akama 2015). For example, one of the most frequently used algorithms for modern crypto systems is the RSA algorithm. Proposed in 1978, the RSA algorithm is an example of an asymmetric encryption algorithm that uses a public and private key for encryption and decryption. The public key is an encryption key that need not be kept secret and can be sent in the clear, while the private key must be kept secret. In order to compromise the integrity and security of the RSA encryption algorithm, an eavesdropper needs to compute the private key used in a secure data exchange. Once the private key can be computed, the eavesdropper can decipher encrypted data sent between two points using an RSA-encrypted link. This private key is generated on the basis of prime factorization, and since it is currently computationally infeasible to compute the private key given only the public key, it is therefore a secure, trusted and widely used method for providing privacy and confidentiality (Menezes et al. 1997). This implication can be appreciated by considering the theory of the RSA encryption algorithm (Akama 2015). First, a set of two very large integers, p and q, are selected and their product, n, is computed such that

$$n = pq. \tag{5.43}$$

With the values of p and q kept secret, the Euler function is computed to find the number pair (e, d) that satisfies the following conditions

$$\varphi(n) = (p-1)(q-1) \tag{5.44}$$

$$\gcd(d, \varphi(n)) = 1 \tag{5.45}$$

$$ed \equiv 1 \bmod \varphi(n) \tag{5.46}$$

where,

d is thus coprime with $\varphi(n)$ and can be computed by using the extended Euclid algorithm

e is the inverse of d

The number pair (n, e) forms the public key and the number d forms the secret key. Data being transmitted between two points are encrypted by using the encryption function defined as

$$E_e = m^e \pmod{n}. \tag{5.47}$$

Upon reception, the encrypted data need to be decrypted. Decryption takes place by using the decryption function

$$D_d = c^d \left(\mod n \right). \tag{5.48}$$

The backbone of the RSA encryption algorithm (and many other encryption algorithms) is the fact that while computing the multiplication of integers at high speed is easy to perform, the factoring of large integers at high speed is nearly impossible. Shor's algorithm uses a completely different approach: it replaces the problem of factorization with period finding. The problem is then solved by using the quantum Fourier transform (QFT). We will now briefly consider the outlines of Shor's algorithm (Akama 2015). Consider two integers, a and b, and their greatest common divisor gcd(a, b). The factorization problem is then computed as follows:

- First, an integer x satisfying $x < N$ is found.
- Compute $f = \gcd(x, N)$. If $f \neq 1$, it implies that f is a factor of N. If this is not the case, proceed to the next step.
- Find the smallest integer r such that $x^r \mod N = 1$.
- If either $\gcd(x^{r/2} - 1, N)$ or $\gcd(x^{r/2} + 1, N)$ does not equal 1, then it is the factor. If this is not the case, start again at step 1 with a different value.

Given the outline of the Shor quantum algorithm above, the following example illustrates how Shor's quantum algorithm works. Suppose we have a number $N = 15$, which needs to be factorized. Step 1 of the Shor algorithm is to select an integer x that satisfies $x < N$. We will select $x = 7$ for this example. In step 2 of the Shor algorithm, $f = \gcd(x, N) = \gcd(7, 15)$ is computed. This calculation results in $f = 1$. According to the description of the algorithm conditions, we need to continue to step 3, since the condition $f \neq 1$ has not been met. It is now required to find the smallest integer r that would satisfy the condition $x^r \mod N = 1$. Thus, it is required to find an integer value for r that would result in a remainder of 1 when x^r is divided by N. Starting with $r = 1$, we compute $x^r \mod N$ until a remainder of 1 is achieved. Therefore, this is mathematically given by

$$r = 1: \quad 7^1 / 15 = 0$$

which results in a remainder of 3, and

$$r = 2: \quad 7^2 / 15 = 3$$

resulting in a remainder of 4, and

$$r = 3: \quad 7^3 / 15 = 22$$

that results in a remainder of 13, and

$$r = 4: \quad 7^4 / 15 = 160$$

that results in a remainder of 1. Thus, the value $r=4$ gives us a remainder of 1 after division. This value of r represents the order (also called the period) of f. The last step is to compute $\gcd(x^{r/2}-1, N)$ and $\gcd(x^{r/2}+1, N)$, which results in

$$\gcd\left(x^{\frac{r}{2}}-1, N\right) = \gcd\left(7^{\frac{4}{2}}-1, 15\right) = \gcd(48,15) = 3$$

and

$$\gcd\left(x^{\frac{r}{2}}+1, N\right) = \gcd\left(7^{\frac{4}{2}}+1, 15\right) = \gcd(50,15) = 5.$$

Since neither of the above two results result in a value of 1, they give us the factors of N, which are 3 and 5.

The procedure outlined in the third step of Shor's algorithm presents an interesting feature of the algorithm. In this step, finding the required value of r can be accomplished by means of the QFT (Akama 2015). Similar to the discrete Fourier transform (DFT), the QFT is a linear transformation on quantum bits in a unitary nature. It can thus be noted that the third step of Shor's algorithm (which utilizes the QFT) is arguably the most prominent part of the algorithm. However, the QFT is actually a very prominent procedure in quantum computing, since it is used in various quantum algorithms other than Shor's. The QFT is defined by the expression shown by (Akama 2015):

$$A_q : |a\rangle \rightarrow \frac{1}{\sqrt{q}} \sum_{c=0}^{q-1} |c\rangle e^{2\pi i a c / q}. \tag{5.49}$$

In (5.49), q and a are integers satisfying the condition $0 \le a < q$. The value of $a(\mathrm{mod}\ q)$ is calculated as a superposition state, resulting in the quantum state

$$\frac{1}{\sqrt{q}} \sum_{a=0}^{q-1} |a,0\rangle. \tag{5.50}$$

The resulting value is stored in a register, which can be named register_1. Similar to the calculation in (5.50), the value of $x^a(\mathrm{mod}\ n)$ is calculated and stored in another register, named register_2. Also, since the value of a is stored in register_1, the calculation is performed in reverse, yielding the state represented by (5.50). Applying the QFT to register_1 results in the state represented by

$$\frac{1}{\sqrt{q}} \sum_{a=0}^{q-1} |a, x^a (\mathrm{mod}\ n)\rangle \tag{5.51}$$

and

$$\frac{1}{\sqrt{q}} \sum_{a=0}^{q-1} \sum_{c=0}^{q-1} e^{2\pi i a c / q} |c, x^a (\mathrm{mod}\ n)\rangle. \tag{5.52}$$

Finally, the state of register_1 is measured. Assuming $0 \leq k < r$, it is possible to calculate the probability that $\left| c, x^k (\bmod n) \right\rangle$ with

$$\left| \frac{1}{q} \sum_{a: x^a \equiv x^k}^{q-1} e^{2\pi i a c / q} \right|^2 . \tag{5.53}$$

Thus, from the above probability, $r = x^k (\bmod n)$ can be found. In summary, Shor's algorithm provides an efficient and fast means of factoring a large prime N in the order log$N$. In particular, the fast computation of the order (or period) r can be performed in polynomial time, which has previously been regarded as an impossible task (Akama 2015). It is thus clear why Shor's quantum algorithm poses such a big threat to the integrity of cryptosystems. Shor's quantum algorithm makes high-speed prime factorization in polynomial time possible. The safety of the RSA encryption algorithm, among other encryption algorithms, will thus be compromised if quantum computers can be practically implemented.

Other Quantum Algorithms

Several other quantum algorithms have been proposed in addition to Shor's algorithm. Some of these notable quantum algorithms will be briefly discussed here.

The Hidden Subgroup (HS) algorithm, which is essentially a generalized version of Shor's algorithm, finds application in determining a subgroup from a particular group. Since the HS problem is closely related to other problems such as prime factorization, it is an important problem in the fields of mathematics and computer sciences. The HS problem is defined as follows: given a group G, a subgroup $H \leq G$ and a set X, the function $f : G \to X$ is said to hide the subgroup H if $g_1, g_2 \in G, f(g_1) = f(g_2)$ and if and only if $g_1 H = g_2 H$ for the cosets of H. The HS algorithm generally makes use of the QFT, although it is currently being researched whether the HS algorithm can be optimized to waive the need for the QFT (Akama 2015).

Grover's algorithm has been proposed as a solution for searching for data in an unstructured database, where data are not sorted according to any particular structure. Since there is no set structure in which data are sorted in an unstructured database, search functionality within an unstructured database is significantly more computationally expensive than in a structured database. Therefore, the need arises for an efficient and fast search algorithm in unstructured databases. Although a classical algorithm for linear searching within an unstructured database exists, its complexity with data size n is (n). Conversely, Grover's algorithm offers a reduction in complexity to $O\left(\sqrt{n}\right)$. Despite the fact that Grover's algorithm does not offer an exponential improvement in comparison to the classical algorithm, it does offer a quadratic improvement. Furthermore, upon increasing the data size n, especially for very large values of n, the improvement offered by Grover's algorithm over the classical algorithm becomes extremely significant (Akama 2015).

A quantum walk is the quantum version of the random walk computed with classical algorithms. A random walk is typically applied in the context of simulation, e.g. where successive movements based on given movements are determined probabilistically. A typical example of such a probabilistic movement model is the random motion of particles resulting from collision with other particles, a phenomenon called Brownian motion. Random walks can be either discreet or continuous, depending on the structure of time in which the movement is considered. While classical random walks describe particle position as definitive and certain, quantum walks describe particle position as a superposition. Quantum walks can have many advanced applications in solving computational problems in the fields of mathematics, quantum physics and computer sciences (Akama 2015).

QUANTUM CODES AND QUANTUM CRYPTOGRAPHY

In the discussion of Shor's quantum algorithm, it was clearly noted that the safety of cryptographic systems and cryptographic code structures could be compromised with the realization of quantum computers. This fact poses great uncertainty about the future security and integrity of the internet and communications. However, the field of scientific research has repeatedly proven that new problems encountered with the introduction of new technologies only motivate even more research and advancement to more advanced solutions to these problems. The field of cryptography has been no exception (Akama 2015).

Quantum code is a new concept to accommodate the situation of classical cryptography safety being compromised. Classical cryptography codes guarantee their safety by the fact that they require inconceivable and impossible amounts of classical computing power in order to decode them, making it virtually impossible for an eavesdropper to tap into any communication link. Since quantum computers' computing abilities will compromise this safety barrier in principle, quantum codes guarantee their safety by a completely different method based in the heart of quantum physics – the Heisenberg uncertainty principle. In short, the Heisenberg uncertainty principle denotes that the measurement of two physical values with precision is impossible. This is because the more certainty that is gained on one physical parameter of, for example, a particle, the less certainty there is with regard to the other physical parameter. Thus, tapping a secure communications link with encryption that utilizes quantum codes becomes a seriously non-trivial matter. This is the basis of the motivation for quantum codes and quantum cryptography. Furthermore, classical cryptography is distinguished from quantum cryptography by the fact that classical information can be copied with little or no trace, while quantum information is impossible to copy or clone. This property of quantum information is known as the no-cloning theorem (Bergou and Hillery 2013; Akama 2015).

Recalling the fact that the RSA crypto algorithm involves the public exchange of keys and the fact that the introductions of algorithms like that of Shor can easily compromise the safety of such algorithms, a new approach to public key distribution is necessary. In 1984, a new theory for quantum key distribution was proposed, which is now known as the BB84 protocol. In the BB84 protocol, two sets of orthonormal bases are used to establish a shared key between two communicating parties. These two base sets are called the z basis (with states $\{ |0\rangle, |1\rangle \}$), and the x basis (with states $\{ |+x\rangle, |-x\rangle \}$). Furthermore, the x basis states $|\pm x\rangle = (|0\rangle \pm |1\rangle)/\sqrt{2}$.

Note that these particular states correspond to the superposition states encountered in the Hadamard gate, as discussed earlier. In order to map these four states to bit values, the states $|0\rangle$ and $|+x\rangle$ correspond to a bit value of 0 and the states $|1\rangle$ and $|-x\rangle$ correspond to a bit value of 1 (Akama 2015; Bergou and Hillery 2013).

Traditionally, the two names Alice and Bob are used to explain how communication sequences and exchanges work between two parties using cryptography. Upon the establishment of a shared key, Alice sends Bob qubits of which the states are chosen from the two sets of orthonormal bases at random. When Bob receives a qubit, he measures it in either one of the two bases, also chosen at random. Thereafter, Bob announces publicly only which basis he used to measure the qubit's state (note that he does not announce the result of his measurement). Alice responds by indicating whether he used the same basis as she used. If they are in agreement regarding the basis used, the bit value corresponding to that particular qubit is stored, otherwise it is discarded. If Bob is using the same basis that Alice uses, he will measure the same qubit state as the one that Alice sent. However, if he doesn't choose the correct basis, the results that Bob measures will be random (Bergou and Hillery 2013; Akama 2015).

At first glance, the exchange of qubits between Alice and Bob and the measurement of these states might seem superfluous. However, when an eavesdropper (Eve) enters the exchange, Eve must first capture the qubit that Alice has sent to Bob, measure it in one of the two bases and prepare a new qubit based on that result to be transmitted to Bob. The complication that Eve faces is that there is no way for her to know with certainty which basis to use to measure Alice's qubit. Eve is thus left with no choice but to guess which basis is the correct one. In the event that Eve guesses the correct basis and transmits the correct qubits between Alice and Bob, her eavesdropping goes undetected. However, with a 50% probability that she will guess the wrong basis and another 50% probability that the recipient also obtains an incorrect measurement result, Eve's activity will introduce errors in the exchange and her presence will be revealed. An additional layer of security is added when Alice and Bob detect these errors by publicly comparing a particular subset of bits, of which the same bases were chosen by both of them. In the event that no errors are found in this comparison, it means that an eavesdropper was not present. Conversely, if any error is detected, it indicates the presence of an eavesdropper, in which case all exchanged bits are discarded and the key exchange starts over (Akama 2015; Bergou and Hillery 2013).

The key feature that makes the quantum code secure is the fact that qubits that are in superposition states cannot be measured (or observed) without destroying the superposition. Consequently, it is impossible to transform the qubits into their original superposition states without prior knowledge of these states. Since an eavesdropper does not have prior knowledge of the qubits' original states, it is impossible to reproduce (or clone) the qubits with 100% precision (Bergou and Hillery 2013; Akama 2015).

Although the applications of a quantum computer sound promising, a substantial amount of research and development still needs to be conducted. Several quantum physics properties pose unique theoretical and technical challenges that first need to be overcome before any advancement toward realizing a quantum computer can be achieved. Of particular concern is the concept of de-coherence, which refers to the effect of noise on qubits and which could result in unintended and erroneous state and phase changes. Furthermore, there is the inherent problem of destroying the superposition state of a qubit upon direct observation. The means to an efficient workaround to this problem is an aspect that still requires much research and development (DiVincenzo 2009).

It is also important to keep in mind that quantum computing does not act as a new technique for extending or extrapolating the curve of Moore's law. Furthermore, the concept of a quantum computer does not belong to the field of nanotechnology, although it could benefit from the advancements made in nanotechnology that could aid in realizing a quantum computer. The reason for this is that the principles on which a quantum computer would function are fundamentally different from those of classical computers and their use of semiconductor devices. In fact, in some aspects, quantum computers do not necessarily present big advantages over classical computers. However, it is clear from the new possibilities in problem-solving through computing that the advantages of the quantum computer over the classical computer will far outweigh the disadvantages (DiVincenzo 2009).

CONCLUSION

This chapter reviews the technologies that will propel Moore's law for near-future-generation semiconductor components, since traditional material scaling is showing signs of slowing down. The identified technologies are graphene-based electronic circuits, optoelectronics, molecular electronics, spintronics and solid-state computing. The chapter briefly introduces each of these technologies in the introductory section, followed by a detailed description of each.

Future-generation technologies not only ensure constant scaling, they also drive innovation in low-power, high-performance components for a generation of applications that are implemented for the IoT and mobile computing. These applications are pushing the semiconductor industry to invent new ways to increase computing power and decrease power consumption for higher complexity portable data processing.

REFERENCES

Akama, S. (2015). *Elements of Quantum Computing: History, Theories and Engineering Applications*. Switzerland: Springer International Publishing. ISBN 978-3-319-08283-7.

Appenzeller, J. (2008). Carbon nanotubes for high-performance electronics – Progress and prospect. *Proceedings of the IEEE*, 96, 201–211.

Bakshi, V., Lebert, R., Jaegle, B., Wies, C., Stamm, U., Kleinschmidt, J., Schriever, G., Ziener, C., Corthout, M., Pankert, J., Bergmann, K., Neff, W., Egbert, A., Gustafson, D. (2007). Status report on EUV source development and EUV source applications in EUVL. *Mask and Lithography Conference (EMLC), 23rd European*, 22–26 January, Grenoble, France, VDE, 1–11.

Banerjee, S. (2007). New materials and structures for transistors based on spin, charge and wavefunction phase control. *AIP Conference Proceedings*, 931, 445–448.

Bergou, J. A., Hillery, M. (2013). *Introduction to the Theory of Quantum Information Processing*. New York: Springer.

Bodermann, B., Wurm, M., Diener, A., Scholze, F., Gross, H. (2009). EUV and DUV scatterometry for CD and edge profile metrology on EUV masks. *Mask and Lithography Conference (EMLC), 2009 25th European*, 1–12.

Bret, T., Hofmann, T., Edinger, K. (2014). Industrial perspective on focused electron beam-induced processes. *Applied Physics A*. 117(12), 1607–1614.

Choi, H., Mody, C. C. M. (2009). The long history of molecular electronics: Microelectronics origins of nanotechnology. *Social Studies of Science*, 39(2), 11–50.

Chung, K., Karle, T. J., Rab, M., Greentree, A. D., Tomljenovic-Hanic, S. (2012). Broadband and robust optical waveguide devices using coherent tunneling adiabatic passage. *Optics Express*, 20, 23108–23116.

Cui, A., Dong, H., Hu, W. (2015). Nanogap electrodes towards solid state single-molecule transistors. *Small*, 11, 6115–6141.

Czerniak, M. (2015). What lies beneath? 50 years of enabling Moore's law. *Solid State Technology*, 58, 25–28.

De Heer, W. A., Berger, C., Conrad, E., First, P., Murali, R., Meindl, J. (2007). Pionics: The emerging science and technology of graphene-based nanoelectronics. *2007 IEEE International Electron Devices Meeting*, 199–202.

DeHon, A. (2009). Sublithographic architecture: Shifting the responsibility for perfection. *Into the Nano Era*, 1st ed., H. Huff, Ed. Berlin Heidelberg: Springer .

DiVincenzo, D. (2009). Quantum computing. *Into the Nano Era*, 1st ed., H. Huff, Ed. Berlin Heidelberg: Springer .

Dröscher, S., Molitor, F., Ihn, T., Ensslin, K. (2014). Graphene constrictions. *Physics of Graphene*, H. Aoki, M. S. Dresselhaus, Eds. Springer International Publishing .

Filipenko, O., Donskov, O., Chala, O. (2015). The influence of geometric characteristic on a bandwidth of the photonic crystal waveguide. *Problems of Info Communications Science and Technology (PIC S&T)*, 2015 Second International Scientific-Practical Conference, 93–94.

Fiori, G., Iannaccone, G. (2013). Multiscale modeling for graphene-based nanoscale transistors. *Proceedings of the IEEE*, 101(7), 1653–1669.

Franklin, A. D., Luisier, M., Han, S., Tulevski, G., Breslin, C. M., Gignac, L., Lundstrom, M. S., Haensch, W. (2012). Sub-10 nm carbon nanotube transistor. *Nano Letters*, 12, 758–762.

Ganguly, S., Camsari, K. Y., Datta, S. (2017). Evaluating spintronic devices using the modular approach. *IEEE Journal on Exploratory Solid-State Computational Devices and Circuits*, 2, 51–60.

Heath, J. R., Ratner, M. A. (2003). Molecular electronics. *Physics Today*, 43–49.

Heebner, J., Grover, R., Ibrahim, T. (2008). Optical dielectric waveguides. *Optical Microresonators: Theory, Fabrication, and Applications*, J. Heebner, R. Grover, T. Ibrahim, Eds. New York, NY: Springer, 9–70.

Hristova, H. S., Rangelov, A. A., Montemezzani, G., Vitanov, N. V. 2016. Adiabatic three-waveguide coupler. *Physical Review A - Atomic, Molecular, and Optical Physics*, 93, 033802-1.

Hutcheson, G. (2009). The economic implications of Moore's law. *Into the Nano Era*, 1st ed., H. Huff, Ed. Berlin Heidelberg: Springer .

Joannopoulos, J. D., Johnson, S. G., Winn, J. N., Meade, R. D. (2011). *Photonic Crystals: Molding the Flow of Light*. 2nd ed. Massachusetts: Princeton University Press.

Kawahira, H., Hayashi, N., Hamada, H. (1999). PMJ 99 panel discussion review: OPC mask technology for KrF lithography. *Proceedings of SPIE 3873, 19th Annual Symposium on Photomask Technology*, 318.

Kulkarni, S. K. (2015). *Nanotechnology: Principles and Practices*. New Dehli: Springer International Publishing. ISBN 978-3-319-09170-9.

Lin, H., Lin, J. C., Chiu, C. S., Wang, Y., Yen, A. (1999). Sub-0.18-µm line/space lithography using 248-nm scanners and assisting feature OPC masks. *Proceedings of SPIE Vol. 3873, 19th Annual Symposium on Photomask Technology*, 307.

Mack, C. (2015). The multiple lives of Moore's law. *IEEE Spectrum*, 52(4), 31–31.

Mack, C. A. (2011). Fifty years of Moore's law. *IEEE Transactions on Semiconductor Manufacturing*, 24(5), 202–207.

Menezes, A. J., van Oorschot, P. C., Vanstone, S. A. (1997). *Handbook of Applied Cryptography*. Boca Raton: CRC Press.

Mesawich, M., Sevegney, M., Gotlinsky, B., Reyes, S., Abbott, P., Marzani, J., Rivera, M. (2009). Microbridge and e-test opens defectivity reduction via improved filtration of photolithography fluids. *Proceedings of SPIE Vol. 7273, Advances in Resist Materials and Processing Technology XXVI*, 72730O.

Miller, D. A. B. (2009). Device requirements for optical interconnects to silicon chips. *Proceedings of the IEEE*, 97, 1166–1185.

Mohanram, K., Yang, X. (2010). Chapter 10: Graphene transistors and circuits. *Nanoelectronic Circuit Design*, N. K. Jha, D. Chen, Eds. New York, NY: Springer Science & Business Media, Springer, 349–376.

Moore, G. (1995). Lithography and the future of Moore's law. *Proceedings of SPIE 2440, Optical/Laser Microlithography VIII*, Santa Clara, CA, 2.

Orcutt, J., Ram, R., Stojanović, V. (2013). Chapter 12 – CMOS photonics for high performance interconnects. *Optical Fiber Telecommunications (6th ed.)*. I. P. Kaminow, T. Li, A. E. Willner, Eds. Boston: Academic Press, 419–460.

Ruffieux, P., Wang, S., Yang, B., Sanchez-Sanchez, C., Liu, J., Dienel, T., Talirz, L., Shinde, P., Pignedoli, C. A., Passerone, D., Dumslaff, T., Feng, X., Mullen, K., Fasel, R. (2016). On-surface synthesis of graphene nanoribbons with zigzag edge topology. *Nature*, 531, 489–492.

Service, R. F. (2015). Beyond graphene. *Science*, 348, 490–492.

Singh, R., Kumar, D., Tripathi, C. C. (2015). Graphene: Potential material for nanoelectronics applications. *Indian Journal of Pure & Applied Physics*, 53, 501–513.

Skuse, B. (2016). The trouble with quantum computing. *Engineering and Technology*, 11(11), 54–57.

Stan, M., Rose, G., Ziegler, M. (2009). Hybrid CMOS/molecular integrated circuits. *Into the Nano Era*, 1st ed., H. Huff, Ed. Berlin Heidelberg: Springer .

Stoffer, R., Hoekstra, H. J. W. M., De Ridder, R. M., Van Groesen, E., Van Beckum, F. P. H. (2000). Numerical studies of 2D photonic crystals: Waveguides, coupling between waveguides and filters. *Optical and* Quantum *Electronics*, 32, 947–961.

Sullivan, R. F. (2007). The impact of Moore's law on the total cost of computing and how inefficiencies in the data center increase these costs. *ASHRAE Transactions*, 113(5), 457–461.

Tong, X. C. (2014a). Optoelectronic devices integrated with optical waveguides. *Advanced Materials for Integrated Optical Waveguides*, X. C. Tong, Ed. Cham: Springer International Publishing, 103–160.

Tong, X. C. (2014b). Silicon-on-insulator waveguides. *Advanced Materials for Integrated Optical Waveguides*, X. C. Tong, Ed. Cham: Springer International Publishing, 253–287.

Tong, X. C. 2015. Fundamentals and design guides for optical waveguides. *Advanced Materials for Integrated Optical Waveguides*, X. C. Tong, Ed. Cham: Springer International Publishing, 1–51.

Tummala, R. (2001). *Fundamentals of Microsystems Packaging*. New York: McGraw-Hill.

Vardi, M. Y. (2014). Moore's law and the sand-heap paradox. *Communications ACM*, 57(5), 5.

Verma, S., Kulkarni, A. A., Kaushik, B. K. (2016). Spintronics-based devices to circuits: Perspectives and challenges. *IEEE Nanotechnology Magazine*, 10(4), 13–28.

Wang, X., Yin, C., Cao, Z. (2016). Basic analysis on optical waveguides. *Progress in Planar Optical Waveguides*, X. Wang, C. Yin, Z. Cao, Eds. Berlin: Springer, 1–16.

Xiying, M., Weixia, G., Jiaoyan, S., Yunhai, T. (2012). Investigation of electronic properties of graphene/Si field-effect transistor. *Nano Express: Nanoscale Research Letters*, 7, 677.

Yanjun, S., Lianhe, D., Yanbing, L. (2009). Study on fabrication technology of silicon-based silica array waveguide grating. *Proceedings of SPIE 7282, 4th International Symposium on Advanced Optical Manufacturing and Testing Technologies: Advanced Optical Manufacturing Technologies*, 20 May 2009, 72821R.

Yoshikawa, S., Fujii, N., Kanno, K., Imai, H., Hayano, K., Miyashita, H., Shida, S., Murakawa, T., Kuribara, M., Matsumoto, J., Nakamura, T., Matsushita, S., Hara, D., Pang, L. (2015). Study of defect verification based on lithography simulation with a SEM system. *Proceedings of SPIE Vol. 9658, Photomask Japan 2015: Photomask and Next-Generation Lithography Mask Technology XXII*, 96580V.

Zahn, P. (2007). Spintronics: Transport phenomena in magnetic nanostructures. *Materials for Tomorrow*, 1st ed., S. Gemming, M. Schreiber, J. Suck, Eds. Berlin Heidelberg: Springer, 59.

Zhang, C., Katsuki, S., Horta, H., Imamura, H., Akiyama, H. (2008). High-power EUV source for lithography using tin target. *Industry Applications Society Annual Meeting, 2008. IAS '08*. IEEE, 1–4.

Zhao, J., Liu, L., Li, F. (2015). Application of GO in electronics and optics. *Graphene Oxide: Physics and Applications*. Berlin: Springer Berlin Heidelberg, 57.

Zheng, Q., Kim, J. (2015). Application of graphene-based transparent conductors (TCs). *Graphene for Transparent Conductors*. New York: Springer , pp. 179.

Zimmermann, H. (2010). *Integrated Silicon Optoelectronics*. Berlin: Springer Berlin Heidelberg.

6 Microelectronic Circuit Thermal Constrictions Resulting from Moore's Law

Wynand Lambrechts and Saurabh Sinha

INTRODUCTION

Transistors heat up since they use electrons, subatomic elements that have an electric charge, to execute calculations and transmit information (Airhart 2015). As electrons begin to flow through a conductive element such as metal, they collide with particles called atoms and give off heat. The more electrons that collide (because of, for example, an increased transistor count) and the higher the rate of collisions (because of, for example, an increased operating frequency), the more heat that is generated in electronic circuits. In digital computing, to perform useful computations, distinguishable states of memory cells must be irreversibly changed. The thermodynamic entropy S to change memory cells within its potential states is typically described by

$$S = k \ln(W) \tag{6.1}$$

where,

k is Boltzmann's constant (1.38×10^{-23} J/K)
W is a count of the number of microstates

For n memory cells within m states, the number of microstates is determined by

$$W = m^n. \tag{6.2}$$

Derived from the second fundamental law of thermodynamics, the variation in entropy dS is given by

$$dS = \frac{\delta Q}{T} \tag{6.3}$$

where,

    δQ is the energy spent to perform the operations

    T is the operating temperature in units of kelvin

The energy required to write information into a single binary memory bit (E_{bit}) is equal to

$$E_{\text{bit}} = kT \ln(2) \tag{6.4}$$

which is referred to as Landauer's Principle. The Landauer's boundary is therefore the smallest possible quantity of energy necessary to transform one bit of information (bit). Therefore, at 300 K, with a constant value for k, the minimum energy needed to change one bit of information is equal to approximately 2.87×10^{-21} J. The energy can be converted to be represented in eV by applying the conversion coefficient, such that

$$E_{(eV)} = E_{(J)} \times 6.241509 \times 10^{18} \tag{6.5}$$

which results in approximately 0.017 eV (Kumar 2015) for a single bit. Thermal limitations in binary computing dictate the rise in temperature that a semiconductor material can withstand and the thermal efficiency of the techniques employed to remove heat from the source – typically by using moving air or a cooling liquid. The material used governs the heat transfer coefficient, determined by factors such as

- The specific heat.
- Viscosity.
- Thermal conductivity.
- The heat capacity of the material.

The geometry of the cooling structure also plays a significant role in the heat transfer coefficient. In thermodynamics and in mechanics, the heat transfer coefficient h is simply defined by

$$h = \frac{q}{\Delta T} \tag{6.6}$$

where,

    q is the intrinsically generated heat transported away from the source, referred to as the heat flux and specified in W/m^2

    ΔT is the difference in temperature between the material surface area and the immediate environment to which the heat is being transferred (sunk)

Importantly, the heat transfer coefficient is not only a property of the material, but depends on the entire *ecosystem* around the heat source, where the statistical behavior of thermodynamic systems that are not in equilibrium is typically described by

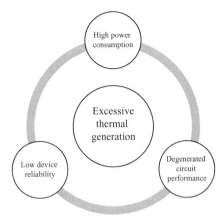

FIGURE 6.1   Ramifications of excessive heat generation in electronic devices.

theorems such as the Boltzmann transport equation. The design of optimal thermal and mechanical microelectronic circuits requires accurate information about the thermal conductivities of the materials used to construct these components (Paul and Baltes 1993). The evolution of Moore's law has become increasingly dependent on efficient means to remove heat from active semiconductor devices. An exponential increase in the number of active devices per unit area and a scaling limitation inducing historically neglected leakage currents have forced researchers and design engineers to consider heat-generation factors traditionally not considered critical. Thermal runaway of electronic components has detrimental effects on device power consumption, performance and reliability, as outlined in Figure 6.1.

The effects of excessive thermal generation as outlined in Figure 6.1 require complex and typically cumbersome analysis of the heat sources and the ambient environment in an integrated circuit to model temperature distribution and thermal effects. Numerical thermal analysis approaches, for instance, the finite element method (FEM) and computational fluid dynamics (CFD)-based finite volume method (FVM), are time-consuming and resource-intensive, especially for transistor-level computations. More practical approaches with relatively small trade-offs in terms of accuracy and computational effort are to build compact thermal models (CTMs) to predict temperature profiles using classical heat transfer equations such as the first law of thermodynamics in a solid domain, with boundary conditions. A typical issue of modeling heat dissipation from transistors to the ambient environment is the development of models inclusive of transistor-level and package-level heat flow. Because of large-scale differences among transistor devices (of a few nanometers in length), interconnects and packaging types, effective heat-transfer simulations are complex. For layered structures, as is typically the case with semiconductor devices (on transistor and package level), several mechanical and emissivity properties of the materials are required to model heat transfer. These properties include

- Density in kg/m^3.
- Volumetric heat capacity in J/kgK.

- Thermal conductivity typically specified in W/mK.
- Material emissivity to define radiation properties.

Thermal conduction in solids from a relatively warmer to a colder body is governed by the relationship

$$\rho C_p \frac{\partial T}{\partial t} - \nabla (k \nabla T) = Q \qquad (6.7)$$

where,

ρ defines the density of the material
C_p refers to its heat capacity
k is the thermal conductivity (in this equation, k therefore does not denote Boltzmann's constant)
T is the temperature
t represents time
Q is the power of the heat source

The transient heat flow at an interface also depends on the heat transfer coefficient h and is used to determine the heat profile in terms of time, such that

$$\frac{T - T_f}{T_i - T_f} = \exp\left[\frac{-hAt}{\rho C_p V}\right] \qquad (6.8)$$

where,

T is the current temperature
T_f is the temperature of the system in its final state as it reaches equilibrium
T_i is the temperature of the system measured during its initial state
A is the area of the interface

In CMOS processing, thermal characterization means that the thermal conductivities of several layers of silicon dioxide, polysilicon, metal and passivation layers as well as the interaction among these layers are required to model thermal transfer accurately. The microstructure of thin films often differs greatly from that of bulk samples and influences their thermal properties considerably (Paul and Baltes 1993). The maximum operating temperature in electronic circuits has implications for both chip and package level reliability and performance (Gurrum et al. 2004). On chip level, enhanced electro-migration is caused by higher current densities resulting from an increase in the temperature of circuit interconnects. The temperature increase at chip level is not uniform either, which leads to signal skewing, especially for high-frequency clock signals. Off-chip, high-temperature operation, beyond the rated operating conditions, rapidly leads to various mechanisms of system failure, leading to both short- and long-term fatigue and failures. There are three mechanisms that govern heat transport, namely

- Conduction – a simplistic mechanism that describes the transfer of heat between two surfaces of different temperature.

- Convection – a complex mechanism that occurs if a solid surface interacts with a gas or with a liquid of a dissimilar temperature.
- Radiation – a complex mechanism where heat is both received and emitted from a source.

Thermally aware electronic circuit simulations can be performed to model and illustrate thermal inertia during operation by implementing thermodynamic and electronic parameter analogies. Figure 6.2 depicts the central thermodynamic and electrical relationships used to represent an equivalent thermal circuit analogy for analyzing heat transfer.

As shown in Figure 6.2, the elementary definitions in the domains of electrical and thermal analysis are similar. In both the thermal and the electrical domain, there are *through* and *across* variables, which describe the presence of energy movement. The *through* variable is a factor that typically flows from a specific reference point toward another; in the electrical domain, this parameter is the flow of current, and in the thermal domain, this parameter is power flow between two or more terminals. The *across* variable forces the flow of current or heat through a resistive element; in the electrical domain, it is voltage, and in the thermal domain, it is a difference in temperature across terminals. The analogies and simplified relationships between thermodynamic and electric parameters used to model and define thermo-electric heat transfer are listed in Table 6.1.

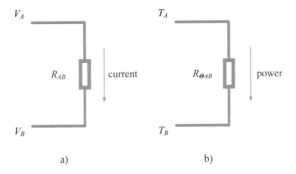

a)   b)

FIGURE 6.2 Central thermodynamic and electrical relationships of (a) the electrical domain and (b) the thermal domain.

TABLE 6.1
Electrical and Thermodynamic Parameter Analogy

Thermodynamic Parameter/Symbol	Unit	Electrical Parameter/Symbol	Unit
Temperature, T	K	Voltage, V	V
Power, P	W	Current, I	A
Heat, Q	J	Charge, q	C
Thermal resistance, R_Θ	K/W	Resistance, R	V/A
Thermal capacitance, C_Θ	J/K	Capacitance, C	C/V

The analogies presented in Table 6.1 are based on the principle that, for electrical characterization, Ohm's law completely defines circuit behavior, whereas for the thermal domain, Fourier's law describes the behavior in a similar fashion. Mathematically, these similarities can be described by, firstly, the derivation from Ohm's law,

$$\Delta V_{AB} = V_A - V_B = I \times R_{AB} \qquad (6.9)$$

and secondly, the derivation from Fourier's law,

$$\Delta T_{AB} = T_A - T_B = P_D \times R_{\Theta AB} \qquad (6.10)$$

where,

V_A and V_B	are the voltages at two terminals
R_{AB}	is the resistance to the flow of current between these terminals
T_A and T_B	are the temperatures at two points
P_D	is the power/energy in the circuit
$R_{\Theta AB}$	is the thermal resistance between the two points

One example of a single active device (here, it is the transistor) thermal coupling circuit is presented in Figure 6.3. This circuit does not, therefore, make provision for thermal coupling between adjacent transistors and the environment, as is practical in most electronic circuit definitions. For higher-order circuit modeling, the power-dissipating elements should be replicated for each heat-generating device.

As shown in Figure 6.3, a single transistor in the electrical domain is modeled in the thermal domain by a power source (*through* variable). Each interface, the device itself, the device-to-heatsink and the heatsink-to-ambient temperature are modeled separately as a thermal resistance between the interfaces and a thermal capacitance to the reference terminal. The ambient temperature is modeled as an *across* variable, which sinks heat generated in the system. To extend the thermal circuit for multiple devices, a method to assess device behavior is given in Figure 6.4.

The method presented in Figure 6.4 does not include transient conditions and therefore does not include thermal capacitors to the system reference. The individual power-dissipating devices are independently analyzed and assigned a junction-to-ambient thermal resistance. The internal air temperature increase beyond the ambient temperature is estimated based on the physical size and other thermally significant features of the ecosystem. The device junction temperature is estimated from the individual thermal resistances with respect to the power dissipated and the internal temperature of the individual component, referred to as the ambient temperature for each power-generating device. This model provides a relatively good approximation of thermal coupling in a system when considering its reasonably low complexity, although it does not take into account thermal coupling among devices within its proximity and assumes a constant internal air temperature.

Thermal device design has become an integral component of microelectronic engineering with CMOS transistor gate lengths scaling below 45 nm (Pop and Goodson 2006). Shrinking dimensions leads to submicron-scale hot spots in the

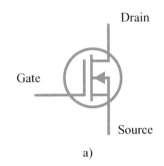

a)

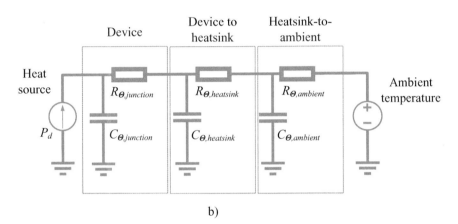

b)

FIGURE 6.3 A single transistor thermal coupling circuit in (a) the electrical and (b) the thermal domain, used to model transient heat exchange between the source (transistor) and the ambient environment.

drain of CMOS transistors, leading to increases in the series resistance of the drain as well as the source injection electrical resistance (Pop and Goodson 2006). In a transistor, an applied electric field such as a voltage results in a lateral electrical field that has its highest value close to the drain of the device – accelerating the charge carriers. The charge carriers, for example conduction band electrons in n-channel metal-oxide semiconductor field-effect transistors (MOSFETs), will inevitably gain substantial energy and therefore experience a rise in temperature. Temperature is directly proportional to molecular speed (by the square rule); therefore, higher-velocity molecules lead to increases in temperature. Within the atomic structure of the material, the electrons are able to scatter with one another and with lattice vibrations (phonons), and to collide and scatter with interfaces and imperfections, as well as with impurity atoms. Scattering with phonons leads to increased temperature within the internal lattice, whereas other scattering mechanisms typically merely disturb the momentum of electrons. The heating of the lattice due to phonon scattering is referred to as Joule heating and is subject to optical phonon emission. Optical phonons are relatively slow-moving atoms and fundamentally do not contribute to heat transfer away from the heat source (Pop and Goodson 2006). Various novel but

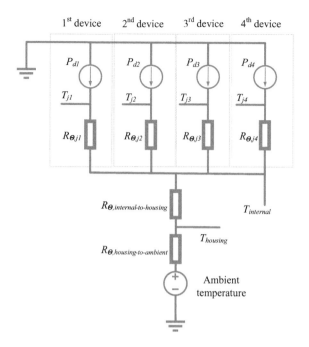

FIGURE 6.4 A multiple device thermal coupling circuit in the thermal domain, used to model transient heat exchange between the sources (transistors) and the ambient environment.

complicated device geometries have been proposed to remove heat from the source of electronic circuits; therefore, the active devices, as well as different material processing, have also been introduced to lower the thermal conductivity of bulk silicon. The thermal conductivity of various materials used in device fabrication, which aims to re-engineer the characteristics of bulk silicon, is given in Table 6.2, adapted from Pop and Goodson (2006).

TABLE 6.2

Thermal Conductivities for Commonly used Microelectronic Device Fabrication Materials

Material	Thermal Conductivity (W/mK)
Silicon (bulk)	148
Germanium (bulk)	60
Silicide	40
Silicon (10 nm thin film)	13
Silicon-germanium ($Si_{0.7}Ge_{0.3}$)	8
Silicon dioxide (SiO_2)	1.4

Source: Adapted from Pop, E. and Goodson, K. E., *Transactions of the ASME*, 128, 102–108, 2006.

The most notable materials that are typically incorporated in bulk transistor designs include Si, SiO_2, and silicide contacts, where the high thermal conductivity of Si, as seen in Table 6.2, facilitates heat transfer from the channel of the transistor to the rear of the chip. This heat is then traditionally removed from the immediate environment using a heat sink. The physical and thermal properties of the non-traditional materials that have been introduced are listed in Table 6.2 and are considered in advanced device manufacturing, although the thermal conductivities of these materials are typically much lower than bulk silicon; however, these materials do introduce additional advantages such as higher mobility and lower capacitance to increase switching speeds. As more non-traditional materials, for example high-κ dielectrics, Ge and various silicide combinations are introduced to traditional semiconductor processing techniques, the "magnitude of boundary thermal resistance and its significance in future-generation nanoscale device behavior is becoming more critical to understand and model" (Pop and Goodson 2006). Active and passive power dissipation in CMOS electronic circuits are the dominant sources of heat generation, and the following section highlights the significance of these forms of power dissipation.

CMOS ACTIVE AND PASSIVE POWER DISSIPATION

For CMOS electronic circuits, the total power dissipation P_{TOT} includes active and passive components during operation. Figure 6.5 summarizes the overall power consumption of these active and passive contributors in CMOS electronic circuits, adapted from Butzen and Ribas (2006).

According to Figure 6.5, first taking into account the active power dissipation in CMOS devices, there are two primary components that contribute to this active power dissipation, namely

*   The dynamic switching power as the load capacitor is charged and discharged.
*   The short-circuit power from non-ideal input waveforms.

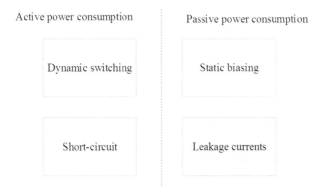

FIGURE 6.5 Active and passive power consumption contributors in CMOS circuits, highlighting the contribution by total leakage currents. (Adapted from Butzen, P. F. and Ribas, R. P., *Universidade Federal do Rio Grande do Sul*, 1–28, 2006.)

A common electronic circuit used to describe the phenomenon of dynamic switching power consumption is the CMOS inverter circuit. The CMOS inverter combines both PMOS and NMOS transistors placed at its input, and drives an output signal through a specified load capacitance C_L. The representation of a CMOS inverter circuit with its in-depth derivation of its circuit operation is provided in Chapter 1 of this book. This section will only aim to highlight the dynamic switching power consumption in the CMOS inverter and compare it with the passive power consumption mechanisms, applicable to semiconductor device processing and presented in this chapter. To present the contributing factors of dynamic switching power consumption P_{DS}, consider its expression given by

$$P_{DS} = fC_L V_{dd}^2 \qquad (6.11)$$

where,

  f is the operation frequency
  C_L is the value of the load capacitor in farads
  V_{dd} is the external voltage from the power supply

Evident from this expression, dynamic power dissipation increases with an increase in operational frequency, load capacitance and supply voltage – all three being external stimuli. These three external gauges have been at the forefront of driving technology advances and key factors of quantifying Moore's law for many decades. In recent device scaling advances, physical device limitations have introduced intrinsic device features that degrade system performance and have slowed down the obsession with increases in frequency and the lowering of supply voltage.

The second active component, with reference to Figure 6.5, is the short-circuit power dissipation, which occurs during a change in the input signal from a low to a high, or from a high to a low, where, for a relatively short period, both the transistors (again in an inverter configuration, for example) will conduct simultaneously. During this time where both transistors conduct a flow of current, a direct path exists where this current can flow between the positive supply and the grounded terminal. The magnitude of the flowing current is proportional to three factors: the slope (input signal voltage ramp) of the input signal, the characteristics of the output load and the physical sizing (geometry) of the transistors. The short-circuit power dissipation P_{sc}, also described in Chapter 1 of this book, can be expressed by

$$P_{sc} = K \left(V_{dd} - 2V_{th} \right)^3 \tau f \qquad (6.12)$$

where,

  K (a constant value) depends on the geometry of the transistor
  V_{th} is the threshold voltage of the transistor
  τ is the rise or fall time of the input signal
  f is the operational frequency

In a static CMOS inverter circuit (therefore, if no switching is occurring), if the value of the supplied input voltage is higher relative to V_{th} for the NMOS device, and lower than V_{th} for the PMOS device, short-circuit power consumption can still occur. Typically, short-circuit power consumption contributes less than a fifth of the active power consumption, where dynamic switching power consumption contributes the largest factor.

Passive power consumption refers to the dissipation of energy in a circuit during periods when no switching of transistor gate terminals is taking place. As shown in Figure 6.5, passive power consumption is categorized into static biasing power consumption and leakage currents. Static biasing power consumption occurs in special cases where CMOS transistors are used as pass-through switches to drive output stages. The threshold voltage drop across these driver transistors reduces the available supply voltage on the output stage and weakens the on-state of these transistors, consequently presenting a static biasing current from the power supply to the ground nodes. Leakage currents, however, have various mechanisms that influence the magnitude and significance in a CMOS circuit. These leakage currents are typically prescribed to device-specific attributes, such as their physical construction and the geometry of their active areas. The following section describes CMOS leakage currents specific to recent aggressively scaled geometries, where ultrathin passivation layers contribute more to leakage currents compared with historically relatively large devices.

AGGRESSIVELY SCALED CMOS LEAKAGE CURRENTS

A more non-specific phenomenon (as opposed to static biasing power consumption under the passive consumption category described in the previous section), which has become much more dominant with aggressive device scaling, is leakage currents between positive or negative power supply terminals and ground nodes, where degraded inputs are not the primary cause, as with static biasing consumption. As presented in Butzen and Ribas (2006), there are three main mechanisms that contribute to leakage currents in CMOS transistors, namely

- Subthreshold leakage current (SLC).
- Gate oxide leakage current.
- Reverse-bias *pn*-junction leakage current.

Additional mechanisms that have a lesser – but non-zero – effect on leakage current are

- When hot carriers are injected that originates in the substrate and toward the gate oxide.
- Gate-induced drain leakage current.

Adapted from Roy et al. (2003), Figure 6.6 depicts leakage current mechanisms that occur in deep-submicron transistors.

As shown in Figure 6.6, several mechanisms cause leakage currents in CMOS transistor geometries, where each of these mechanisms are dependent on various characteristics such as device scaling, layer thickness, applied electric field and

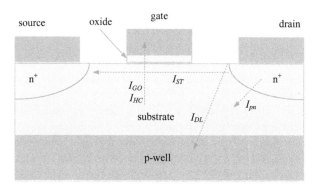

FIGURE 6.6   Graphical representation of transistor leakage current mechanisms. (Adapted from Roy, K., et al., *Proceedings of the IEEE: Contributed Paper*, 91(2), 2003.)

material properties of the semiconductor. These mechanisms are listed with reference to Figure 6.6, with a brief description of the physical occurrence of each:

- *pn*-junction reverse-bias leakage currents between the drain/source and the well of the transistor, represented by I_{pn}.
- SLCs flowing laterally between the drain and source terminals, represented by I_{ST}.
- Tunneling current into as well as through the gate oxide, specifically prominent in current-generation devices with extremely thin oxide, represented by I_{GO}.
- Injected hot carriers from the bulk or substrate in the direction of the gate oxide, represented by I_{HC}.
- Leakage current through the drain terminal induced by an accumulation layer in the gate terminal, represented by I_{DL}.

The principle of each mechanism is described in the following sections. The first leakage current mechanism that is contributed and discussed is the subthreshold current, which transpires laterally amid the drain and source terminals if operated in weak inversion.

SUBTHRESHOLD LEAKAGE CURRENT

SLC transpires amid the drain and source of a CMOS transistor. This phenomenon occurs if the potential on the gate terminal is biased lower than V_{th}. This state of operation is referred to as subthreshold or weak inversion operation and results in a high variation in output current with relatively small variations in the gate voltage of the transistor and, therefore, a large transconductance- (g_m) to-current ratio (slope). The transconductance in weak inversion is given by

$$g_m = \frac{I_{D,\text{weak}}}{nV_T} \tag{6.13}$$

where,

$I_{D,\text{weak}}$ is the drain current in weak inversion

V_T is the thermal voltage (kT/q)

n is the subthreshold swing coefficient (slope factor or body effect coefficient) (Roy et al. 2003), which is discussed in further detail in this section

In weak inversion, "the minority carrier concentration is small, but not zero" (Roy et al. 2003). Subthreshold current is dominated by the longitudinal diffusion current of the minority carriers (Oza and Kadam 2014) and has an exponential relationship with V_{th} and the voltage on the gate of the transistor. Figure 6.7 depicts a NMOS transistor in weak inversion and shows how the minority carriers in the channel (along its length) vary. Figure 6.7 is adapted from Roy et al. (2003).

The exponential correlation concerning the gate voltage and V_{th} on the SLC is expressed by

$$I_{D,\text{weak}} \propto \exp\left(\frac{V_{gs}}{nV_T}\right) \tag{6.14}$$

where n is the slope factor. The magnitude of the SLC is also dependent on temperature, power supply characteristics, geometry and intrinsic process characteristics. Assuming a long-channel uniformly doped active device, the subthreshold swing coefficient can be determined by

$$n = 1 + \frac{C_b}{C_g} \tag{6.15}$$

where,

the base capacitance C_b is given by ε_s/w_d

ε_s is the dielectric constant of the material

w_d is the width of the depletion region

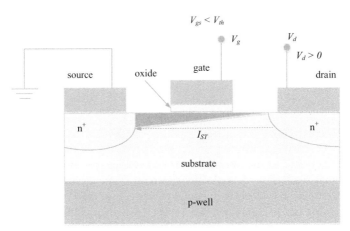

FIGURE 6.7 Graphical representation of an NMOS transistor in weak inversion showing how the minority carriers in the channel vary. (Adapted from Roy, K., et al., *Proceedings of the IEEE: Contributed Paper*, 91(2), 2003.)

The gate oxide capacitance C_g is expressed as ε_{ox}/t_{ox}, where ε_{ox} is the dielectric constant of the gate oxide and t_{ox} is the gate oxide thickness. The subthreshold slope, which is the reciprocal value of the subthreshold swing, characterizes the transition time between off and on (low and high current) states. It is therefore a fundamental function described by the discrepancy of the current in the drain to the gate-source potential of the transistor, such that the swing S is defined as

$$S = \ln(10)\frac{dV_{gs}}{d\ln I_D}. \tag{6.16}$$

The electron density therefore decreases exponentially with increasing energy, which leads to the fundamental relationship between subthreshold swing as a function of temperature $S(T)$, and kT/q, such that

$$S(T) = \ln(10)\frac{kT}{q}. \tag{6.17}$$

Therefore, at room temperature operation (300 K), the fundamental intrinsic theoretical limitation for MOSFET subthreshold swing is approximately 60 mV/dec (Chueng 2010). To overcome this fundamental limitation, since power supply voltages are decreasing to values close to the transistor threshold voltage, current-controlling techniques such as tunneling (tunneling field-effect transistors or TFETs) and impact ionization have been introduced. TFETs are showing promising results for next-generation low-power circuits, and are discussed in further detail in this chapter. In Butzen and Ribas (2006), the SLC $I_{D,weak}$, which is the transistor drain current expression in weak inversion, is expressed as

$$I_{D,weak} = I_s e^{\frac{V_{gs}-V_{th}}{nV_T}}\left[1 - e^{-\frac{V_{ds}}{V_T}}\right] \tag{6.18}$$

where,

V_{ds} is the voltage across the drain-source junction
V_{gs} is the potential across the gate-source junction

the transistor specific current I_s can be expressed by

$$I_s = 2n\beta V_T^2 \tag{6.19}$$

where β is the intrinsic current gain. Evident from the derivations of SLC is the dependence not only on the intrinsic threshold voltage, but also on the supply voltage. Continual scaling of microelectronic circuit supply voltages to lower electric fields on junctions and decrease dynamic power consumption in order to maintain device reliability has led to an increase in SLC. Roy et al. (2003) discuss additional descriptions on drain-induced barrier lowering, the body effect, the effects of channel length and V_{th} roll-off and the effects of temperature with reference to SLC.

Decreasing V_{th} therefore increases the SLC exponentially, and reducing the gate length also increases leakage current due to higher electric fields per available area.

Therefore, in an integrated circuit, because of process variations, the transistors with smaller threshold voltage and/or the shortest gate length contribute more to overall leakage currents, and this non-uniform distribution can lead to unpredictable behavior and failure. SLC also increases with a rise in temperature, which in certain cases can have a snowball effect and lead to device failure.

Kao et al. (2002) present several circuit-design practices to manage SLCs. These practices can reduce SLCs in both active (switching) and passive (standby) phases of circuit operation to reduce the overall power consumption. In general, there are two key methodologies to manage SLCs in circuit topologies, namely source biasing, which biases the source of an *off* transistor to decrease the leakage current, and direct V_T manipulation by using a "multiple threshold voltage process" (Kao et al. 2002), which allows switching between "high-performance and low-leakage components" (Kao et al. 2002). Kao et al. (2002) present circuit techniques to alleviate SLC, such as

- Source biasing.
- Transistor stacking.
- Dual V_T partitioning.
- Multi-threshold CMOS dual V_T techniques.
- Variable threshold CMOS.
- Optimal V_{dd}/V_T operating points.

to reduce leakage currents in CMOS circuits. Kao et al. (2002) therefore quantify the effect of SLCs on total power consumption and propose circuit methodologies to reduce this impact and maintaining acceptable circuit performance.

The following section reviews reverse-bias *pn*-junction leakage currents between the drain and source-to-bulk junctions in CMOS transistors.

REVERSE-BIAS *PN*-JUNCTION LEAKAGE CURRENT

In CMOS transistors, the drain and source-to-bulk are usually reverse-biased, leading to a reverse-bias *pn*-junction between these nodes. A leakage current can occur at these junctions (also called band-to-band tunneling (BTBT)), typically due to either "minority carrier diffusion or drift near the edge of the depletion region, or due to electron-hole pair generation in die depletion region of the reverse-biased junction" (De and Borkar 1999). BTBT is a strong function of the area of the junction as well as the doping concentration. For CMOS transistors that are heavily doped in both the *p*- and *n*-regions, and with a high electric field across the reverse-biased junctions, the *pn*-junction is dominated by BTBT leakage currents between the valence band of the *p*-region and the conduction band of the *n*-region. In this case, and for electron tunneling to transpire, the overall potential drop through the junction should be higher than the semiconductor band gap (Roy et al. 2003). Semiconductor devices with a low band gap and with high source and drain resistance typically present the largest BTBT current characteristics. Band gap engineering of the semiconductor devices through choice of material, or by varying doping characteristics and material composition, can alleviate the BTBT current, whereas the drain and source resistance can be lowered through varied diffusion and annealing techniques during processing. To characterize and quantify BTBT leakage current, the electric field that is applied across the transistor junction is first considered. The electric field at the leakage junction is typically described by

$$E = \sqrt{\frac{2qN_aN_d\left(V_R + V_{bi}\right)}{\varepsilon_s\left(N_a + N_d\right)}}$$
(6.20)

where,

N_a is the acceptor doping concentration in the p-region
N_d is the donor doping concentration in the n-region
V_R is the applied reverse-bias voltage
V_{bi} is the built-in potential of the pn-junction
ε_s is the permittivity of the semiconductor material

The BTBT current density J_{btb} is given in Roy et al. (2003) as

$$J_{btb} = A\frac{EV_R}{E_g^{1/2}}\exp\left(-B\frac{E_g^{3/2}}{E}\right)$$
(6.21)

where E_g is the energy band gap of the material, and the coefficients A and B are given by

$$A = \frac{\sqrt{2m^*}q^3}{4\pi^3\hbar^2}$$
(6.22)

and

$$B = \frac{4\sqrt{2m^*}}{3q\hbar}$$
(6.23)

where,

m^* is the effective mass of the electron
$\hbar$ is Planck's reduced constant (essentially $h/2\pi$)

The probability of tunneling, P, can be approximated by the Wentzel-Kramers-Brillioun method, which simply states that

$$P = |T|^2 \approx \exp\left(-2\int_{x_v}^{x_c}|k(x)|\,dx\right)$$
(6.24)

where,

T is the transmission coefficient through the semiconductor medium
x_v is the point where the electron energy equals the valence band edge energy
x_c is the point where the electron energy equals the conduction band edge energy
$k(x)$ is the wave vector (Van Engelen et al. 2015)

BTBT leakage current has also been modeled numerous times in photonic devices, since such devices are typically reverse-biased during operation. Schenk (1993) provides an in-depth review on the theory and a simplified model of the BTBT in silicon semiconductors. Leakage currents also occur at the gate oxide interface and with continuous scaling of semiconductor devices, this interface is becoming extremely thin and these leakage currents are becoming a substantial contributing factor to high power consumption in switching circuits. Gate oxide tunneling leakage currents are described in the following section.

GATE OXIDE TUNNELING LEAKAGE CURRENTS

Thin oxide films are predominantly used as insulating materials in a wide range of CMOS electronic devices (Haris et al. 2016). Continuous scaling, governed by Moore's law, dictates that the thickness of oxide materials is commonly only a few nanometers in total. Oxide thickness (t_{ox}) limits to the amount they can be scaled have been well defined by comparing the gate tunneling current to the subthreshold source-to-drain leakage current (Bowman et al. 2000; Hirose et al. 2000). The aggressively driven recent downscaling of t_{ox} improves current drivability in MOSFET transistors, while tunnel leakage current in these extremely thin gate oxides is a significant limitation of semiconductor integration (Hirose et al. 2000). For optimum performance with respect to the thickness of t_{ox}, Bowman et al. (2000) present plots of the power-area product caused by a performance-limited power area for a variety of t_{ox} thicknesses. Bowman et al. (2000) use a "tunneling current model for the oxide thickness, while the electrical effective oxide thickness is calculated by including the gate-depletion effect". Choi et al. (1999) present a compact direct tunneling current model for circuit simulation to predict the CMOS circuit performance of thin gate oxides (less than 2 nm).

Although the properties of SiO_2 with its interface at Si are nearly ideal, a continual decrease in the gate oxide thickness as part of the scaling of transistors to improve speed and lower operating voltages has led to prominent gate oxide tunneling currents. The dominant leakage current in conventional CMOS devices results mainly from short channel effects due to drain-induced barrier lowering (DIBL). Leakage currents through thin gate oxides reach values comparable to the current flowing through devices in the on state, rendering such devices unusable because of unwanted and high static power consumption. Increases in switching speed have alleviated this effect somewhat, since the devices are in standby mode for very short periods of time, but the overall effect cannot be neglected. These quantum mechanical effects (QMEs) are therefore an important consideration in modern nanoscale CMOS transistor modeling. If a MOSFET is heavily doped in its channel region and the oxide is very thin, very large values of the electric field in the oxide occur, typically in the MV/cm range. Electrons can then tunnel from the channel through the oxide barrier relatively easily because of its decreased thickness. Direct gate tunneling leakage is often the principal source of leakage current, which can lead to improper operation of the designed circuitry and to high power consumption, which is especially troublesome in mobile devices and within the IoT revolution. Quantum mechanical tunneling (QMT) is the phenomenon of electrons breaking the oxide barrier because of large electric fields, and can occur as two primary mechanisms, depending on the magnitude of the electric field. These two categories are quantum mechanical direct tunneling (QMDT) and Fowler-Nordheim (FN) tunneling. FN tunneling occurs in lower (with respect to QMDT) gate voltages applied to the device and with this mechanism, electrons are able to trickle into the conduction band created by the oxide layer. The dissimilarity between these two constituents of current is evident from the direction of the electric field responsible for accelerating electrons. The direct tunneling of electrons into the oxide layer is typically modeled using only quantum mechanical theory, which plays a significant part in defining the tunneling mechanism in MOSFETs. The mechanisms of direct tunneling include electron tunneling in the

- Conduction band (ECB).
- Valence band (EVB).
- Hole tunneling in the valence band (HVB).

The total leakage current in a gate oxide where electrons are tunneling through QMT is a summation of FN tunneling and QMDT, but FN is typically much lower than QMDT and can be disregarded. The energy band diagram for a typical metal-oxide-p-type semiconductor device that does not have an applied bias voltage is shown in Figure 6.8, adapted from Ranuárez et al. (2006).

In Figure 6.8, the conduction band denoted by E_c, E_i represents the intrinsic Fermi level, E_{fs} represents the Fermi level in the substrate and the Fermi level in the metal is given by E_{fm}. If a positive supply is applied to the metal layer in relation to the p-type substrate in Figure 6.8, the band diagram on the metal side of Figure 6.8 is decreased

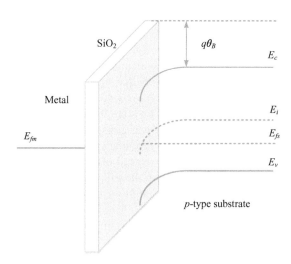

FIGURE 6.8 Energy band diagram for a metal-oxide-p-type semiconductor lacking an applied bias voltage. (Adapted from Ranuárez, J. C., et al., *Microelectronics Reliability*, 46, 1939–1956, 2006.)

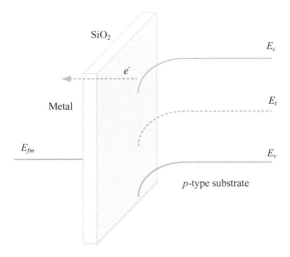

FIGURE 6.9   Energy band diagram of the FN tunneling mechanism.

(lowered) and tunneling of electrons from the conduction band of the substrate material into the conduction band of the oxide can follow. This nearly triangular barrier, as adapted from Ranuárez et al. (2006), is shown in Figure 6.9.

The current density of the FN tunneling mechanism J_{FN} can be described by the current versus the applied voltage, which has been derived in Lezlinger and Snow (1969) and also presented by Ranuárez et al. (2006), such that

$$J_{FN} = \frac{q^3}{16\pi^2\hbar\phi_b} E_{ox}^2 \times \exp\left[\frac{-4\sqrt{2m_{ox}^*}\,\phi_b^{3/2}}{3\hbar q E_{ox}}\right] \qquad (6.25)$$

where,

q is the elemental electron charge
$\hbar$ is Planck's reduced constant ($h/2\pi$)
m_{ox}^* is the effective mass of the electron in the insulator
ϕ_b is the barrier height at the semiconductor-oxide interface
E_{ox} is the electric field across the oxide (Ranuárez et al. 2006)

As seen in (6.1), for a specified voltage on and thickness of the oxide, the tunneling current component depends simply on two internal factors, the

- m_{ox}^*
- ϕ_b.

The following sub-section briefly describes the Shottky barrier height, which is used for current density calculations in this chapter.

THE SCHOTTKY BARRIER HEIGHT

The Schottky barrier height is defined independently for the cases of n-type and p-type materials, practically determined as starting at the conduction band edge and the valence band edge, respectively. The Schottky barrier heights of n-type and p-type materials are ideally related to each other by

$$\phi_b^{(n)} + \phi_b^{(p)} = E_g \tag{6.26}$$

where $\phi_b^{(n)}$ and $\phi_b^{(p)}$ are the barrier heights for n-type and p-type materials, respectively. The calculation of the barrier height for an n-type material is given as

$$\phi_b^{(n)} = \phi_M - \chi \tag{6.27}$$

where,

ϕ_M is the work function of the metal with $\phi_M = (E_0 - E_f)/q$

χ is the electron affinity in eV (or kJ/mol) with $\chi = (E_0 - E_c)/q)$, and for a p-type material, the barrier height is given by

$$\phi_b^{(p)} = \frac{E_g}{q} + \chi - \phi_M. \tag{6.28}$$

The measured barrier heights for selected metal-semiconductor junctions are listed in Table 6.3.

A convenient method to determine the Schottky barrier height is through capacitance-voltage (CV) measurements of the semiconductor-metal interface. The CV data can be applied to calculate the total barrier height ϕ_B, through the definition of the potential barrier, which electrons in the semiconductor must pass over in order to get into the semiconductor, defined as

$$q\phi_{bi} = q\phi_B - \left(E_c - E_f\right)_{\text{bulk}} \tag{6.29}$$

TABLE 6.3
Measured Barrier Height for Selected Metal-Semiconductor Junctions

	Ag	Al	Au	Cr	Ni	Pt	W
ϕ_M	4.3	4.25	4.8	4.5	4.5	5.3	4.6
n-Si	0.78	0.72	0.8	0.61	0.61	0.9	0.67
p-Si	0.54	0.58	0.34	0.5	0.51		0.45
n-Ge	0.54	0.48	0.59		0.59		0.48
p-Ge	0.5		0.3				
n-GaAs	0.88	0.8	0.9			0.86	0.8
p-GaAs	0.63		0.42				

where ϕ_{bi} causes electrons to transfer away from the boundary. Electrons close to the boundary/interface are at a higher average energy compared with the ones in the bulk because of the electric field, thus the presence of ϕ_{bi}. This equation can be rewritten as

$$q\phi_{bi} = q\phi_B - kT \ln \frac{N_a}{N_d} \tag{6.30}$$

where N_a and N_d are the acceptor and donor carrier concentrations, respectively. Considering the width of the depletion region w_d of a semiconductor material, which is a function of the applied voltage (V) and the material concentration, and is given by the square root function

$$w_d = \sqrt{\frac{2\varepsilon_s \left(\phi_{bi} + V \right)}{qN_d}} \tag{6.31}$$

where ε_s is the relative permittivity of the semiconductor material, and the capacitance C of the junction is given by

$$C = \frac{\varepsilon_s A}{w_d} \tag{6.32}$$

where A is the area of the junction, it follows that a plot of the linear function $1/C^2$, which is defined by Equation (6.33), can be used to determine the total barrier height of the semiconductor-metal interface.

$$\frac{1}{C^2} = \frac{2\left(\phi_{bi} + V \right)}{qN_d\varepsilon_s A^2} \tag{6.33}$$

Once the built-in potential in (6.33) is known, the barrier height can be determined through (6.30).

Returning to the current density of the FN tunneling mechanism J_{FN}, a typical solution to (6.33) is to graphically plot the function of $\log(J_{FN}/E_{ox}^2)$ versus $1/E_{ox}$, which is known as the FN plot, and is also described in Lezlinger and Snow (1969). This plot yields a straight line, which is independent of oxide thickness, and works well with datasets of experimental data compared with the theoretical model (Ranuárez et al. 2006). However, with more recent technology advances and aggressive device scaling, leading to oxide thicknesses of only a few nanometers, (6.33) becomes less relevant to use owing to effects such as (adapted from Ranuárez et al. 2006)

- Direct tunneling in a small electric field.
- Carrier quantization initiated in the inversion and the accumulation layer.
- Additional tunneling mechanisms (apart from ECB) such as EVB and HVB.
- The effects of determinate temperature on the disposal of carriers leading to tunneling.
- The depletion effects on the polysilicon gate (Ranuárez et al. 2006).

DIRECT TUNNELING CURRENT

A classical approach to modeling the direct tunneling of electrons through a thin gate oxide layer is presented in various papers and books, including Chaudry (2013), Choi et al. (1999) and Lee and Hu (2001), and is briefly adapted here to highlight the significance of this phenomenon in semiconductor power consumption. Consider, first, the simplified cross-section of a MOSFET transistor, in this instance a *p*-type substrate with *n*-type diffused source and drain channels, presented in Figure 6.10.

Figure 6.10 shows the typical cross-section of an *n*-type MOSFET. By increasing the gate bias voltage, a depletion layer of electrons is formed at the interface of the oxide and the body, which forms a passage for the transport of electrons between the source and the drain. In order to improve the performance of the device, the channel length (*L*) can be decreased; however, aggressive scaling and reduction of the channel length also leads to DIBL and velocity saturation. DIBL has an effect on the conductance at the output of the transistor configuration and its threshold voltage. With a reduction in the length of its channel, the electric field in the direction of transport (drain to source) increases and the control on the gate deteriorates, allowing electrons to flow freely between the drain and source, notwithstanding the control imposed by the gate voltage. One method to counteract this mechanism is to increase the width of the gate oxide and use high-κ dielectric material; however, as highlighted in this chapter, modern transistors already push the limits of the gate oxide thickness. Velocity saturation occurs when there is no further current increase as the channel length declines. The mechanisms that cause this phenomenon are "the increasing carrier scattering in high electric fields due to the short channel length and the carrier group velocity, which no longer increases as the channel length decreases" (Khanna 2016). To counteract these mechanisms, novel materials with greater mobility of carriers, for example III–V compound semiconductor materials such as GaAs, GaN and InAs as well as carbon-based compounds and nanostructure materials, have been developed. Despite these advances, oxide tunneling limitations still exist and require alternative methods to improve performance.

Figure 6.11 represents the same cross-section of a typical *n*-type MOSFET with a large electric field applied across the gate terminal.

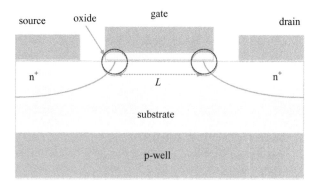

FIGURE 6.10   Cross-section of a MOSFET transistor with *p*-type substrate and *n*-type diffused source and drain channels.

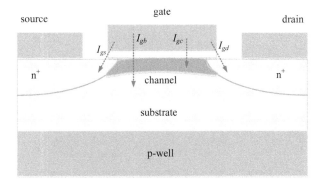

FIGURE 6.11 Cross-section of a MOSFET transistor with *p*-type substrate and *n*-type diffused source and drain channels with a large electric field applied across the gate terminal of the device.

As shown in Figure 6.11, several current mechanisms can be recognized in the MOS transistor. If the device is operating in the inversion regime, a tunneling current is induced between the gate of the transistor and the conducting channel, designated I_{gc}. In addition, in both accumulation and inversion, a current flow, again between the gate of the transistor and, in this case, the bulk (I_{gb}), exists in MOS transistors. For all operating regimes, particularly significant in short-channel devices, currents also flow among the sections where the gate terminal overlaps the drain (I_{gd}) and the source (I_{gs}). Cao et al. (2000), as well as Ranuárez et al. (2006), provide a convenient summary of the principal current mechanisms for each tunneling component, adapted and presented in Table 6.4.

Direct tunneling of electrons into the conduction band of the electrodes is much more likely with an oxide thickness of a few nanometers. Traditional tunneling current models are becoming less ideal to accurately define the tunneling of electrons across the potential barrier in the silicon-oxide interface (Ranuárez et al. 2006). Electron tunneling at lower electric fields cannot be neglected, and these barriers are not triangular but are approximately trapezoidal. Figure 6.12 represents the energy band diagram of the direct tunneling mechanism across a trapezoidal barrier into the conduction band of the insulating material, adapted from Ranuárez et al. (2006).

TABLE 6.4
Dominant Leakage Current Mechanisms for MOSFET Transistors

Current Component	I_{gc}	I_{gb}		I_{gs}, I_{gd}
Operating region	Inversion	$V_g > 0$	$V_g < 0$	All
PMOS	HVB	ECB	EVB	HVB
NMOS	ECB	EVB	ECB	ECB

Source: Adapted from Cao, K. M., et al., *International Electron Devices Meeting 2000 (IEDM)*, 815–818, 2000 and Ranuárez, J. C., et al., *Microelectronics Reliability*, 46, 1939–1956, 2006.

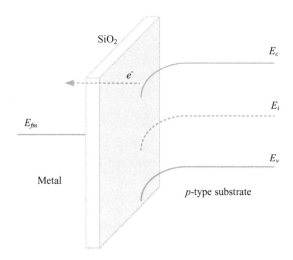

FIGURE 6.12 Energy band diagram of the direct tunneling mechanism across a trapezoidal barrier.

The semi-empirical model current density of the direct tunneling mechanisms J_{DT} can be described by the current that has been derived in Ranuárez et al. (2006) and Lee and Hu (2001). This current is a function of the applied voltage, such that

$$J_{DT} = \frac{q^3}{16\pi^2 \hbar \phi_b} C\left(V_g, V_{ox}, t_{ox}, \phi_b\right) E_{ox}^2$$

$$\times \exp\left[\frac{-4\sqrt{2m_{ox}^*}\,\phi_b^{3/2}}{3\hbar q |E_{ox}|}\left[1 - \sqrt{1 - \frac{qV_{ox}}{\phi_b}}\right]^{3/2}\right] \tag{6.34}$$

where,

V_{ox} is the voltage over the oxide

$C(V_g, V_{ox}, t_{ox}, \phi_b)$ is a correction function given by Ranuárez et al. (2006)

$$C\left(V_g, V_{ox}, t_{ox}, \phi_b\right) = \exp\left[\frac{20}{\phi_b}\left(\frac{|V_{ox}| - \phi_b}{\phi_{bo}} + 1\right)^{\alpha}\left(1 - \frac{|V_{ox}|}{\phi_b}\right)\right]\frac{V_g}{t_{ox}} N \tag{6.35}$$

where α is a fitting parameter. The density of carriers, N, injected into the surface is described by Equation (6.36) is applicable to ECB and HVB in inversion and accumulation.

$$N = \frac{\varepsilon_{ox}}{t_{ox}}\left\{n_{inv}V_t \ln\left[1 + \exp\left(\frac{V_{ge} - V_{th}}{n_{inv}V_t}\right)\right] + n_{acc}V_t \ln\left[1 + \exp\left(-\frac{V_g - V_{fb}}{n_{acc}V_t}\right)\right]\right\} \tag{6.36}$$

The parameter V_{fb} is referred to as the flat-band voltage, the gate-depletion voltage subtracted from the gate voltage is given by V_{ge}, V_g is the actual potential on the gate and η_{inv} and η_{acc} are fitting parameters (Ranuárez et al. 2006). Similarly, the density of carriers N for EVB is described by

$$N = \frac{\varepsilon_{ox}}{t_{ox}} n_{\text{inv}} V_t \ln\left[1 + \exp\left(\frac{|V_{ox}| - E_g / q}{n_{\text{EVB}} V_t}\right)\right] \qquad (6.37)$$

where,

$\quad \eta_{\text{EVB}}$ is a fitting parameter

$\quad E_g$ is the semiconductor band gap (Ranuárez et al. 2006)

These approximations and derivations show that the direct tunneling leakage current density of thin oxides presents relatively involved and complex models to quantify the amount of current leaking through the oxide barrier. Ranuárez et al. (2006) provide further a derivation on modeling gate oxide tunneling current using BSIM* modeling. TFETs are *steep-slope* devices that show significant prospects for ultra-low-power electronic devices. A fundamental characteristic of the TFET is the likelihood of a reverse sub-threshold gradient beneath the 60 mV/dec limit of normal MOSFETs. TFETs are described and analyzed in the following section.

TUNNELING FIELD-EFFECT TRANSISTORS

TFETs have proven to be promising candidates for next-generation low-power integrated circuits since they provide a subthreshold swing lower than the imposed kT/q limitation (sub-thermal) (Cao et al. 2015). The significance of a lower subthreshold swing is that these transistors allow power supply voltage scaling while maintaining a high *on-off* current ratio (Cao et al. 2015). Figure 6.13 depicts the cross-section comparison of a traditional *n*-type MOSFET device and a *n*-type TFET.

Note that, from Figure 6.13, for the TFET, the traditional *n*-type region at the source terminal is replaced by a *p*-type dopant and the *p*-type substrate is replaced by an intrinsic layer. TFETs have demonstrated the ability to "overcome the 60 mV/dec subthreshold swing limit of conventional MOSFETs" (Alper et al. 2015; Lu and Seabaugh 2014). Extremely accurate electrostatic control and the use of very thin semiconductor layers where quantum effects become dominant are required to maximize the BTBT in these devices (Alper et al. 2015). To recognize the minimized subthreshold swing, as described in (6.16) and (6.17), consider the drain current flowing through a "degenerately-doped p^+-n^+ tunnel junction, with transport mechanism described by the Zener tunnel direction (reverse direction)", adapted from Zhang et al. (2006) and given as

$$I_D = a V_{\text{eff}} E \exp\left(-\frac{b}{E}\right) \qquad (6.38)$$

---

* The Berkeley Short-Channel IGFET Model (BSIM) is a transistor model intended to accurately reflect the device operation of continuously scaled geometries.

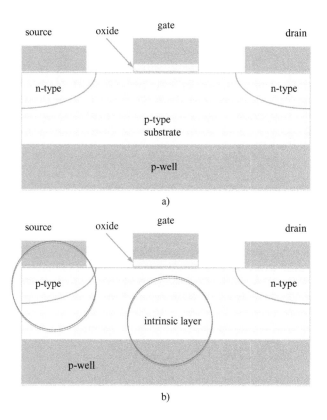

FIGURE 6.13   Cross-sectional view of (a) the traditional MOSFET transistor and (b) the low subthreshold swing TFET.

where,

V_{eff}	is the tunnel-junction bias voltage
E	is the electric field
a and b	are constants depending on parameters of the semiconductor material and the cross-sectional area of the component

The derivations of a and b are given in further detail in Zhang et al. (2006). The derivative (therefore the slope) of the tunneling current equation in relation to the gate-source voltage can be implemented to represent the equation for subthreshold fluctuations (swing) of a TFET, such that

$$S = \ln(10) \frac{dV_{gs}}{d \ln I_D} \tag{6.39}$$

which results in (adapted from Zhang et al. 2006)

$$S = \ln(10) \sqrt{\frac{1}{V_{eff}} \frac{dV_{eff}}{dV_{gs}} + \frac{E+b}{E^2} \frac{dE}{dV_{gs}}} \tag{6.40}$$

showing two individual terms in the denominator within the square root that can be capitalized on to realize a small subthreshold swing not limited by the thermal voltage, kT/q. As stated in Zhang et al. (2006), the first term in (6.40) indicates that the transistor should be constructed such that "the gate-source voltage directly modulates the tunnel-junction bias voltage V_{eff}". Related to transistor physical geometry, this means that the transistor must have a thin or a high-κ gate dielectric and a very thin body, which will ensure that the gate electric field modulates the channel directly. Assuming a very thin oxide thickness, within the low-nanometer range, the term dV_{eff}/dV_{gs} approaches a value of 1, which means that the first term in (6.40) simplifies to $1/V_{\text{eff}}$. Therefore, the subthreshold swing increases proportionally with the gate-source voltage.

Focusing on the second term in (6.40), the subthreshold swing can be lowered by maximizing the derivative (slope) of the gate-source voltage. The magnitude of this field therefore varies the tunneling width, although it is not possible to individually and mutually exclusively engineer the tunnel-junction voltage together with the junction electric field. Zhang et al. (2006) provide a low subthreshold swing device design by exploring transistor geometry through two-dimensional Poisson derivation. In Jin et al. (2017), the BTBT *off*-current characteristics of BiCMOS SiGe transistor channels are explored, considering the promising characteristics of this technology for sub-5 nm nodes.

A mechanism that increases V_{th}, degrades performance and causes heating of the transistors is discussed in the following section.

HOT-CARRIER INJECTION FROM THE SUBSTRATE TO THE GATE OXIDE

Hot-carrier injection leads to degradation in MOSFETs, of which the key basis is heating within the channel of the MOSFET. The heated active transporters can be introduced into the gate oxide and degrade the characteristics of the transistor, including an increase in V_{th}, lowering the carrier mobility (the combination leading to a lower drain current) and, therefore, lowering the transconductance. As a result, oxide and interface damage is initiated by the addition of holes and electrons into the gate oxide, through

- Traps being created within the oxide.
- Electrons and holes remaining trapped within the oxide layer.
- The creation of interface states at the semiconductor-oxide interface.
- Electrons and holes getting trapped by the interface states (Sugiharto et al. 1998).

If it is assumed that the gate and drain of a transistor are connected to V_{dd}, carriers pick up energy as they exchange across the channel, and these carriers come to be *hot* and are attracted to the gate node. These *hot* carriers may be injected into the gate oxide where they become trapped and cause a shift in the thermal voltage. The "high electric field near the thin Si-SiO$_2$ boundary of short-channel transistors induce a sufficient gain in energy into carriers to cross the interface potential barrier and enter the oxide layer" (Roy et al. 2003). The barrier height energy for holes is 4.5 eV, whereas for electrons it is 3.1 eV; therefore, the hot-carrier injection of electrons from the Si to the SiO_2 is more likely to occur than that of holes being injected through the barrier. Gate current in n-type MOSFETs under stress conditions is due to holes and electrons; in p-type MOSFETs, the gate current is approximately 1000 times as large and solely generated by electrons (Rosenbaum et al. 1991). A possible mechanism to control this leakage current is the scaling down of the supply voltage with the device dimension, although this practice introduces various other limitations, and recent advances have stepped away from the further scaling of V_{dd}.

The two most important hot-carrier injection types (in n-type MOSFET transistors) are the

- Drain avalanche hot-carrier (DAHC) injection.
- Channel hot-electron (CHE) injection (Sugiharto et al. 1998).

In DAHC injection, a lateral field speeds up carriers to induce impact ionization due to impacts with the semiconductor material in the vicinity of the drain. The highest horizontal electric field, a function of the gate and drain voltages and transistor values such as the depth of the junction, the thickness of the oxide and the effective length of the channel, can be described according to Sugiharto et al. (1998):

$$E_{\text{peak}} = \frac{\left(V_D - V_{D,\text{sat}}\right)}{l} \tag{6.41}$$

where,

l is the effective length of the lateral electric field (pinch-off length adjacent to the drain)

V_D is the drain voltage

$V_{D,\text{sat}}$ is the saturation voltage

Determining the electric field across a junction and the temperature of the electrons directly is difficult to achieve in practice, therefore, in most instances, the bulk and gate currents are observed. The relationship between the substrate and gate currents and the lateral electric field is given in Sugiharto et al. (1998).

The preliminary electron-hole pair formation can lead to additional impact ionization, resulting in an avalanche effect (Sugiharto et al. 1998). CHE injection occurs if the produced carriers drift out of the bulk and generate a hole-current within the substrate in n-type MOSFETs and electron current in p-type MOSFETs. When the electron in the channel has enough energy to advance over the Si-SiO_2 interface, it can be conveyed and introduced inside the gate oxide (Sugiharto et al. 1998). Additional hot-carrier injection mechanisms, depending on the operating conditions of the transistor, include

- Minority carriers induced by secondary impact ionization.
- Radiative electrons within the large electric field area of the substrate (Sugiharto et al. 1998).

As a result of device degradation due to gate oxide currents from hot-carrier injection, a reliability assessment for both transistor-level as well as on-circuit level must be defined to ensure the long-term reliability of integrated circuits. Techniques to reduce the effects of hot-carrier injection include supply voltage downscaling in conjunction with device dimensions, although this practice introduces various other limitations and recent advances have stepped away from the further scaling of V_{dd}. In addition, suitable transistor drain engineering and incorporation of a higher quality of the gate oxide boundary can alleviate some of these hot-carrier effects. This procedure typically requires enhanced cleaning procedures and temperature control during growth of the gate oxide.

Another mechanism that causes additional power consumption/leakage currents in submicron transistors is gate-induced drain leakage (GIDL), discussed in the following section.

GATE-INDUCED DRAIN LEAKAGE

In deep-submicron MOSFETs, the concentration of doping is increased and the junction profile becomes increasingly rapid, which leads to an intensification in the electric field. This induces a leakage current, called GIDL, which has become a significant leakage current component in modern semiconductor active devices (Yuan et al. 2008). "If a MOSFET is biased in the accumulation region, a significant drain leakage current can be observed if the drain bias is much lower than the breakdown voltage of the device" (Yuan et al. 2008). GIDL currents are caused by the electron BTBT current in the reverse-biased channel-to-drain *pn*-junction. The band-to-band reverse-biased *pn*-junction current density equation is given by (6.21) earlier in this chapter. In Rafhay et al. (2011), a schematic representation of the impact of GIDL on the transistor drain current versus gate voltage is provided, adapted and presented in Figure 6.14.

As shown in Figure 6.14, GIDL leakage current mechanisms primarily have an impact on device performance by limiting the *off*-state current to a minimum value (Rafhay et al. 2011). From a physical perspective, if the gate of the MOSFET is subjected to a bias voltage to produce an accumulation layer at the semiconductor

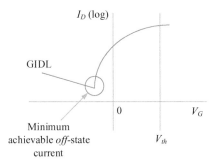

FIGURE 6.14 The effects of GIDL on the gate voltage versus drain current characteristics of a MOSFET.

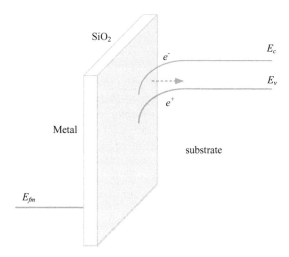

FIGURE 6.15 Band diagram of the GIDL regime shown in the perpendicular direction. (Adapted from Rafhay, Q., et al., *2011 Ultimate Integration on Silicon*, 1–4, 2011.)

surface, the area below the gate terminal takes on a potential virtually similar to that of the *p*-type substrate. "Accumulated holes at the surface therefore behave as a *p*-type region that is more heavily doped than the substrate, which causes the depletion layer at the surface to be much narrower than elsewhere" (Roy et al. 2003). This narrowing of the surface leads to an intensification in the immediate electric field and the effects of the higher field at these surfaces increase the possibility of leakage currents between the drain and gate. A large negative gate voltage depletes the *n*-type drain section below the gate and causes additional peak electric field increases that lead to avalanche multiplication and BTBT (Roy et al. 2003). The band diagram of the GIDL regime is given in Figure 6.15.

Several recent works on the modeling of GIDL in modern devices such as vertically stacked nanowire FETs, SiGe pFETs and SOI FinFETs are described in Fan et al. (2013), Hur et al. (2016) and Tiwari et al. (2014). In Ravi et al. (2014), an electronic circuit is proposed to reduce the GIDL currents in output buffers without compromising the speed of operation of such circuits.

CONCLUSION

High current consumption in electrical circuits has various detrimental effects on circuit reliability, invoking arbitrary operating conditions and heat generation. In submicron electronic components, circuits and systems, thermal constrictions confine advances in components driven by Moore's law. Although it is known that Moore's law is slowing down (as of 2017), thermal limitation incurs further restrictions, which can be mitigated if the mechanisms are defined and recognized. CMOS devices are approaching their atomistic and quantum mechanical physics limitations and leakage currents are proving to be a major hurdle to overcome because of aggressive scaling. In Haron and Hamdioui (2008), five primary categories of challenges of CMOS scaling are given, which are listed as

- Physical challenges due to tunneling and leakage currents, which have an impact on the performance and functionality of the devices.
- Material challenges from the inability of the dielectric and conductive materials to provide sufficient and reliable insulation and conduction.
- An increasing number of transistors per chip (per area), which leads to high power consumption and thermal dissipation, defined as power-thermal challenges.
- Technological challenges, which primarily relate to the limitations and challenges of developing photolithographic tools capable of operating reliably at extremely short wavelengths.
- Economic challenges from the exponential increase in fabrication cost, typically associated with the cost of processing equipment and testing facilities.

This chapter focuses on the first of these five categories, namely the physical challenges associated with increased leakage currents and electron tunneling through the gate oxide layer. The scaling of CMOS devices brings about equivalent scaling factors for the dielectric oxide layers, which have reached a point where focused electric fields at device junctions are enabling electrons to move right through the oxide layer and affect device performance and operation. The decrease in device dimensions therefore induces physical limitations not only in the gate oxide thickness, but also in introducing short-channel effects, lowering power and threshold voltages and increasing channel doping, which leads to reduced carrier mobility and BTBT.

This chapter introduces the mechanisms present in CMOS active and passive power consumption that contribute to the total power consumption in analogue and digital electronic circuits. The aggressively scaled CMOS leakage current mechanisms that are reviewed in this chapter are

- SLCs.
- Reverse-biased *pn*-junction leakage currents.
- Tunneling leakage currents.
- Hot-carrier injection from the bulk into the oxide of the gate.
- GIDL.

and the chapter provides a physical and mathematical description of each of these mechanisms. The aim is to reduce power consumption in submicron electronic circuits by first understanding the mechanisms that cause severe device integrity compromises when the scaling of transistors leads to parasitic effects during operation. The chapter also briefly introduces the notion of TFETs, which are steep-slope transistors capable of ultra-low power consumption through their inverse subthreshold slope beneath the traditional 60 mV/dec limit for MOSFETs. The TFET is simply a gated pin-diode, which is operated under reverse-biased conditions and utilizes BTBT as opposed to the thermal injection used by traditional MOSFETs as the source of the carrier injection mechanism. TFETs have received increased popularity, especially with foundries moving toward sub-20 nm FinFET transistors.

REFERENCES

Airhart, M. (2015). Hot chips: Managing Moore's law. Retrieved 10 March 2017 from http://news.utexas.edu

Alper, C., Palestri, P., Padilla, J. L., Gnudi, A., Grassi, R., Gnani, E., Luisier, M., Ionescu, A. M. (2015). Efficient quantum mechanical simulation of band-to-band tunneling. *2015 Joint International EUROSOI Workshop and International Conference on Ultimate Integration on Silicon*, 141–144.

Bowman, K. A., Wang, L., Tang, X., Meindl, J. D. (2000). Oxide thickness scaling limit for optimum CMOS logic circuit performance. *30th European Solid-State Device Research Conference*, 300–303, .

Butzen, P. F., Ribas, R. P. (2006). Leakage current in sub-micrometer CMOS gates. *Universidade Federal do Rio Grande do Sul*, 1–28.

Cao, K. M., Lee, W. C., Liu, W., Jin, X., Su, P., Fung, S. K. H., An, J. X., Yu, B., Hu, C. (2000). BSIM4 gate leakage model including source-drain partition. *International Electron Devices Meeting 2000 (IEDM)*, 815–818.

Cao, W., Jiang, J., Kang, J., Sarkar, D., Liu, W., Banerjee, K. (2015). Designing band-to-band tunneling field-effect transistors with 2D semiconductors for next-generation low-power VLSI. *2015 IEEE International Electron Devices Meeting (EDM)*, 12.3.1–12.3.4

Chaudhry, A. (2013). Fundamentals of nanoscaled field effect transistors. *Springer Science and Business Media*. Chapter 2. New York, April 2013.

Choi, C. H., Oh, K. H., Goo, J. S., Yu, Z., Dutton, R. W. (1999). Direct tunneling current model for circuit simulation. *Electron Devices Meeting*. IEEE Technical Digest, 735–738.

Chueng, K. P. (2010). On the 60 mV/dec @ 300 K limit for MOSFET subthreshold swing. *Proceedings of 2010 International Symposium on VLSI Technology, System and Application*, 72–73.

De, V., Borkar, S. (1999). Technology and design challenges for low power and high performance. *International Symposium for Low Power Electronics and Design*, 163–168.

Fan, J., Li, M., Xu, X., Huang, R. (2013). New observation on gate-induced drain leakage in silicon nanowire transistors with epi-free CMOS compatible technology on SOI substrate. *2013 IEEE SOI-3D-Subthreshold Microelectronics Technology Unified Conference*, 1–2.

Gurrum, S. P., Suman, S. K., Joshi, Y. K., Fedorov, A. G. (2004). Thermal issues in next-generation integrated circuits. *IEEE Transactions on Device and Materials Reliability*, 4(4), 709–714.

Haris, A., Iris, K., Anel, T., Adnan, M., Senad, H. (2016). Modelling the generation of Joule heating in defective thin oxide films. *International Symposium on Industrial Electronics (INDEL)*, 1–4, 3–5 November 2016.

Haron, N. Z., Hamdioui, S. (2008). Why is CMOS scaling coming to an end? *3rd International Design and Test Workshop*, 98–103.

Hirose, M., Koh, M., Mizubayashi, W., Murakami, H., Shibahara, K., Miyazaki, S. (2000). Fundamental limit of gate oxide thickness scaling in advanced MOSFETs. *Semiconductor Science and Technology*, 15, 485–490.

Hur, J., Lee, B., Kang, M., Ahn, D., Bang, T., Jeon, S., Choi, Y. (2016). Comprehensive analysis of gate-induced drain leakage in vertically stacked nanowire FETs: Inversion-mode versus junctionless mode. *IEEE Electron Device Letters,* 37(5), 541–544.

Jin, S., Park, H., Luisier, M., Choi, W., Kim, J., Lee, K. (2017). Band-to-band tunneling in SiGe: Influence of alloy scattering. *IEEE Electron Device Letters*, 38(4), 422–425.

Kao, J., Narendra, S., Chandrakasan, A. (2002). Subthreshold leakage modeling and reduction techniques. *IEEE*, 141–148.

Khanna V.K. (2016). Short-channel effects in MOSFETs. *Integrated Nanoelectronics. NanoScience and Technology*. New Delhi: Springer.

Kumar, S. (2015). Fundamental limits to Moore's law. Retrieved 10 March 2017 from http://arxiv.org

Lee, W., Hu, C. (2001). Modeling CMOS tunneling currents through ultrathin gate oxide due to conduction- and valence-band electron and hole tunneling. *IEEE Transactions on Electron Devices*, 48(7), 1366–1373.

Lezlinger, M., Snow, E. H. (1969). Fowler-Nordheim tunneling into thermally grown SiO2. *Journal of Applied Physics*, 40(1), 278–283.

Lu, H., Seabaugh, A. (2014). Tunnel field-effect transistors: State-of-the-art. *IEEE Journal of the Electron Devices Society*, 2, 44–49.

Oza, A., Kadam, P. (2014). Techniques for sub-threshold leakage reduction in low power CMOS circuit designs. *International Journal of Computer Applications*, 97(15), 10–13.

Paul, O., Baltes, H. (1993). Thermal conductivity of CMOS materials for the optimization of microsensors. *Journal of Micromechanics and Microengineering*, 3(3), 110–112,.

Pop, E., Goodson, K. E. (2006). Thermal phenomena in nanoscale transistors. *Transactions of the ASME*, 128, 102–108.

Rafhay, Q., Xu, C., Batude, P., Mouis, M., Vinet, M., Ghibaudo, G. (2011). Revisited approach for the characterization of gate induced drain leakage. *2011 Ultimate Integration on Silicon*, 1–4.

Ranuárez, J. C., Deen, M. J., Chen, C. (2006). A review of gate tunneling current in MOS devices. *Microelectronics Reliability*, 46, 1939–1956.

Ravi, H., Goel, M., Bhilawadi, P. (2014). Circuit to reduce gate induced drain leakage in CMOS output buffers. *22nd International Conference on Very Large Scale Integration*, 1–5.

Rosenbaum, E., Rofan, R., Hu, C. (1991). Effect of hot-carrier injection on n- and pMOSFET gate oxide integrity. *IEEE Electron Device Letters*, 12(11), 599–601.

Roy, K., Mukhopadhyay, S., Mahmoodi-Meimand, H. (2003). Leakage current mechanisms and leakage reduction techniques in deep-submicrometer CMOS circuits. *Proceedings of the IEEE: Contributed Paper*, 91(2).

Schenk, A. (1993). Rigorous theory and simplified model of the band-to-band tunneling in silicon. *Solid-State Electronics*, 36(1), 19–34.

Sugiharto, D. S., Yang, C. Y., Le, H., Chung, J. E. (1998). Beating the heat [CMOS hot-carrier reliability]. *IEEE Circuits and Devices Magazine*, 14(5), 43–51.

Tiwari, V. A., Scholze, A., Divakaruni, R., Nair, D. R. (2014). Modeling of gate-induced drain leakage mechanisms in silicon-germanium channel pFET. *IEEE 2nd International Conference on Emerging Electronics*, 1–5.

Van Engelen, J. P., Shen, L., van der Tol, J. J. G. M., Smit, M. K. (2015). Modelling band-to-band tunneling current in InP-based heterostructure photonic devices. Proceedings of the 20th Annual Symposium of the IEEE Photonics Benelux Chapter. Brussels, Belgium, 27–30 November 2015, 26–27.

Yuan, X., Park, J., Wang, J., Zhao, E., Ahlgren, D. C., Hook, T., Yuan, J., Chan, V. W. C., Shang, H., Liang, C., Lindsay, R., Park, S., Choo, H. (2008). Gate-induced-drain-leakage current in 45-nm CMOS technology. *IEEE Transactions on Device and Materials Reliability*, 8(3), 501–508.

Zhang, Q., Zhao, W., Seabaugh, A. (2006). Low-subthreshold-swing tunnel transistors. *IEEE Electron Device Letters*, 27(4), 297–300, April 2006.

7 Microelectronic Circuit Enhancements and Design Methodologies to Facilitate Moore's Law – Part I

Wynand Lambrechts and Saurabh Sinha

INTRODUCTION

ICs have always had a direct association with their cost per area, with a technical priority to increase the transistors per area while maintaining a steady improvement in the cost per area. In parallel, semiconductor manufacturers are aiming to increase the yield of microelectronic circuits, therefore reducing the number of defects per area. There are multiple ways of reducing transistor size while maintaining as few defects as possible, either through process enhancements or effective electronic circuit schematic and layout designs. Eliminating a single transistor in an electronic circuit could have vast effects on the final footprint of the IC, where building blocks, or cells, are often multiplied several thousands or even millions of times to achieve the desired performance. Eliminating passive components such as resistors, inductors or capacitors from an electronic circuit has a much larger effect on the overall area. Eliminating active or passive components, however, should not be implemented at the expense of circuit performance, but rather to its advantage, or should have a negligible effect. The design parameters of microelectronic circuits determine various aspects of the final product. These aspects of the final circuit can be described in terms of

- Performance (operation speed, power dissipation, functionality, and flexibility to account for tolerances in device manufacturing).
- The physical size of the die and, therefore, the cost of production.
- The time to design, which directly influences the cost of engineering and the production schedule.
- Ease of manufacturability and testability, again influencing the overall cost of production.

As with almost any design, be it mechanical or electrical, certain trade-offs must be weighed against one another to determine if a design alteration or optimization achieves the required operation. This chapter primarily focuses on simplified techniques and generalized suggestions to take advantage of the available area for semiconductor ICs more effectively and reduce the cost per die associated with the semiconductor industry. In order to clearly understand the requirements to save space on a wafer, not only a precise understanding of the circuit design is needed, but also a methodology for schematic design and schematic-to-layout generation must be followed. The first of these requirements, a thorough understanding of circuit behavior, is often a more difficult approach in view of the vast number of circuit topologies and applications that exist for each of these circuits. Designers must consider all trade-offs in lieu of the desired application, and decide which components may have larger tolerances, or can be omitted completely. Providing a guideline for the implementation of all possible circuits with respect to all possible applications is not a practical approach. However, a methodology and considerations of circuit performance versus device area are useful tools to achieve high-performance and small-footprint designs. The design description for an IC can be described in terms of three primary domains. In each of these domains, various design decisions are made to solve a particular problem based on its required performance and its application. These domains can be hierarchically divided into levels of design abstraction, referred to as the abstraction hierarchy. The relationships between these domains and abstraction hierarchy levels have been described by the Gajski-Kuhn Y-diagram (Gajski and Kuhn 1983), where three radial lines represent the three description domains, and along each line are the types of associated objects in that domain, an expansion of the abstraction hierarchy levels. The Gajski-Kuhn Y-diagram is presented in Figure 7.1.

The Gajski-Kuhn Y-diagram can essentially be used to describe function representations, modeling means and structural elements within an electronic circuit (hardware) design methodology. From the Gajski-Kuhn Y-diagram, the creation of a physical description from a structural analysis is achieved through circuit layout synthesis. As seen in Figure 7.1, the stages of development of ICs, described by the various perspectives, follow a methodology of three domains and five hierarchy levels, defined within the five concentric circles. The three domains related to the design process of a microelectronic circuit, with reference to Figure 7.1, are the

- Behavioral domain, the domain in which a component is described by defining its input and output response, and therefore its temporal and functional behavior.
- Structural domain, a domain in which a component is described in terms of an inter-connection of more primitive components, thus a set of subsystems that form the building blocks of the entire system.
- Physical or geometrical domain, which specifies how to build a structure that has the required connectivity to implement the prescribed behavior. This domain therefore describes the geometric properties of the system and its subsystems, with regard to information on its size, shape and physical placement.

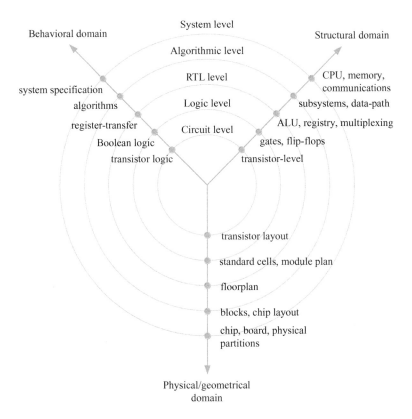

FIGURE 7.1   The Gajski-Kuhn Y-diagram, where the circles represent the abstraction levels and the three straight lines forming the Y represent the aspect from which these layers are examined. (Form Gajski, D. D. and Kuhn, R. H., *Computer*, 16(12), 11–14, 1983.)

Traditionally, and still relevant in nano-electronic device design, the design abstraction levels defined by the five concentric circles in Figure 7.1 have included some form of

- An architecture, functional or system level.
- An algorithmic level.
- A register-transfer level (RTL).
- A logic level.
- A circuit level (Gajski and Kuhn 1983).

To describe the usage of the Gajski-Kuhn Y-diagram, the system level abstraction layer for the behavioral, structural and physical/geometrical domains is presented as an example. At system level, the basic properties of the electronic system are determined, as described by the problem definition – the functions that the electronic circuit should be able to perform. With respect to the behavioral description of the system, block diagrams are typically used to generate abstractions of the internal signals and their time and/or frequency response. The system level behavior should therefore completely describe the desired output response for a given input signal. Within the structural domain, the requirements needed to implement the desired circuitry must be defined in terms of a user specification, such as defining the processing power, memory usage and communication protocols that are required to process and represent the input signal as a desired output. From the perspective of the physical or geometrical domain, at system level, the size, shape and connection requirements to use the top-level system are to be described. These requirements define to the user how and where to integrate the functional system with the highest level of integration. Chip size, printed circuit board or integration-substrate size, or supporting circuitry requirements are specified at this abstraction layer in the physical or geometrical domain. A similar methodology is followed to completely define and describe the abstraction level consideration of the algorithmic, RTL, logic and circuit levels in all three domains. Since hardware design, especially electronic circuit design, is typically a top-down approach, the Gajski-Kuhn Y-diagram perceives the three domains as the *top* design consideration, with the more detailed abstraction layers representing the *bottom* design considerations. The specific sequence used in a design methodology does not necessarily have to be defined by the Gajski-Kuhn Y-diagram (therefore its radial approach), and variations of this methodology are frequently implemented, with the principle of the diagram aiding the design process.

Semiconductor IC design flow can be constructed with reference to the domains and abstraction layers presented by the Gajski-Kuhn Y-diagram. The various levels of design include the design specifications, which abstractly describe the functionality, interface and architecture of the desired IC. A simplified design flow for a very large-scale integration (VLSI) system is presented in Figure 7.2.

According to Figure 7.2, the top-down approach of a simplified IC design flow starts with a design specification to define the function of the circuit. This step is typically followed by the preliminary schematic design of the integrated subsystems, simulated and tested as individual cells using computer-aided design (CAD) software, such as SPICE simulators. In practice, a substantial amount of time and effort is spent on simulation and circuit optimization at this stage to ensure that the sub-circuit operates adequately, given its pre-defined input signals. Designers and researchers seldom proceed beyond this point if each sub-circuit does not function according to the exact specifications provided. Symbols are then created for each sub-circuit and interconnected to form a full architecture of the required operation, where a full-circuit simulation should ideally only be used as a final check that each interconnect provides the correct signal levels and bias to the subsequent sub-circuit. If discrepancies are noticed at this point, the design flow is halted and returned to the schematic design of the individual (problematic) circuits. The process is repeated up to the point where the full system simulation satisfies the initial design specifications.

At this point in the design flow, a completely different discipline of the IC manufacturing process is entered, namely the layout of the components. Data on the layout of the components will be provided to the manufacturer (foundry) that fabricates the devices; this is a crucial stage in the development process. This information relates directly to the processes used in the foundry, such as the photolithography, wet/dry chemical etching or metallization techniques, and limitations incurred owing to the equipment used. Any errors that may occur at this stage are typically more

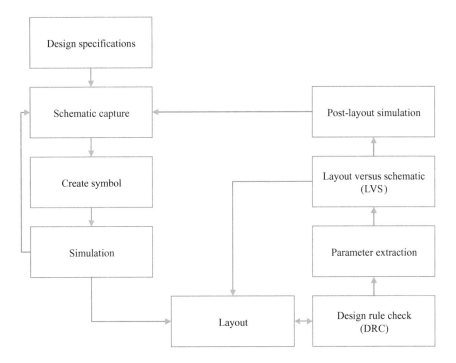

FIGURE 7.2   The design flow emanating from the Gajski-Kuhn Y-diagram for an electronic IC from the specifications stage through to the post-layout simulations to verify system level operation.

time-consuming and resource-intensive to identify and could take significant time and effort in the overall design process to complete successfully. It is relatively commonplace for schematic designers to hand over a design to layout designers, with frequent inputs provided by the schematic designer. The layout procedure is followed by a DRC, which is a set of rules provided by manufacturers that define the placement and layer limitations of a specific process. For example, a DRC will determine whether the placement of two metal lines (interconnects) are at least a pre-defined distance apart. However, the DRC does not guarantee that the layout adheres to the circuit operation as defined in the schematic layout, and is purely a geometric check of the placement of layers. If any errors are identified, typically geometric limitations that have been exceeded, the layout process is revised and the relevant corrections are made. An extraction process that reads the information of the layout and re-constructs a SPICE circuit-level representation of the devices defined by the DRC is required. With this information, a layout versus schematic (LVS) process is run on the schematic and layout extracted data and compared electrically. This ensures that the circuit operation that the schematic designer intended matches the layout design and adheres to the physical placement rules and limitations of the manufacturer. If there are any discrepancies between the schematic operation and the extracted layout data, the layout process must be revised and corrected where applicable. Only if the LVS process passes without any errors (or, in certain cases, waiving non-critical warnings) can the post-layout simulations be performed. The post-layout simulations take into account all geometric placements, including the effects of parasitic parameters (such as stray inductance or capacitance), interconnects and measurement pads, and determines if the desired output signal(s) still matches the output signals of the schematic design in the second step of the design flow. At this point, the ideal result would be a perfect (or within acceptable tolerances) match between the post-layout simulation and desired schematic output, whereas if any discrepancies are identified, the design flow is, in practice, restarted to determine the parasitic (not explicitly designed for) parameters in the circuit layout.

According to the design flow of even the simplest microelectronic circuit, this process can be extremely time-consuming and complex. The cost of device manufacturing is typically high, depending on the process node designed for and, typically, primarily because of photolithography mask manufacture and equipment setup; efficient circuit design techniques are therefore crucial to decrease the development time of an IC. As Moore's law is pushing the geometric scales of semiconductor processes, photolithography masks, process optimization and equipment setup are becoming more expensive with decreasing node size. Aspects such as efficiency, yield and components per area are becoming even more crucial for commercial institutions and scientific researchers, and understanding the importance of circuit enhancements and design methodologies to achieve these aspects is undeniably beneficial.

A structured design approach to CMOS ICs warrants efficient, successful and optimized use of the available semiconductor area for circuits. A structural approach includes considerations of

- Design, using a set hierarchy of circuit components and primitive devices.
- Regularity, where the reuse of modules and standard cell libraries is encouraged.
- Modularity, which allows modules to be treated as black boxes, meaning that each module is optimized individually and available to be used at any interface within the circuit.
- Physical and temporal locality, which is the ability to reuse elements and modules within a relatively short time as well as in relatively close proximity.

The hierarchical approach to implementing sub-modules from larger circuit models is commonly referred to as the divide-and-conquer technique, the process of repeating sub-module division until the complexity of the circuit is manageable and can be relatively easily optimized in the layout. Dividing a design into submodules, following the hierarchical approach, can be complemented by incorporating regularity in the approach. In IC design, the use of iteration to form arrays of identical cells illustrates the use of regularity combined with hierarchical design. Furthermore, following a modularity methodology in IC design, the hierarchy and regularity principles are expanded by having well-defined functions as well as interfaces among sub-modules. It is important to ensure that not only the circuit (sub-module) operates efficiently and is well defined, but also that the interfaces among adjacent modules add a modular design, which introduces fewer unknown variables among cascading modules. In performing these techniques, the complexity and optimizations of each circuit module and sub-module are not apparent to the designer implementing the design, and this reduces the number of global variables that should be accounted for in an IC design, effectively reducing the complexity of the full circuit design.

The following section reviews the commonly used microelectronic circuit components, with regard to the associated footprints of each of these primitive devices. The footprint of each device plays a substantial role in determining the effective space that is available on a die, which can be used to improve the performance of the circuit.

MICROELECTRONIC CIRCUIT COMPONENT FOOTPRINTS

Only a handful of microelectronic circuit components are commonly used in IC design. The associated footprint of these devices, along with careful placement, correct sizing in terms of their electrical characteristics and – often – required symmetry, directly influences the overall performance of the system in its entirety. The most commonly used microelectronic circuit components (primitive devices) include

- The transistor, which is an active component completely defined by its four terminals: the gate, usually formed by polycrystalline silicon, the source and drain regions, and the substrate or bulk region.
- The diode, a two-terminal semiconductor device, which primarily conducts the flow of current in a single direction and has an ideally low resistance to the flow of current in the opposite direction.
- The resistor, which is a passive two-terminal device that restricts the flow of current in both directions, depending on its internal resistance.
- The capacitor, also a passive component that is used to store energy in an electric field between two conducting plates and a dielectric material located in between these plates.
- The inductor, another passive component used to store electrical energy in a magnetic field when an electric current is passed through its two terminals.

The reasoning behind the description of these components as either active of passive devices, with reference to the information provided in this chapter, relates to the layout geometry of each of these devices and the placement consideration of the devices in the vicinity of one another. It is commonly accepted that passive devices are large, whereas active devices are typically much smaller in comparison. Common practice in microelectronic design often aims to avoid the use of passive components as much as possible, or to place these devices off-chip to allow more active devices to fit on the allocated die size.

This chapter reviews the transistor with respect to common practices in IC design, layout considerations and circuit enhancements for various examples, which can improve circuit performance through space saving and the concepts of electronic design. The next chapter reviews the remaining passive components and provides reliability and yield optimization for each. Transistor optimization is arguably the most common technique used in modern IC design, since an IC could contain millions and even billions of transistors. This review is focused on MOSFETs, but similar analogies and approaches can be followed for bipolar transistors.

TRANSISTOR OPTIMIZATIONS

The fundamental operation of a transistor is not covered in this book, which assumes that the reader has an understanding of its operation; this book aims to highlight design concerns and layout techniques to minimize die size in microelectronic circuit design. For this discussion, the CMOS transistor is used as an example, where similar derivations for bipolar junction transistors can be applied. Two types of CMOS transistors exist: the n-channel MOS transistor, or NMOS, and the p-channel MOS transistor, or PMOS. Both these transistor variants have four terminals that completely define the transistor: the drain, gate, source and bulk or substrate of the transistor. In an NMOS transistor, the bulk is typically connected to the ground node of the circuit, whereas for a PMOS transistor, the bulk is connected to the positive supply. The operation point and performance of the transistor, in terms of its power capabilities, are defined through three parameters to which the designer has access (and therefore are not only process-dependent): the channel length L, the channel width W and the drain current I_d flowing through the transistor. As a brief revision, the drain current of the transistor is defined as

$$I_D = a\frac{W}{L}\left(V_{gs} - V_t\right)^2\left(1 + \lambda V_{ds}\right) \tag{7.1}$$

where,

  a is a process variable defined by the oxide capacitance and carrier mobility
  V_{gs} is the gate-source voltage
  V_t is the threshold voltage
  λ is process-dependent
  V_{ds} is the drain-source voltage

There is therefore a direct (linear) correlation between the maximum drain current that can flow through the transistor and its W/L ratio. Depending on the required drain current and gain, the W/L ratio is typically minimized to decrease the physical size of the transistor. The sizing of the transistor is, however, not a *random* process involving a choice of a geometry that best fits the layout requirements of the designer. The size of the transistor involves considering the exact behavior of the component within the circuit, based on pre-defined specifications for each application. Most methods used to size MOS transistors, typically in amplifier applications, assume that the transistor is operating in strong inversion and use the transistor gate voltage overdrive (V_{ov}) as the key parameter. The overdrive voltage for a MOS transistor is quantified by the voltage over the gate of the transistor and its source voltage (V_{gs}) in surplus of V_t (the minimum gate-source voltage needed to switch the transistor on; the threshold voltage). Mathematically, the overdrive voltage is therefore determined by

$$V_{ov} = V_{gs} - V_t. \tag{7.2}$$

As a brief summary, the three behavioral operating regions of a transistor are defined as

- The region of strong inversion where the gate-to-source voltage V_{gs} is greater than roughly 100 mV of V_t and the inversion channel is strong and the drift current is dominant.
- The region of weak inversion where the gate-to-source voltage is far below that of the threshold voltage and the inversion channel is weak owing to a small number of free charges. In the weak inversion region, the dominant method of conduction is diffusion current and I_D has an exponential relationship to V_{gs}.
- The region of moderate inversion, if V_{gs} is approximately equal to V_t and both diffusion and drift currents are evident.

For submicron technology, weak inversion models are commonly exploited to precisely determine the transistor geometry. A method to synthesize the MOS transistor with respect to all the regions of operation is the transconductance (g_m)-over-drain current I_D ratio versus I_D, normalized with respect to the W/L ratio of the transistor. The relevance of the g_m-over-drain current technique is due to four primary reasons, namely that

- In analog circuits, this technique is strongly associated with performance.
- It links design variables (transconductance, cut-off frequency and drain current) to the specifications of the circuit, such as its operating bandwidth and power consumption.
- The operating region of the device can be determined through this technique.
- The dimensions of the transistor can be calculated through this technique.

Through the small-signal definition of a MOS transistor, the transconductance-over-drain current can be described in terms of the overdrive voltage, also referred to as the transconductance efficiency, by

$$\frac{g_m}{I_D} = \frac{2}{V_{ov}} \tag{7.3}$$

where, furthermore, the W/L ratio can be defined in terms of the transconductance, carrier mobility and oxide capacitance through

$$\frac{W}{L} = \frac{g_m}{\mu C_{ox} V_{ov}} \tag{7.4}$$

where, for a fixed g_m and L, a smaller V_{ov} translates into a physically bigger (in this case, wider, since L is kept constant) transistor, thus presenting a larger gate-to-source capacitance, C_{gs}, adversely affecting the operating bandwidth of the device. The cut-off frequency of the transistor is defined in terms of the gate-to-source capacitance as

$$\omega_T = \frac{g_m}{C_{gs}} \tag{7.5}$$

where the gate-to-source capacitance varies with the size of the transistor. Therefore, only applying the V_{ov} as a design parameter in analog circuits is not an effective method to size the transistor efficiently. The coefficient g_m/I_D is the slope of the I_D curve plotted against the gate voltage V_g on a logarithmic scale, described by the relationship

$$\frac{g_m}{I_D} = \frac{\delta I_D / \delta V_g}{\delta V_g} \tag{7.6}$$

which can be rewritten as

$$\frac{g_m}{I_D} = \frac{\delta \log (I_D)}{\delta V_g} \tag{7.7}$$

where the maximum slope of this curve, therefore the maximum g_m/I_D ratio, appears if the transistor is operating in weak inversion. This slope decreases until it reaches the strong inversion region, where the slope is at its minimum. In the weak and moderate inversion regions, the value of the gate voltage and the over-drive voltage are low and this makes these regions optimal for low power supply voltage operation. A representation of I_D versus the gate-to-source voltage V_{gs} is given in Figure 7.3.

From Figure 7.3, the variation in slope of I_D versus V_{gs} can be seen within the three operating regions: weak inversion, moderate inversion and strong inversion. Note the indication in Figure 7.3 of the threshold voltage V_t. The g_m/I_D ratio is also a measure of the efficiency of translating I_D into transconductance; a larger g_m/I_D ratio therefore means that higher transconductance can be reached for a constant drain current in relation to a normalized current, I_N, given by

$$I_N = \frac{I_D}{(W / L)}. \tag{7.8}$$

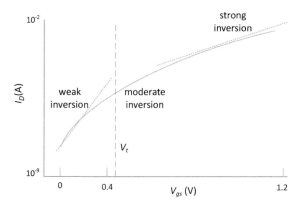

FIGURE 7.3   A representation of I_D versus the gate-to-source voltage V_{gs} for an NMOS transistor with a constant W/L aspect ratio.

In a specific process technology, g_m/I_D is considered a widespread characteristic of transistors within the process technology, and the relationship concerning the g_m/I_D ratio parameter and the operating region is given by

$$\frac{g_m}{I_D} = \frac{1}{I_D}\frac{\delta I_D}{\delta V_{gs}} \tag{7.9}$$

which can be rewritten as

$$\frac{g_m}{I_D} = \frac{\delta(\ln I_D)}{\delta V_{gs}} \tag{7.10}$$

and results in

$$\frac{g_m}{I_D} = \frac{\delta\left(\ln\dfrac{I_D}{(W/L)}\right)}{\delta V_{gs}} \tag{7.11}$$

where the universal curve of g_m/I_D versus I_N is used to find the aspect ratio of the transistor, typically the W parameter since L is kept constant. Considering a fixed transistor length, the result in (7.11) suggests that to increase g_m while maintaining a high value of g_m/I_D, W should be increased, maintaining a constant I_N while increasing I_D. If W is increased, though, the parasitic capacitance increases in the transistor, which reduces f_T, attested to the intrinsic gain-bandwidth product (GBP) of the transistor depending on the operating region. In an early work by Silveira et al. (1996), a g_m/I_D-based methodology for the design of CMOS analog circuits is presented, and provides a detailed description of transistor characteristics such as gain and bandwidth with respect to the geometry and g_m/I_D ratio.

Although the transistor voltage-to-current gain and bandwidth are typical design parameters that determine the size of the transistor and thus its W/L ratio, another important consideration in transistor design is the effect of device scaling on fluctuations in the current within the transistor, leading to noise in the component and inevitably amplified in the circuit. The following section reviews the impact of device geometry on generated noise.

THE IMPACT OF TRANSISTOR SCALING AND GEOMETRY ON NOISE

In a MOS transistor, a resistive channel under the gate is modulated by the gate-to-source voltage V_{gs}, such that it controls the drain current I_D. As a result, the resistive channel under the gate generates thermal noise, which can be embodied in the small-signal equivalent model of the transistor by a noise-current generator from the drain to the source. This noise-current generator in the MOS transistor drain to its source is shown in Figure 7.4.

The thermal noise generated by the transistor can be represented by

$$i_d^2 = 4kT\left(\frac{2}{3}g_m\right)\Delta f \tag{7.12}$$

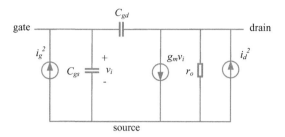

FIGURE 7.4 Noise-current generator between the drain and the source of a MOS transistor represented by the small-signal equivalent model. (From Gray, P. R. et al., *Analysis and Design of Analog Integrated Circuits*, John Wiley and Sons, Inc, New York, 2001.)

where,

k is Boltzmann's constant

T is the operating temperature

Δf is the noise bandwidth

In addition to the thermal noise generated by the resistive channel under the gate of the MOS transistor, the conducting current near the surface of the semiconductor material in the (lateral) MOS device acts as a surface state trap that captures and releases current carriers, leading to flicker noise ($1/_f$ noise) in the transistor. The flicker noise generated by the flowing current can be represented by

$$i_d^2 = K \frac{I_D^a}{f} \Delta f \tag{7.13}$$

where,

K is constant for a specific device

a is a technology-constant typically between 0.5 and 2

For long-channel devices with channel lengths longer than approximately 1 μm, the noise current generated in the transistor is a sum of the thermal and flicker noise, independently of each other, and given as

$$i_d^2 = 4kT\left(\frac{2}{3}g_m\right)\Delta f + K\frac{I_D^a}{f}\Delta f \tag{7.14}$$

where the thermal noise component in short-channel devices is typically two to five times larger than in long-channel lengths. The gate leakage current is another source of noise in the MOS transistor, referred to as shot noise, and given by

$$i_g^2 = 2qI_G\Delta f \tag{7.15}$$

where,

q is the elemental electron charge

I_G is the gate leakage current

Since the gate leakage current is typically very small, in the order of 1 fA, the shot noise in MOS transistors is generally neglected. At very high frequencies, typically seen in high-frequency amplifiers, a fourth source of noise starts to dominate the noise generation in the transistor. At an arbitrary point in the channel, the gate-to-channel voltage exhibits a random component due to fluctuations along the channel caused by thermal noise. These fluctuations generate noisy AC gate currents, i_g, due to the capacitance between the gate and the channel. For a long-channel device in the active region, this noise-current is described by

$$i_g^2 = \frac{16}{15} kT \omega^2 C_{gs}^2 \Delta f \qquad (7.16)$$

where C_{gs} can be rewritten as

$$C_{gs} = \frac{2}{3} C_{ox} WL, \qquad (7.17)$$

which clearly shows the effect of the width and length of the transistor on the noise source. As the device dimensions carry on reducing with each new generation of MOS processing technology, the consequence of a discrete deficiency on device performance becomes more prominent (Tsai and Ma 1994). In submicron channel lengths and widths, higher-order noise effects have been experienced, especially at low bias currents, as random telegraph noise (RTS), which is categorized by "discrete switching events of the channel current, trapping and de-trapping conduction carriers by individual interfacial defects" (Tsai and Ma 1994). As flicker noise has been deliberated on and studied for many years, the combination now leads to the two sources complementing each other to better understand RTS noise (Tsai and Ma 1994). In Tsai and Ma (1994), the channel length dependence of RTS noise is derived from the random current variations resulting from the trapping and de-trapping of carriers; these random fluctuations in the drain current, assuming negligible variations in carrier mobility across the channel, are defined in Tsai and Ma (1994) by

$$\Delta I_D = \left(\frac{1}{L^2} \right) q \mu V_d \qquad (7.18)$$

for a minor voltage on the drain V_d. This result suggests that if RTS is measured at a fixed average carrier density Q_s, where

$$Q_s = \frac{qN}{WL} \qquad (7.19)$$

where,
    N is the total number of carriers without trapping events in the channel
    ΔI_D scales inversely with L^2

In addition, in Tsai and Ma (1994), the channel length dependence on $1/f$ noise is also investigated. For a low drain voltage V_d and additional assumptions defined in Tsai and Ma (1994), the $1/f$ noise power spectral density $S_{ID}(f)$ is used to determine the dependence of $^1/_f$ noise on channel length. As a result, the power spectral density in a MOSFET is defined as

$$S_{I_D(f)} = \frac{M}{L^3} \qquad (7.20)$$

where M is defined as

$$M = \frac{kTW}{\gamma f}(\mu V_d)^2 (q + Q_s \alpha \mu)^2 N_t \left(E_{fn}\right) \qquad (7.21)$$

where,

    k being Boltzmann's constant
    T the absolute temperature
    γ an attenuation coefficient of the electron wave function in the oxide
    f the frequency of the noise spectrum
    α the scattering coefficient associated with the trapped charge
    N_t the occupied trap density per unit area
    E_{fn} the quasi Fermi level (Tsai and Ma 1994)

From (7.20), it is seen that there is a channel length dependence of $^1/L^3$ with respect to the power spectral density of $^1/_f$ noise in the MOSFET, assuming that all constants in M are held constant.

Transistor layout efficiency can only be achieved if the physical layout of the layered structure that defines a transistor on semiconductor material is implicitly defined. The following section reviews the physical layout properties of the MOSFET and aims to provide optimization techniques with respect to its geometrical attributes and electrical performance.

PHYSICAL LAYOUT PROPERTIES OF THE MOSFET

To lay out a MOS transistor in a CMOS or bipolar-CMOS (BiCMOS) process, several layers/masks are required to define the structure. The combination of these layers defines the transistor for CAD software to recognize it as a transistor and to perform LVS and post-layout simulations based on the geometries of each layer (Binkley et al. 2003). The DRC defines the minimum and maximum size, overlap and distance among adjacent layers. These layers, which define a transistor, include the n-diffusion, p-diffusion, polysilicon, n-type or p-type well, and metal layers that are isolated from one another by an intermediate oxide (typically SiO_2) layer, which is also used to form vias between layers. A layout of a pre-defined NMOS transistor is shown in Figure 7.5 to demonstrate the layer functions of a transistor.

From Figure 7.5, the following conclusions regarding the transistor, which relate to effective device patterning and layout techniques to optimize its physical size with respect to the desired performance, can be drawn:

- The active area of the transistor is defined in the diffusion region.
- The channel with W of the device is 6 µm.

- The channel length L of the transistor is 0.6 μm (600 nm).
- The aspect ratio of the transistor, its W/L ratio, is 10.
- The source and drain diffusion should be packed with the maximum amount of contact holes to decrease the resistance of the path to the diffusion and maximize the current flowing through the contacts.
- The polysilicon gate should overlap the active area by a process-specific minimum distance.

The minimum line width that can be used in a process is defined by the process node, a quantity that is constantly scaled to drive Moore's law, achieve smaller

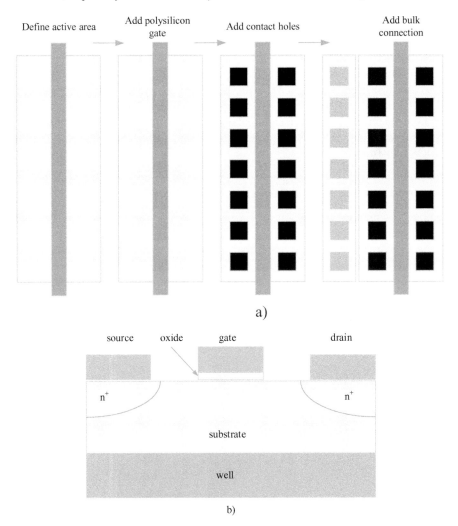

FIGURE 7.5 A simplified representation of an NMOS transistor (a) top view and (b) side view, which demonstrates the layer functions of a defined transistor in a semiconductor material.

device geometries and increase the total transistor count, which typically leads to an improvement in device performance, depending on what figure of merit is used to define its performance. A higher transistor count in low-noise circuits, for example, might lead to higher noise generation and therefore decreased system performance, if noise is the figure of merit. As presented in Jeon and Kang (2016), the gate parasitic capacitance between the polysilicon and the metal interconnects mainly affects the degradation of high-frequency and noise performance. The transistor layout as well as the interconnect effects are the two dominant factors that influence transistor performance, especially at high frequencies in the high-GHz range. These effects are typically only identified in post-layout simulations, or, in the worst case, during its real-world performance owing to external parasitic components introduced without the knowledge of the schematic or layout designer. It is therefore crucial to treat each component in the design independently and to determine its optimal geometry and placement.

Returning to Figure 7.5, from a visual perspective it is noticeable that the geometry of the transistor seems somewhat un-optimized. Visual inspection of component geometry and placement is often a first step to optimize the device, without specific regard to its physical operation and fundamental definitions. Possibly the most common transistor optimization is the use of a multiple-finger approach to design the CMOS transistor.

MULTIPLE-FINGER TRANSISTOR GEOMETRIES

Figure 7.6 represents a single-finger versus a multiple-finger approach to transistor layout.

As seen in Figure 7.5a and 7.6a, the single-finger transistor layout with a high W/L ratio (10 in this instance) is replaced by a multiple-finger transistor, as shown in Figure 7.6b. This technique has several advantages, which could have a substantial effect on device performance and total die size. There are several advantages as well

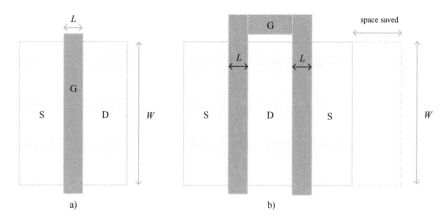

FIGURE 7.6 CMOS transistor with (a) a single-finger design with a high W/L aspect ratio versus (b) a CMOS transistor with a similar aspect ratio, but using multiple (two) fingers.

as disadvantages in using a multiple-finger layout geometry. These are discussed in the following sections.

Reduced Polysilicon Gate Resistance

The first advantage of using a multiple-finger approach is that it effectively decreases the resistance of the polysilicon gate electrode, R_g. The total gate width of the transistor (6 μm in this example) is defined as

$$W_{total} = W_f \times N \tag{7.22}$$

where,
- W_f is the width of a gate in a fingered structure
- N is the number of fingers in the multiple-finger layout

Therefore, if a single-finger transistor is used, $N = 1$, and the total width of the gate is similar to that of the desired width (6 μm × 1 = 6 μm). To achieve an equivalent total width of the gate electrode, as per the specified W/L aspect ratio, with L constant, the number of fingers can be increased, which means that the width of each gate is reduced to ensure a constant total width. By doubling the number of fingers from $N = 1$ to $N = 2$, the following relationship holds:

$$6\mu m = W_f \times 2 \tag{7.23}$$

therefore

$$W_f = \frac{6\mu m}{2} = 3\mu m \tag{7.24}$$

which gives a gate length of $W_f = 3$ μm for a two-finger design. Furthermore, there are now two polysilicon gates in parallel, which decreases the resistance factor by 2, from the identity

$$R_1 \parallel R_2 = \left(\frac{1}{R_1} + \frac{1}{R_2} \right)^{-1} \tag{7.25}$$

and if $R_1 = R_2$ (since the parallel polysilicon gates are of equal dimension),

$$R_1 \parallel R_1 = \left(\frac{2}{R_1} \right)^{-1} = \frac{R_1}{2}. \tag{7.26}$$

In total, the gate electrode effective length has therefore been reduced fourfold from its original value, that is, from 6 μm to 1.5 μm, which substantially decreases its electrical resistance. The gate electrode resistance can be described in further detail by

$$R_g = \frac{\rho W}{12 T_g L_g N} \tag{7.27}$$

where,

ρ is the gate resistivity of the gate material
T_g is the gate thickness

The factor 12 is derived from the current flowing profile through the width of the gate. Therefore, assuming a constant ρ, T_g and L_g, a decrease in the width W of the transistor finger will proportionally also decrease the value of the gate electrode resistance. Thus, a fourfold reduction in the original width decreases the gate electrode resistance by the same factor. Since the input resistance of the transistor is a combination of the gate electrode resistance and the intrinsic input impedance (the resistance path of the gate current between the source and the drain terminals), a lower R_g results in a lower input impedance of the transistor. A lower input impedance of the transistor effectively leads to lower power consumption and higher frequency operation, since the input impedance and the input capacitance create an additional pole, which limits the cut-off frequency f_T and the maximum operating frequency f_{max} of the transistor.

Reduced Drain and Source Area and Perimeter

A multiple-finger transistor layout also decreases the drain and source area and the perimeter, which effectively reduces the gate capacitance of the transistor. With respect to the single-finger transistor layout in Figure 7.6a and the multiple-finger layout in Figure 7.6b, consider the effective and total width of the polysilicon gate for each configuration. The drain and source area (A_{D_x} and A_{S_x}, where x represents the number of fingers) and the perimeter of the single-finger transistor configuration in Figure 7.6a can be described by

$$A_{D\_1} = A_{S\_1} = W \times W_D \tag{7.28}$$

where W and W_D are shown in Figure 7.6a and b. The perimeter of the drain and source is given by

$$P_{D\_1} = P_{S\_1} = W + 2W_D \tag{7.29}$$

where both the drain and source perimeters are equal. From Figure 7.6b, the area of the drain of the multiple-finger transistor ($N=2$) can be derived as

$$A_{D\_2} = \frac{1}{2} W \times W_D \tag{7.30}$$

and the area of the source is derived as

$$A_{S\_2} = \frac{1}{2} W \times W_D \times 2 = W \times W_D \tag{7.31}$$

where factor 2 comes from the fact that there are now two source connections present in the configuration. The perimeter of the drain is therefore

$$P_{D\_2} = W + 2W_D \tag{7.32}$$

and the perimeter of the source is equal to

$$P_{S\_2} = 2\left(\frac{1}{2}W + 2W_D\right) = 4W_D + W \tag{7.33}$$

where, again, two source regions exist for the configuration presented in Figure 7.6b. To generalize the area and perimeter of the drain and source region for all even numbers, an integer variable i can be introduced, such that the number of fingers N is equal to $2i$, which represents an even number. In this case, Table 7.1 summarizes the area of the drain and source region with respect to the number of fingers, where W_f is the width of a single finger, represented by

$$W_f = \frac{W}{2i} \tag{7.34}$$

for an even number of fingers, and

$$W_f = \frac{W}{2i+1} \tag{7.35}$$

for an odd number of fingers.

From Table 7.1 and the above equations, for a two-finger transistor (which is further derived to represent all even and odd numbers of fingers), it can be seen that the area of the drain region of a two-finger approach, A_{D_2}, is half that of the area of a single finger, A_{D_1}. The area of the source region of a two-finger transistor is similar to that of the single finger for a two-finger configuration, but decreases when $N > 2$. The perimeter of the drain of the two-finger transistor is equal to that of the single-finger $N > 2$, whereas the perimeter of the source region of the two-finger transistor is also equal to that of the single-finger transistor, but decreases with $N > 2$.

TABLE 7.1

The Area and Perimeter of the Single-Finger Transistor Compared with Multiple-Finger Transistors, where the Number of Fingers is an Even Number for $N = 2i$ and Odd for $N = 2i+1$

Number of Fingers	Area of the Drain (A_D)	Area of the Source (A_S)	Perimeter of the Drain (P_D)	Perimeter of the Source (P_S)
1	$W \times W_D$ (A_{D_1})	$W \times W_D$ (A_{S_1})	$W + 2W_D$ (P_{D_1})	$W + 2W_D$ (P_{S_1})
$2i$ (even)	$0.5 A_{D\_1}$	$\dfrac{i+1}{2i} A_{S\_1}$	$2i \times W_D$	$2(i+1)W_D + 2W_f$
$2i+1$ (odd)	$\dfrac{i+1}{2i+1} A_{D\_1}$	$\dfrac{i+1}{2i+1} A_{S\_1}$	$2(i+1)W_D + W_f$	$2(i+1)W_D + W_f$

Improved MOS Transistor Frequency Response

The intrinsic parasitic diffusion capacitances of a MOS transistor are also related to the junction and sidewall components of the source and drain region. These junction and sidewall capacitances are determined by the area and perimeter of the source and drain regions. Since the area and the perimeter of multiple-finger transistors decrease for an increased number of fingers, the junction and sidewall capacitances will also decrease for a higher number of fingers. The frequency response, and therefore the 3dB cut-off frequency (f_T), of a MOS transistor is defined as

$$f_T = \frac{1}{2\pi} \frac{g_m}{C_{gs} + C_{gb} + C_{gd}} \tag{7.36}$$

where,

C_{gs} is the gate-to-source capacitance
C_{gb} is the gate-to-substrate or bulk capacitance
C_{gd} is the gate-to-drain capacitance
g_m is the top gate transconductance of the transistor, defined as

$$g_m = \sqrt{2I_D \mu C_{ox} \frac{W}{L}} \tag{7.37}$$

where,

μ is the carrier mobility
C_{ox} is the oxide capacitance

From the 3 dB cut-off frequency, it can therefore be seen that the intrinsic parasitic capacitances between the gate and the drain/source have a significant effect on the high-frequency operation of the transistor and should therefore be minimized, ideally through multiple-finger transistor configurations. The presence of the oxide capacitance, defined as

$$C_{ox} = \frac{\varepsilon_{ox}}{t_{ox}} \tag{7.38}$$

where,

ε_{ox} is the permittivity of the oxide material
t_{ox} is the thickness of the oxide layer

This shows that a decrease in oxide thickness, which is typically the case with decreasing node size, increases the performance of the high-frequency operation of the transistor, although it presents additional challenges of higher leakage currents and therefore higher power consumption, as shown in Chapter 6 of this book.

Improved Transistor Matching

Another advantage of using a multiple-finger transistor layout is based purely on the geometric advantages from a layout perspective. It is generally easier to match transistors physically on a layout with techniques such as the common centroid and interdigitated layouts, if the transistors have a more square geometry compared with a long and thin topology. In semiconductor processing, a practical mismatch will always exist among components on the same wafer, because of process tolerances over large surfaces. Therefore, creating a transistor layout that is as compact as possible will ensure that there are many evenly distributed parasitic characteristics over the entire transistor. A long and thin transistor might exhibit a slight variation in intrinsic impedance, intrinsic capacitance, contact-hole variations and variable current density along its width, which could lead to mismatches and unexpected circuit behavior.

TRANSISTOR LAYOUT MATCHING

The multiple-finger geometry supports improved transistor matching in the circuit layout. Transistor matching is an important aspect in circuits where identical transistors are required, since it is practically impossible to manufacture exactly identical devices because of small process variations. Transistor matching aims to ensure that any variations in the transistors are mirrored symmetrically so that each transistor experiences similar systematic and random variations. Circuits where transistor matching is crucial occur, for example, when an input pair of transistors is used in operational amplifiers, or in current mirror circuits. Matching techniques are applied to CMOS and BJT technologies and depend on the application of each circuit.

In the work presented by Pelgrom et al. (1989), a parameter P is defined, which describes the properties of apparent mismatches in MOS transistors from varying process parameters. Sources of random fluctuations include line edge roughness, random dopant fluctuations and variability in gate oxide thickness. The parameter P is used to describe the summation of all variations, since these variations are very small, and consists of correlation distances much smaller than the dimensions of the transistor (Pelgrom et al. 1989). P is a normally distributed parameter and its variance σ^2 is related directly to the W and L of the transistor (Pelgrom et al. 1989), such that

$$\sigma^2\left(P\right) \propto \frac{1}{WL} \tag{7.39}$$

which shows that for large values of W and L, the variance of P increases. Various electrical and technological parameters are accounted for with P, including carrier mobility, the substrate factor, threshold voltage and the gate charge. The complete definition of the variance of P is further described in Pelgrom et al. (1989) as

$$\sigma^2\left(P\right) = \frac{A_P^2}{WL} + S_P^2 D^2 \tag{7.40}$$

where,
  A_P is the area probability constant
  S_P describes the variation of P with spacing
  D is the distance among transistors

The proportionality constants can be measured and used to predict the mismatch variance of the circuit (Pelgrom et al. 1989). Long-channel MOS transistor matching properties can be determined by the drain current flowing through the transistor, defined by

$$I_D = \frac{\beta}{2}\left\{\frac{(V_{GS}-V_t)^2}{1+\theta(V_{GS}-V_t)}\right\} \tag{7.41}$$

where,

θ　is determined by the mobility reduction effect and the source resistance

β　is defined by the oxide capacitance, carrier mobility and geometry of the transistor, such that

$$\beta = \frac{C_{ox}\mu W}{L} \tag{7.42}$$

where C_{ox} is given in (7.38). Furthermore, in Pelgrom et al. (1989), the mismatch factors for the threshold voltage at zero bias, the substrate factor and the current gain are derived as

$$\sigma^2\left(V_{T0}\right) = \frac{A_{V_{T0}}^2}{WL} + S_{V_{T0}}^2 D^2 \tag{7.43}$$

$$\sigma^2\left(K\right) = \frac{A_K^2}{WL} + S_K^2 D^2 \tag{7.44}$$

$$\frac{\sigma^2\left(\beta\right)}{\beta^2} = \frac{A_\beta^2}{WL} + S_\beta^2 D^2 \tag{7.45}$$

where the current factor variations account for variations in W and L originating from edge roughness. With respect to the variation factors, CAD software combined with statistical models provided by semiconductor foundry process development kits enables Monte Carlo analysis to evaluate the variations in desired circuit performance, which subsequently allows yield predictions. Matching is a key parameter in CMOS and BiCMOS process development and to predict yield and performance degradation of manufactured devices.

Techniques that are commonly used to match transistors in differential pair configuration, to ensure that each transistor undergoes similar process variation and is matched as closely as possible, are the common centroid and interdigitating matching topologies. These techniques are reviewed in the following paragraph.

COMMON CENTROID AND INTERDIGITATING TRANSISTOR MATCHING

Process variations during the fabrication of semiconductor components may have accuracy and yield limitations and thus influence the desired performance of electronic circuits. These variations are especially pronounced in circuits where exact transistor matching is required, such as in current mirrors and differential pairs. Process variations not only affect the W/L ratio of the transistor, but can also have varying effects on the threshold voltage of these components. Mismatches in circuits such as differential amplifiers could lead to undesirable offsets and a poor common-mode rejection ratio. Consider the example of the MOS Widlar current source, a common implementation used in CMOS circuit design to generate and mirror a bias current. The MOS configuration of the Widlar current source is given in Figure 7.7.

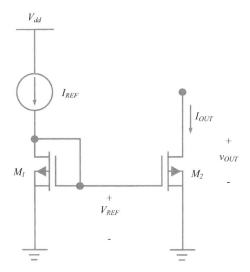

FIGURE 7.7   The Widlar current source using MOS transistors to show the importance of transistor matching for accurate circuit operation.

As shown in Figure 7.7, the Widlar current source comprises two NMOS (or PMOS, depending on the application) transistors, where the current through M_1 is mirrored to the transistor M_2. The mathematical derivation of the circuit shows the significance of having matched transistors for accurate circuit operation. For an input or reference current I_{REF} larger than zero amperes, the transistor M_1 operates in the active region since it is diode-connected, meaning that the gate of the transistor is connected to its drain. Using the transistor geometries $(W/L)_1$ and $(W/L)_2$ as design parameters, an output current should be generated based on the multiplication factor of M_2 (with respect to M_1), as long as M_2 is operating in saturation mode. The drain of M_2 is connected to the load, which does not necessarily have to be a resistor, and typically involves additional MOS transistors. This configuration is typically referred to as a sinking current source, since the output current flows out of the load and into the drain of M_2. For a PMOS current source, the current typically flows into the load, and is referred to as sourcing the current. Importantly, the supplied current does not depend on the output voltage of the circuit, which leads to very stable and well-controlled output currents. The reference current, flowing through M_1, can be described by

$$I_{REF} \approx \frac{1}{2}\left(\frac{W}{L}\right)_1 \mu C_{ox}\left(V_{REF} - V_{t1}\right)^2 \tag{7.46}$$

where,

$(W/L)_1$	is the geometry ratio of M_1
μ	is the carrier mobility
C_{ox}	is the oxide capacitance
V_{REF}	is the reference voltage seen in between the gate of M_1 (and therefore the drain of M_1 and the gate of M_2) and V_{t1} is the threshold voltage of M_1. Similarly, the output current is defined as

$$I_{OUT} \approx \frac{1}{2}\left(\frac{W}{L}\right)_2 \mu C_{ox}\left(V_{REF}-V_{t2}\right)^2 \qquad (7.47)$$

where $(W/L)_2$ is the geometry ratio of M_2. By combining these equations, and assuming that all process parameters (oxide capacitance, mobility, threshold voltage) and geometries are exactly equal, the relationship of I_{OUT} and I_{REF} is described in Equation (7.48).

$$I_{OUT} = I_{REF}\frac{\left(W/L\right)_2}{\left(W/L\right)_1} \qquad (7.48)$$

This shows that the output current scales directly with the reference current and the W/L ratios of both transistors, hence the importance of well-matched transistors. By defining the expected drain current I_D, geometry and threshold voltage as presented in Gray et al. (2001) and adapted here, it follows that

$$I_D = \frac{I_{REF}+I_{OUT}}{2} \qquad (7.49)$$

with

$$\Delta I_D = I_{REF}-I_{OUT} \qquad (7.50)$$

and

$$\frac{W}{L} = \frac{\left(W/L\right)_1+\left(W/L\right)_2}{2} \qquad (7.51)$$

with

$$\Delta\frac{W}{L} = \left(W/L\right)_1-\left(W/L\right)_2 \qquad (7.52)$$

and

$$V_t = \frac{V_{t1}+V_{t2}}{2} \qquad (7.53)$$

where

$$\Delta V_t = V_{t1}-V_{t2}. \qquad (7.54)$$

It follows, in Gray et al. (2001), that

$$\frac{\Delta I_D}{I_D} = \frac{\Delta\frac{W}{L}}{\frac{W}{L}} - \frac{\Delta V_t}{V_{ov}/2} \qquad (7.55)$$

showing the relationship among the variations of the drain current, transistor geometry and threshold voltage, while accounting for the gate-source voltage defined within the overdrive voltage. It can be seen that the current mismatch consists of two components. The first term is a geometric-dependent parameter, which contributes a fraction of the current mismatch and is not proportional to the applied voltage. In the second term, dependence on the threshold voltage mismatch varies with a change in the overdrive voltage, and therefore the gate-to-source voltage applied to the transistor.

Common centroid and interdigitating components are applicable not only to transistors, but also to any component on a semiconductor IC, such as capacitors, resistors and even inductors (or microstrips). If, for example, two components, arbitrarily defined as component A and component B, are to be matched on the layout of the circuit, these components can be matched by first splitting them up into smaller, more manageable geometries, depending on the original size. For this example, components A and B are split into four smaller components, which are exactly one quarter of the intended size of A and B, denoted as A_1–A_4 and B_1–B_4. In the case of transistors, the multiple-finger approach can also be used to modify the W/L aspect ratio and to avoid any long or thin geometries. The common-centroid matching technique is employed by placing these components in such a way that component pairs have the same centroid. Interdigitating these components leads to layout placements such that alternate pairs of each component are placed next to each other; both these techniques are shown in Figure 7.8.

As shown in Figure 7.8b, the common-centroid matching technique compensates for process variations by placing the transistors in such a way that the transistor pairs share a common centroid, as indicated in the figure. The interdigitated geometry in Figure 7.8c differs slightly from the common-centroid layout, in such a way that transistor pairs are placed directly next to each other with as little spacing among these transistors as possible to further reduce the influence of process variation. Another technique that is commonly employed in transistor layout is the placing of dummy components on the sides of the transistors, essentially creating a shorted (its drain and source) transistor. This technique is presented in Figure 7.9.

The dummy devices in Figure 7.9 ensure that all transistors, especially the devices on the sides, see the same material layers on all sides. The shorted dummy devices do not contribute electrically to the operation of the circuit; however, the parasitic contributions of these devices should be taken into account and extracted from post-layout simulations.

If layout optimizations are done to improve transistor yield, size or performance, care should be taken not to introduce additional mechanical stress in the transistor, which could result in non-uniform distribution of stress that affects the transistor characteristics and can change circuit behavior. One such stress-induced mechanism is shallow trench isolation (STI), which is becoming increasingly common with the continuous shrinking of device feature size. STI induces thermal residual stress in active semiconductor materials through post-manufacturing thermal mismatch. "Stress- and strain-based techniques for mobility improvement are starting to dominate traditional geometric scaling to maintain Moore's law trajectories for device performance" (Kahng et al. 2007). Stresses within the transistor could not

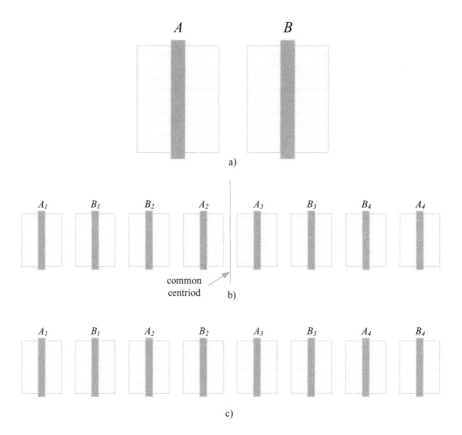

FIGURE 7.8 A representation of matching two components (a) A and B using the (b) common-centroid matching technique and (c) interdigitated devices.

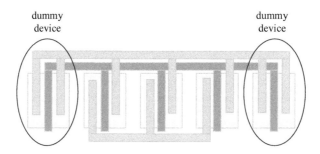

FIGURE 7.9 A representation of dummy components on the sides of a transistor layout, to ensure that all transistors see the same material layers on all sides.

only induce variations in the mobility of the carriers, but also lead to unwanted variations on the thermal voltage. Importantly, not all stresses in transistors have adverse effects on device operation – certain stresses (depending on the direction with respect to the gate of the transistor) could lead to improved device performance. Nevertheless, stress-engineering of transistors requires high-accuracy modeling and three-dimensional simulation to take advantage of STI effectively, and repeatability and consistency across a large number of devices are complex tasks. The following section reviews the mechanisms that lead to STI stresses in MOS transistors.

SHALLOW TRENCH ISOLATION STRESS IN TRANSISTORS

STI is created early during the semiconductor manufacturing process, before the actual transistor is formed. Trenches are etched into the semiconductor material and dielectric material such as SiO_2 is deposited to fill the trenches. The excess dielectric is removed, typically through chemical-mechanical planarization. Because of the different thermal expansion coefficients between the semiconductor material and the STI structures, there is biaxial compressive residual stress in the active region of the transistor after processing. Generally, STI stress increases the current flow through PMOS transistors, hence increasing power consumption, and reduces current flow through NMOS transistors. The STI stress relaxes exponentially with increased distance from the semiconductor and STI structure boundary. The placement of the STI between two NMOS transistors is shown in Figure 7.10.

To lessen the effects of stresses induced by STI, the diffusions within the source and drain can typically be extended if they are placed adjacent to the STI. Adversely, extending the source and drain diffusions increases the parasitic capacitances and may reduce the f_T of the device. To avoid this decrease in device performance due to enlarged diffusion areas, dummy transistors can be placed at the end of each active (operational) transistor in the circuit with a shared diffusion region among these devices. With respect to the impact of the three-dimensional stress and strain effects on CMOS performance, Li et al. (2008) provide the drive current performance results for both NMOS and PMOS transistors. The direction of the compressive stress on the transistor is shown in Figure 7.11.

The performance gains and degradation results obtained in Li et al. (2008) for a decrease in compressive stress along the axis provided in Figure 7.11 are presented in Table 7.2.

As seen in Table 7.2, and as presented originally by Li et al. (2008), both NMOS and PMOS transistors display an improvement or degradation in the drive (drain) current when STI stresses are applied in a certain direction. As seen in Table 7.2, NMOS transistors show an improvement in drive current for decreased compressive stress in the x- and y-direction, with current performance degradation in the z-direction. For PMOS transistors, current performance is degraded with decreased compressive stress in the x-direction, but improved in the y- and z-directions.

TRANSISTOR LAYOUT CHECKLIST

The following list summarizes the typical considerations to take into account when designing a circuit layout, with specific focus on the transistor placement and geometry considerations.

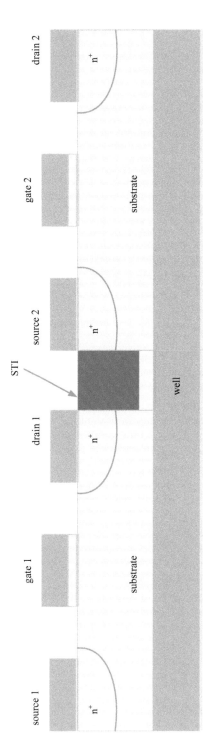

FIGURE 7.10 Placement of the STI between two NMOS transistors.

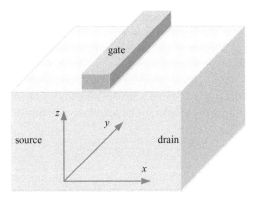

FIGURE 7.11   The Cartesian directions of the compressive stress experimental results provided in Li et al. (2008) on NMOS and PMOS transistors.

TABLE 7.2

Drive Current Performance for Reduced Compressive Stress on NMOS and PMOS Devices

	Drive Current Performance	
Decreased Compressive Stress	**NMOS**	**PMOS**
x-direction	Improved	Degraded
y-direction	Improved	Improved
z-direction	Degraded	Improved

Source: Adapted from Li, Y. et al., *IEEE Transactions on Electron Devices*, 55(4), 2008, with reference to the Cartesian directions indicated in Figure 7.11.

- The first step is to draw a readable and logical transistor diagram.
- Identify all critical elements (such as current mirror transistors and differential pairs), as well as critical nodes in the design, before attempting the layout.
- These critical parameters should be used to identify further components and nodes where minimum parasitic parameters may be introduced and where minimum interference from adjacent components is permitted.
- Mark all transistors that must be exactly matched in the layout clearly, and aim to place these transistors as close to one another as possible.
- Mark axes of potential symmetry in the layout.
- Analyze the sizing of the transistors, especially the W of each transistor, and determine the viability of using multiple-finger geometries.
- Group transistors into stacks and define the expected height and width of the cell containing all circuit elements.

- Sketch a stick diagram representing the full layout of the circuit.
- Finally, define the connection layers (metal layers) for input and output connections and for horizontal and vertical connections to ensure currents flow in parallel.

The layout considerations provided are specifically tailored for transistors in a CMOS process. Additional considerations for passive components are provided in Chapter 8 of this book.

CONCLUSION

Chapters 7 and 8 of this book present microelectronic circuit enhancements and design methodologies to facilitate the continuation of Moore's law with aggressively scaled semiconductor components. This chapter provides considerations for the transistor, a component used very frequently in ICs. This chapter also introduces the design description for an IC, described in terms of three primary domains: the behavioral, structural and physical/geometrical domains. In each of these domains, various design decisions are made to solve a particular problem based on its required performance and its application. These domains can be hierarchically divided into levels of design abstraction, referred to as the abstraction hierarchy, summarized by the Gajski-Kuhn Y-diagram presented in this chapter. Furthermore, semiconductor IC design flow can be constructed with reference to the domains and abstraction layers presented by the Gajski-Kuhn Y-diagram. This design flow is presented in this chapter, with respect to full IC design and not specifically focused on the transistor. A simplified design flow for a VLSI system is presented, with reference to the requirements to design, simulate and test the performance of an IC using CAD tools. Important aspects such as DRC, LVS and post-simulation data extraction are briefly reviewed in this chapter.

Since transistors are used in high volumes of IC design, the microelectronic circuit component footprint section of this chapter, where the geometrical considerations of active components (transistor and diode) and passive components (resistor, capacitor and inductor) are reviewed, focuses only on the transistor. The transistor optimizations are introduced with respect to the drain current equation of the MOSFET. The drain current capacity, gain and geometric relationships of a transistor define the operation of the device for specific applications. The operating regions of the transistor are briefly reviewed, as these have an effect on the drain current definition of the transistor. The relationship between the gain of the transistor, g_m, and the drain current is presented with respect to the size of the transistor; hence the technique of determining the optimal size of a transistor based on its current-carrying requirements for various operating regions. This is achieved by the universal g_m/I_D versus the normalized current I_N of the transistor.

This technique is followed by the impact of transistor scaling and geometrical variations on the noise performance of a MOSFET. A description of the small-signal model of the MOS transistor is provided, with noise-current generators used to model the additive noise in a transistor. Thermal, flicker and shot noise characteristics of the transistor are described, with specific reference to the W and L parameters of the device.

In order to optimize the layout of the MOS transistor, a review of the layout properties that affect the device performance is presented in this chapter. These properties are especially critical in post-layout simulations, where device optimization typically requires thorough knowledge of the physical characteristics of the device, in order to improve it. The most commonly used technique to define transistors with practical and optimal W and L ratios, the multiple-finger layout strategy, is employed. This chapter reviews this strategy, with specific focus on the changes in the electrical characteristics such as the gate resistance and capacitance, due to the implementation of this technique. Transistor layout matching techniques are then presented, in lieu of the optimal structures that can be defined using the multiple-finger approach. Techniques such as the common centroid, interdigitated transistors and dummy device matching are presented.

Finally, a brief introduction to mechanical stresses in transistors, especially prominent in modern, sub 65-nm nodes, as a result of shallow trench isolation is presented, followed by a checklist for transistor design and layout.

REFERENCES

Binkley. D. M., Hopper, C. E., Tucker, S. D., Moss, B. C., Rochelle, J. M., Foty, D. P. (2003). A CAD methodology for optimization transistor current and sizing in analog CMOS design. *IEEE Transactions on Computer-Aided Design of Integrated Circuits and Systems*, 22(2), 225–237.

Gajski, D. D., Kuhn, R. H. (1983). Guest editor's introduction: New VLSI tools. *Computer*, 16(12), 11–14.

Gray, P. R., Hurst, P. J., Lewis, S. H., Meyer, R. G. (2001). *Analysis and Design of Analog Integrated Circuits*. 4th ed. Hoboken: John Wiley and Sons, Inc..

Jeon, J., Kang, M. (2016). Circuit level layout optimization of MOS transistor for RF and noise performance improvements. *IEEE Transactions on Electron Devices*, 63(12), 4674–4677.

Kahng, A. B., Sharma, P., Topaloglu, R. O. (2007). Exploiting STI stress for performance. *Proceedings of the 2007 IEEE/ACM International Conference on Computer Aided Design*, 83–90.

Li, Y., Chen, H., Yu, S., Hwang, J., Yang, F. (2008). Strained CMOS devices with shallow-trench-isolation stress buffer layers. *IEEE Transactions on Electron Devices*, 55(4).

Pelgrom, M. J. M., Duinmaijer, C. J., Welbers, A. P. G. (1989). Matching properties of MOS transistors. *IEEE Journal of Solid-State Circuits*. 24(5), 1433–1440.

Silveira, F., Flandre, D., Jespers, P. G. A. (1996). A g_m/I_D based methodology for the design of CMOS analog circuit and its application to the synthesis of a silicon-on-insulator micropower OTA. *IEEE Journal of Solid-State Circuits*, 31(9), 1314–1319.

Tsai, M., Ma, T. (1994). The impact of device scaling on the current fluctuations in MOSFET's. *IEEE Transactions on Electron Devices*, 41(11), 2061–2068, November 1994.

8 Microelectronic Circuit Enhancements and Design Methodologies to Facilitate Moore's Law – Part II

Wynand Lambrechts and Saurabh Sinha

INTRODUCTION

In microelectronic circuits, especially in applications used for computing data at high rates, such as operations performed by microprocessors and microcomputers, increasing the number of on-chip transistors characteristically relates to higher computing speed, additional memory and more efficient data transfer. However, transistors make up only a finite percentage of the overall chip. The remaining area is consumed by bulky passive components such as resistors, capacitors, inductors, antennas, filters and switches. Therefore, to realize miniaturization, as proposed by Moore's law, efficient layout, circuit enhancements and design methodologies should not only be applied to active components such as the transistor. Passive components need to receive substantial attention as well, in terms of increasing their effective area usage, performance and matching of intrinsic variances in semiconductor processing.

Passive components, however, owing to their relatively large size on a wafer, are highly susceptible to mismatches from process variations across the area they occupy, more so than active transistors, which occupy a much smaller area (per component) on an IC. There are three primary sources of mismatch in semiconductor components, namely

- Systematic mismatches typically introduced by the circuit or layout designer following incorrect or sub-optimal design strategies.
- Random mismatches due to variation in process parameters.
- Gradient mismatches, which are first- or second-order fluctuations over longer lengths across a chip.

Figure 8.1 represents the relationship among the primary mismatch sources: systematic, random and gradient mismatches.

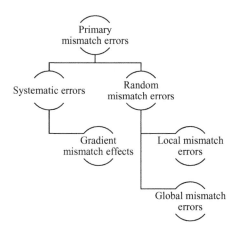

FIGURE 8.1 The relationship among the primary mismatch sources.

Systematic and gradient mismatches, as shown in Figure 8.1, are deterministic trends observed in the mismatch values of the sampled devices on a wafer and can therefore be accurately forecasted, with the process gradients known. Random mismatches, as the name suggests, represent mismatches that are stochastic and cannot be predicted; it is also difficult to consider these errors and mismatches during the circuit and layout design of an IC. Random mismatches are also divided into local and global mismatch errors. Local mismatch errors occur in response to factors such as variations in the metal or the polysilicon grain edge margins, or local variations in

- Etching.
- Implantation.
- Diffusion.
- Gate oxide thickness.
- Permittivity
- Dopant.

Causes of global mismatches are line edge variations, stepper lens aberrations, wafer loading effects and optical proximity effects. In analog electronic design, absolute values are less important when compared with effective ratios. Absolute values are governed by large drifts in parameters, whereas effective ratios are governed by local dissimilarity of parameters. As feature size decreases, guided by Moore's law, the variation between active and passive component values increases.

The concept of mismatch of passive components applies to devices having geometric ratios other than 1:1. The mismatch between any two devices, δ, is the difference between the ratio of their measured values and the ratio of the intended values, divided by the ratio of the intended values, and normalized such that the mismatch becomes independent of the geometric ratio. This is typically described mathematically as

$$\delta = \frac{(x_2 / x_1) - (X_2 / X_1)}{(X_2 / X_1)} \tag{8.1}$$

where,

X_1 and X_2 are the intended values of the two devices

x_1 and x_2 are the actual, measured values of the two devices

This mismatch can be simplified as

$$\delta = \frac{X_1 x_2}{X_2 x_1} - 1 \tag{8.2}$$

where measurements of a large number of devices (pairs) are required to provide a comprehensive random distribution of mismatches. Importantly, the mismatch can be a positive or a negative quantity, and the sign of the mismatch must be retained to determine the average of the mismatches on a wafer. The average, or mean, m_δ, of the measured mismatches can simply be determined by

$$m_\delta = \frac{1}{N} \sum_{i-1}^{N} \delta_i \tag{8.3}$$

where N is the number of samples measured. The mean is a measure of the systematic mismatch, or bias, among matched devices. Finally, the standard deviation of the mismatches, s_δ, can be calculated by

$$s_\delta = \frac{1}{N-1} \sum_{i-1}^{N} (\delta_i - m_\delta)^2 \tag{8.4}$$

which describes the quantity by which the mismatches vary within the total number of samples measured. Layout concerns that have an impact on matching include geometry, since the arbitrary constituent of mismatch increases with increased geometry; proximity, which is the physical separation distance between devices; and the matching orientation, the centerline between matched devices.

This chapter describes the principles of semiconductor integrated passive components, specifically resistors and capacitors, with respect to their fundamental operating principles, intrinsic parameters and, importantly, layout considerations to optimize device performance. Inductors are briefly described; however, a large number of optimization strategies used for resistors and capacitors can be adapted for inductor optimizing. Passive components have a significant influence on the validity of Moore's law and aggressive scaling in semiconductors, since these components occupy such large areas on-chip. It is therefore equally important to consider optimal layout strategies of passive components, along with active transistors, as discussed in Chapter 7 of this book. The first passive component discussed in this chapter, its operating principles and its layout optimizations, is the integrated resistor.

INTEGRATED RESISTOR OPTIMIZATIONS

INTEGRATED RESISTOR LAYOUT

Integrated lumped-element resistors are regularly used in semiconductor IC designs and present several challenges to ensuring accurate values of manufactured devices versus simulated ones. In analog design flow, one of the utmost imperative concerns is to attain precise resistor ratios throughout the layout phase, also referred to as resistor matching (Jiang et al. 2012). Resistors can be realized by depositing thin films of *lossy* material onto a dielectric base, through thick-film technology or by monolithic semiconductor films on a semi-insulating substrate (Bahl 2003). Several characteristics of an integrated resistor should be known and accounted for when implementing these components in integrated designs. These characteristics include, as listed in Bahl (2003), the

- Sheet resistance.
- Thermal resistance.
- Maximum current density (current-handling capacity), maximum working voltage and its power rating, which depends on its area, since a larger area can sink higher dissipated power, as well as on the ambient temperature.
- Variations in the specified resistor values from the same manufactured batch of devices, known as the resistor tolerance.
- Rate of change of the resistor value with an increase or decrease in temperature, referred to as the temperature coefficient of the film.
- Maximum frequency of operation before attenuation due to the occurrence of parasitic components.
- Unwanted random fluctuations generated in the resistor, therefore the intrinsic thermal noise (also known as Johnson or white noise).

Typical integrated resistors are manufactured using several techniques, which include diffused resistors, epitaxial resistors, pinched resistors and thin-film resistors. Figure 8.2 shows an *n*-type IC resistor in a *p*-type substrate, including its mask definitions and a cross-sectional view of the conducting channel.

As seen in Figure 8.2a, the integrated resistor consists of three primary masks to define the components completely. These masks are the oxide mask, the contact mask and the metal or poly mask. In Figure 8.2b, the cross-sectional view shows the conducting channel under the surface of the passivation layer (oxide), specifically in this example, an *n*-type region on a *p*-type substrate. The resistance and sheet resistance are two defining characteristics that define the operation of the device and are reviewed in the following section.

RESISTANCE/SHEET RESISTANCE OF THE INTEGRATED RESISTOR

The sheet resistance (or specific resistance) of an integrated resistor is often computed in terms of pertinent average concentrations and mobility associated with the calculated or estimated concentration averages (Choma 1985). The sheet resistance, R_S, of an integrated resistor is related to its geometry through the integral relationship

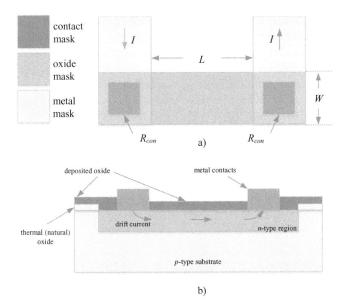

a)

b)

FIGURE 8.2 The physical layout of an *n*-type IC resistor on a *p*-type substrate; (a) top view, showing the typical masks used to define the component, and (b) side view, showing the current components.

of the impurity concentrations and the material mobility. Figure 8.3 shows the geometric characteristics used to determine the sheet resistance.

As shown in Figure 8.3, the thickness of the device is represented by W, the width is represented in the z-direction and the current path length is given by L. The sheet resistance can be derived by

$$\frac{1}{R_S} = q \int_0^W \mu \big[N(x) \big] N(x) dx, \qquad (8.5)$$

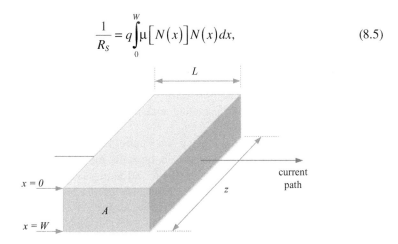

FIGURE 8.3 A simplified representation of the integrated resistor geometry used to derive the sheet resistance of the resistor. (Adapted from Choma, J., *IEEE Transactions on Electron Devices*, 32(4), 845–847, 1985.)

adapted from Choma (1985), where q is the elementary electron charge; $N(x)$ is the impurity concentration of the material, assumed to be spatially variant in only the vertical (x) direction; and $\mu[N(x)]$ is the concentration-dependent mobility of the majority of carriers in the device. The impurity concentration $N(x)$ is determined by the Gaussian function, with reference to Figure 8.2,

$$N(x) = N_0 \exp\left[-\left(\frac{x - X_0}{L_0}\right)^2\right]$$ (8.6)

where the peak concentration of the material is given by N_0 and is evidenced at $x = X_0$, dependent on the implant energy and the background concentration (Choma 1985). Essentially, the sheet resistance of an integrated resistor is the resistivity of the material divided by the thickness (t), given as

$$R_S = \frac{\rho}{t}$$ (8.7)

where the resistivity ρ is given by

$$\rho = R\frac{A}{L}$$ (8.8)

where A is the cross-sectional area of the resistor $(W \times L)$, with reference to Figure 8.3. To determine the resistance of the integrated resistor with respect to the sheet resistance, the following equation is commonly used:

$$R = 2R_{con} + \frac{L}{W}R_s$$ (8.9)

where,
$\quad R_{con}$ is the contact resistance of the metal terminals at the ends of the integrated resistor, as shown in Figure 8.1b
$\quad R_s$ is the sheet resistance specified in Ω/square (or $\Omega/\square$)

Recall that if the resistor geometry is a square, the sheet resistance is specified straightforwardly in Ω. Sheet resistance is typically measured using the four-point probe method, as described in Logan (1967) and Smits (1958). Sheet resistance can be determined theoretically based on the intrinsic properties of the material; however, for certain materials such as InGaAs, InP or InSb, these properties can vary significantly based on physical and electrical external factors such as temperature, power, manufacturing tolerance and uneven distribution through a wafer, and sheet resistance is often measured instead. The following section reviews the techniques used to measure the sheet resistance, R_s, as well as the contact resistance, R_{con}, of integrated resistors.

Test structures are often included in IC design on strategically relevant areas (such as inside the scribe channel) on the wafer to measure characteristics such as sheet resistance, contact resistance, actual feature dimensions, uniformity across the wafer and overlay errors. These test structures are also useful to predict wafer yield and wafer reliability post-production, or in certain cases, during the manufacturing process. A key parameter to measure using these test structures is the resistivity – and hence the sheet resistance and contact resistance – which defines the integrated resistors throughout the IC. The value of the resistance of an integrated resistor is a function of the doping level and the resulting resistivity of the material, the length of the device and the cross-sectional area of the resistor (Baker and Herr 1964). To determine the sheet resistivity and contact resistance of a semiconductor material, three commonly used structures are the Greek cross and the bridge test structure to measure sheet resistance, and the contact resistance is measured using the Kelvin structure. These test structures form part of the van der Pauw structures, and are briefly described in the following paragraphs, with the first discussion presenting the Greek cross to measure sheet resistance.

The Greek Cross (Sheet Resistance)

One of the most widely used test structures is the Greek cross, a structure that yields accurate (up to 1%) resistivity data and is relatively easy to design and lay out on an IC (Walton 1997), incorporating the four-point probe method as described in Logan (1967) and Smits (1958). The simplified layout representation of the Greek cross is given in Figure 8.4.

The Greek cross, which yields the sheet resistance of the material, was described by L. J. van der Pauw in 1958 and proved when current is forced through *B* and exits

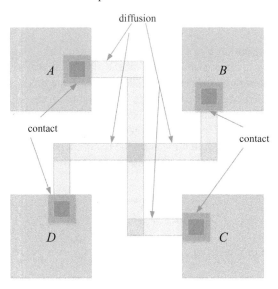

FIGURE 8.4   The layout of the van der Pauw structure, the Greek cross.

C, and the resultant voltage, quantified across A and D are used to determine the impedance $R_{BC,AD}$ of the configuration. This is true if the following relation holds:

$$e^{-i\left(\pi \cdot \frac{d+R_{AB,CD}}{\rho}\right)} + e^{-i\left(\pi \cdot \frac{d+R_{BC,AD}}{\rho}\right)} = 1 \tag{8.10}$$

where,

d	is the thickness of the device
ρ	is the resistivity
$R_{AB,CD}$	is the resistance of the structure when current is forced through contact A and exits through contact B, with the ensuing voltage measured over the contacts labeled C and D in Figure 8.4.

Moreover, the van der Pauw structure resistances are typically small, in the order of a few Ω. Therefore, the ratio of the voltage and current across these devices is small, and more importantly, the voltages are small. Since it is relatively difficult to source an accurate small voltage, these structures encourage forcing a current through the device and measuring the resulting potential difference. From the relation in (8.10), it follows that the sheet resistance is determined by first passing a current through contacts A and B (I_{AB}) and measuring the difference in the potential across contacts D and C (V_{DC}), such that the resistance (not yet the sheet resistance) is given by

$$R = \frac{V_{DC}}{I_{AB}} \tag{8.11}$$

and the sheet resistance, R_s, is determined from this resistance, as

$$R_s = \frac{\pi R}{\ln 2} \tag{8.12}$$

which is specified in units Ω/sq. For a homogenous structure with contact resistances negligibly small in contrast with the rest of the structure, and which is completely symmetrical, only a single differential voltage measurement is required to quantify the sheet resistance. If it is found that there is ambiguity on any of these parameters, an additional technique to improve the accuracy of the measurements further, specifically when the length of the arms is greater or equal to the size of the heart of the cross, is to force a current I_{AB} and measure the resultant potential, and to reverse the measurement and again determine the potential difference. In such a case, the first set of measurement would entail

$$R_{0°} = \frac{V_{DC} - V_{CD}}{I_{AB} - I_{BA}} \tag{8.13}$$

and the second set of measurements would include

$$R_{90°} = \frac{V_{AD} - V_{DA}}{I_{BC} - I_{CB}} \tag{8.14}$$

such that the average measured resistance is given by

$$R = \frac{R_{0°} + R_{90°}}{2} \qquad (8.15)$$

and the accurate sheet resistance can again be equated to (8.12). Importantly, up to eight DC measurements can be taken, averaged and used to determine the sheet resistance with high accuracy, if there is uncertainty about any of the parameters (homogenous, negligible contact resistance and symmetry). The sheet resistance can also be determined by implementing a bridge structure, as discussed in the following paragraph.

Bridge Test Structure (Sheet Resistance)

The bridge structure again assumes a homogenous film thickness as well as negligible sidewall effects due to the applied electric field (which is small, since the applied voltage is also small). The bridge test structure layout is given in Figure 8.5.

As shown in Figure 8.5, the width and the length of the bar structure are denoted by the parameters W and L, respectively. These dimensions should be known to extract the sheet resistance from the bridge structure, unlike the case of the Greek cross, where the dimensions are not required (Enderling et al. 2006). In the bridge test structure, a current is forced through (between) C and D (I_{CD}) and the potential difference across terminals A and B, V_{AB}, is measured. The resistance, R, of the conducting channel is therefore given by

$$R = \frac{V_{AB}}{I_{CD}} \qquad (8.16)$$

where the effective sheet resistance (R_{seff}) can be determined through this measured resistance, R, and the ratio of width and length of the diffusion as shown in Equation 8.18.

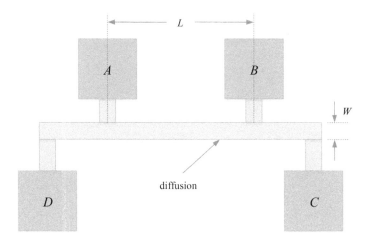

FIGURE 8.5 The layout of the van der Pauw structure, the bridge test structure.

$$R_{\text{seff}} = R\frac{W}{L} \tag{8.17}$$

where W and L are related to the dimensions of the diffusion. The effective sheet resistance will therefore decrease with narrower diffusion, or with increasing length. To determine the actual diffusion sheet resistance, R_s, the following relationship can be used:

$$R_s = R_{\text{seff}}\left(1 + \frac{\Delta W}{W}\right) \tag{8.18}$$

where ΔW is the enlargement of the diffusion region width as a result of lateral diffusion. To measure the resistance of the contacts in a manufactured IC, used throughout the design to provide low-resistance contact to the substrate and the diffusion, the Kelvin structure is commonly used. The following paragraph briefly describes the layout and the measurement technique of the Kelvin structure.

THE KELVIN STRUCTURE (CONTACT RESISTANCE)

The Kelvin structure is another four-point terminal structure, used to measure the contact resistance on a wafer. The layout of the Kelvin structure is provided in Figure 8.6.

The Kelvin structure presented in Figure 8.6 works on the principle that a current is forced through terminals A and C and the resulting potential difference is measured between terminals B and D (Walton 1997). The conduction path between terminal A and the center of the structure (the contact being measured), as well

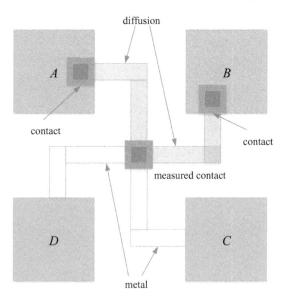

FIGURE 8.6   The layout of the van der Pauw structure, the Kelvin structure.

as between terminal *B* and the contact being measured, is formed by the diffusion layer. The conduction path between *C* and *D* toward the center contact is formed by the metal layer. Therefore, this structure is a three-dimensional structure if studied under a scanning electron microscope. The contact resistance, R_c, is therefore determined by

$$R_c = \frac{V_{BD}}{I_{AC}} \tag{8.19}$$

and the result can be used to determine whether good ohmic contact is achieved during the manufacturing, or if high resistance Schottky behavior is introduced, possibly during incomplete contact etching or imperfect sintering, which aims to enhance properties such as electrical and thermal conductivity. If the contact resistance, R_c, is determined by the Kelvin structure, the specific contact resistivity of the contact can be derived from

$$\rho_c = R_c A_c \tag{8.20}$$

where A_c is the area of the contact, defined by the square of its width, $A_c = (W_c)^2$. To perform optimized layout of integrated resistors, the intrinsic parasitic behavior of these devices should be accounted for, especially the parallel capacitances toward the substrate. The following paragraph reviews the parasitic components of integrated resistors.

INTEGRATED RESISTOR PARASITIC COMPONENTS

Ideally, a resistor is represented by a single resistive component, which defines its ability to oppose the flow of current. In practice, however, resistors have intrinsic, undesired, parasitic characteristics such as inductance, capacitance and additional resistance. These parasitic components could be from metallic contacts inducing inductance and additional resistance, and capacitance from surrounding oxide layers and toward the substrate, as well as fringing fields traversing the length of the resistor. A small-signal AC model (or equivalent circuit) can be used to represent these parasitic components and is typically used to analyze and quantify the high-frequency or DC behavior of the integrated resistor. Several models for the equivalent circuit exist, each with varying complexity; the two most commonly used models are the two-port and one-port element models. The two-port models assume that the resistor is divided into two symmetrical halves, and the two ports of the device are connected in-situ with the operating circuit. The one-port model assumes that the second terminal of the device is grounded and simplifies the analysis. The small-signal equivalent circuit one-port model for integrated resistors is given in Figure 8.7.

As shown in Figure 8.7, several parasitic components are apparent in integrated resistors. These parasitic values include, separately from the ideal resistance, *R*, a series inductance due to metallic conductive lines, represented by *L*, and a parallel capacitance, *C*, across the resistor toward the substrate (ground) on which the device is manufactured. In MOS integrated resistors, the parasitic capacitance to

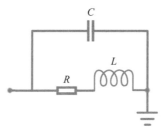

FIGURE 8.7   Simplified one-port model representation of the parasitic components of integrated resistors.

the substrate/well is voltage-dependent. These parasitic components play a significant role in the frequency capabilities of the device and typically limit the maximum operating frequency (bandwidth). To determine the resonance frequency of the
equivalent circuit model presented in Figure 8.7, the input admittance should first be
determined, as adapted from Bahl (2003), such that

$$Y_{in} = \frac{1}{R + j\omega L} + j\omega C \qquad (8.21)$$

where ω is the operating frequency in rad/s ($\omega = 2\pi f$, where f is the frequency in Hz).
By equating the input impedance at resonance, and rewriting with respect to the real
impedance, inductance and capacitance, it follows that the resonance frequency of
the equivalent resistor circuit is given by

$$\omega_0 = \frac{1}{LC} - \frac{R^2}{L^2} \qquad (8.22)$$

where the resistor is primarily inductive below the resonant frequency, and capacitive above the resonant frequency. For more complex resistor shapes, such as mean-
dering designs, a distributed model for thin-film and active semiconductor resistors
can be used, where the resistor is divided into a number of sections (n) as a function
of its wavelength (λ) of operation. A detailed review of the distributed model is presented in Bahl (2003). The meander line resistor is used frequently in IC designs to
conserve area as well as to improve resistor matching (discussed later in this section);
therefore, quantifying its resistance is important. Scalable lumped-element meander-
line resistors allow the accurate creation of components for specific applications, still
taking into consideration parasitic parameters (Murji and Deen 2005). Consider the
meander line resistor as presented in Figure 8.8, adapted from Bahl (2003).

In Figure 8.8, the width of the resistor, which is kept constant for all tracks, is
described by W, l_1 is the length of the track up to its initial bend, l_c is the thickness of
the bend, l_2 is the length of the track vertically and l_3 is the length of the horizontal
section of the track. The total length l is therefore $2(l_1 + l_2) + l_3$ and the resistance of
this structure, according to Bahl (2003), is given by

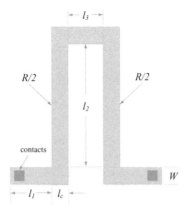

FIGURE 8.8   The meander line resistor often used in IC design to conserve area and improve device matching.

$$R = \frac{\rho}{Wt}\left[l + 4(0.44)l_c\right] \tag{8.23}$$

where,

ρ   is the resistivity of the material
t   is the thickness of the layer

Although it is possible to describe the resistance of meander line resistors mathematically, the complexity of these resistors can become cumbersome, and electromagnetic solvers are often used to determine the resistance, as well as parasitic components within the structure (especially any mutual inductance or capacitance between adjacent tracks). Murji and Deen (2005) provide a relevant analysis of the modeling and parameter extraction of meander-line *n*-well resistors, including all mutual inductance/capacitance parasitic components.

Before continuing to integrated resistor optimization and matching, the figure of merit on the noise performance of integrated resistors is reviewed. Noise characterization and modeling of submicron MOS transistors, as well as the effects of passive devices, have received increased attention, particularly since the higher-than-expected channel thermal noise reported by Jindal (1985) and Chen et al. (2012) due to distributed gate resistance noise, substrate resistance noise, bulk charge effects, substrate current noise and hot carrier channel thermal noise. The following section reviews the noise performance in integrated resistors.

NOISE IN INTEGRATED RESISTORS

In integrated (*n*-well) resistors, low-frequency noise has become crucial to factor in as device scaling advances, driven by Moore's law. Contrary to thermal noise, low-frequency noise is typically evident in 1/*f* noise, with its origin studied using number fluctuation theories (Srinivasan et al. 2010). The effects of process parameters on 1/*f* noise during *n*-well resistor construction can be analyzed by investigating variances

in dose, energy, annealing conditions and shallow-trench isolation depth, characterized by IV-curves and using the noise coefficient as a figure of merit (Srinivasan et
al. 2010). In Srinivasan et al. (2010), the noise coefficient, α, is described by

$$\alpha = WLf S_{I_r} /I_r^2 \tag{8.24}$$

where,
  W is the width of the resistor
  L is its length (therefore, WL defines its area)
  f is the practically determined spot frequency in Hz
  S_{I_r} is the current noise spectral density of the resistor
  I_r is the current flowing through the resistor

Numerous mechanisms could lead to enhanced low-frequency noise. One such
mechanism that is often described is the empirical relationship known as Hooge's
law (Brederlow et al. 2001). According to Hooge's law, the magnitude of the $1/f$ noise
is inversely proportional to the number of free charge carriers, N. Since the noise is
also proportional to the resistance, it follows from Hooge's law that the frequency-
dependent noise spectral density, $S(\omega)$, is proportional to the thickness of the resistor,
such that

$$S(\omega) \propto \frac{1}{t} \tag{8.25}$$

which, as indicated in Dutta and Horn (1981), is consistent with

$$S(\omega) \propto \frac{1}{N} \tag{8.26}$$

and a detailed review of the $1/f$ noise fluctuations is presented in Dutta and Horn
(1981). The relative noise in integrated resistors is assumed to scale with the square
of the sheet resistance (Vandamme and Casier 2004), such that the open-circuit voltage noise, S_V, in a sample can be described by

$$S_V = \frac{KF_R R_s^2 V^2}{WLf} \tag{8.27}$$

where KF_R is a fitting parameter specified in cm²/Ω². In Vandamme and Casier
(2004), the $1/f$ noise is described as a conductance fluctuation that follows Hooge's
empirical relationship, such that the ratio of S_R/R^2 is given by

$$\frac{S_R}{R^2} = \frac{C_{us}}{WLf} \tag{8.28}$$

where C_{us} is the noise expressed for a unit area. By rewriting this relationship, and
including (8.7) and (8.8), Vandamme and Casier (2004) show that the resistor noise
spectral density S_R can be completely described by

$$S_R = \frac{C_{us}R_sL}{W^3 f}$$ (8.29)

which defines the low-frequency noise observations in integrated resistors. The first observation is that there is length, width and area dependence, assuming that the sheet resistance remains constant. Second, for constant resistivity and area, there is noise dependence on the thickness of the resistor. Third, there is noise dependence on the doping concentration of the sheet material, and finally, there is noise dependence on the free carrier mobility in the resistor. As is evident from the above derivations, determining an accurate noise spectral density and, therefore, the noise performance of integrated resistors can become complex, with several factors inducing varying degrees of noise and interference. Thermal and $1/f$ noise have a significant effect on the amplitude and frequency dependence of the intrinsic noise in integrated resistors. Importantly, and considered in the following section with respect to integrated resistor optimization and matching, is the W term in (8.29), which shows that there is inverse cubic proportionality in the resistor noise spectral density to the width of the film, a characteristic that is often used to design integrated resistors with low-noise performance. The following section reviews integrated resistor layout optimization and matching.

RESISTOR LAYOUT OPTIMIZATIONS AND MATCHING

Integrated resistors, as in the case of most active and passive devices on semiconductor ICs, inevitably undergo some form of mismatch for various reasons. These reasons include

- Random statistical fluctuations.
- Process biases.
- Pattern shifts.
- Diffusion interactions.
- Stress gradients.

To match integrated resistors effectively, various techniques and mechanisms can be employed throughout the wafer, specifically on these structures, to reduce the statistical fluctuations between adjacent and sparsely placed devices. Several techniques that alleviate these disparities and are often used to provide improved matching among components include

- The use of structures of equal geometry.
- Placing elements as close as possible to one another, maintaining the technology design rules.
- Placing matched elements to have the same orientation.
- Not designing integrated resistors to be too narrow or too short.
- The use of interdigitated structures.
- The use of common-centroid structures having the same center of gravity.
- Implementing dummy structures to compensate for boundary effects.
- Using thermal effect compensation.
- Incorporating 45° orientations if the device is under excessive stress.

The following paragraphs review the most commonly used integrated resistor matching strategies.

Interdigitated Integrated Resistors

To alleviate the effect of process gradients such as changes in the n-well, $n+$ or $p+$ doping profiles across large areas on the wafer, interdigitated resistor layouts can be implemented. If using this technique, the orientation of the resistors should be consistent among unit cells, therefore these cells should all be placed either vertically or horizontally with respect to adjacent cells. An interdigitated integrated resistor layout is presented in Figure 8.9.

As seen in Figure 8.9, resistors R_A and R_B are divided into equally sized cells, and placed in an $ABAB$ approach. The total resistance of R_A and R_B comprises the series combination of the cells of each of these resistors. By using this approach, each of the cells, and therefore the resistors, has essentially the same parasitic variations, which statistically cancel out to result in matched resistors. However, interdigitated resistor elements, when used in differential circuits, can lead to parasitic interactions during electrostatic discharge events (Voldman 2015). For n-type integrated resistors, it is possible that an internal/parasitic npn bipolar transistor can be formed unaware to the designer amid two resistor segments (Voldman 2015). Figure 8.10 shows the placement of this parasitic npn transistor formed in between two interdigitated resistive elements.

This parasitic npn transistor, as shown in Figure 8.10, can lead to electrostatic failures among signal pins. Similarly, the parasitic npn transistor can be utilized with proper design as electrostatic protection in differential circuits. Additional contacts

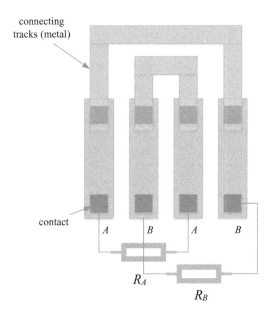

FIGURE 8.9 Interdigitated integrated resistors to alleviate the effects of process gradients across large areas on a wafer.

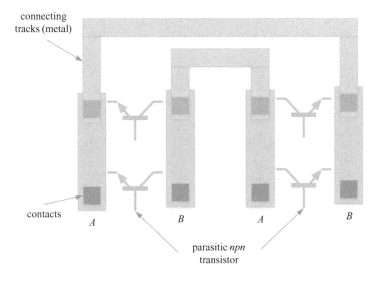

connecting
tracks (metal)

contacts

A   *B*   *A*   *B*

parasitic *npn*
transistor

FIGURE 8.10 Parasitic *npn* transistor formed between two interdigitated resistive elements.

or vias can be added to interdigitated integrated resistor strategies to increase the robustness of passive elements. The following paragraph reviews the common-centroid technique when implemented to match integrated resistors.

Common-Centroid Integrated Resistors

The common-centroid layout, also called the cross-coupled layout, improves device matching between adjacent resistors, but again at the cost of uneven parasitic components between two (or more) elements – similar to the intrinsic *npn* transistor formed among elements in *n*-well interdigitated layouts. The common-centroid layout is somewhat similar to the interdigitated layout; each resistor is again divided into unit value cells. In the common-centroid strategy, each resistor is divided into two elements. The common-centroid technique is presented in Figure 8.10, with reference to the interdigitated structure in Figure 8.9.

Starting from the center of the common centroid, as depicted in Figure 8.11, variations in process parameters in both the *x*- and *y*-directions are always *seen* by one of these elements of each resistor. Therefore, each element will undergo equal deterioration from the process variations in each direction (Fayed and Ismail 2006). If the number of elements in Figure 8.11 is even, further accuracy can be achieved in matching by interdigitating each resistive element, as described in the previous paragraph. Dummy elements, in addition, increase the accuracy of matching devices, and are described in the following paragraph.

Dummy Elements

The use of dummy elements is another strategy to improve matching between two or more elements of integrated resistors. Figure 8.12 represents two resistive elements, divided into smaller sections of equal geometry, with dummy elements placed on the sides of the first and last element.

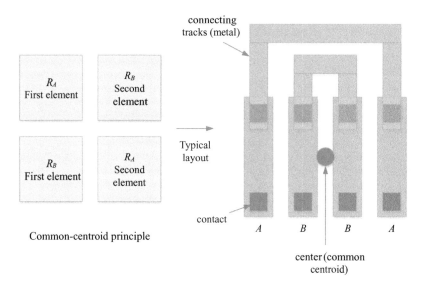

FIGURE 8.11   Four matched common-centroid integrated resistors to alleviate the effects of process gradients across large areas on a wafer. (Adapted from Fayed, A. and Ismail, M., *Adaptive Techniques for Mixed Signal System on Chip*, Springer-Verlag US, New York 2006.)

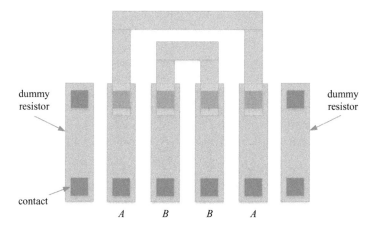

FIGURE 8.12   Two resistive elements, divided into smaller sections of equal geometry, with dummy elements placed on the sides of the first and last element. (Adapted from Baker, R. J., *CMOS: Circuit Design, Layout, and Simulation*. Wiley-IEEE Press, Hoboken, NJ, 2011.)

In Figure 8.12, the final amount of diffusion, the section forming the resistive channel in *n*-well resistors, is different under the photoresist, at the edges and between the inner and outer unit cells. This is a common characteristic due to varying dopant concentrations at these regions on the wafer, a typical manufacturing tolerance. These variations result in a mismatch among resistors, which can be compensated for by using dummy elements in the design. Dummy elements can be added to interdigitated as well as to common-centroid layout strategies, and ensure that the unit resistors *see* similar adjacent structures. Dummy elements are typically tied to the ground (such as the substrate) or connected to the positive supply potential.

Another passive component that significantly increases the total area of on-chip components, typically much larger than transistors, for example, is the MOS capacitor. Capacitive components are often implemented off-chip, purely to decrease the overall area used on-chip; however, for small capacitances (in the pF and fF range), on-chip MOS capacitors are still used. The proper design and matching of these components are crucial, since the available area on a chip dictates the number of active components that can fit on a single die, and Moore's law can only continue to thrive if layout optimizations of passive components are considered during the layout design phase, essentially to make space for the active components (transistors). The MOS capacitor optimization strategies are reviewed in the following section, which include first-principle mathematically derived parameters and characteristics of these components.

MOS CAPACITOR OPTIMIZATION

INTEGRATED CAPACITOR LAYOUT

The MOS capacitor is constructed to consist of a semiconductor body, also referred to as the substrate; an insulator film, typically using material such as SiO_2, Si_3N_4 or high-dielectric materials in a modern and very thin dielectric construction; and a metal electrode, referred to as the gate of the device. The simplified construction of a MOS capacitor is given in Figure 8.13.

The capacitance of the MOS capacitor in Figure 8.13 is simply defined by the well-known capacitance equation,

$$C = \frac{\varepsilon_0 A}{d} \tag{8.30}$$

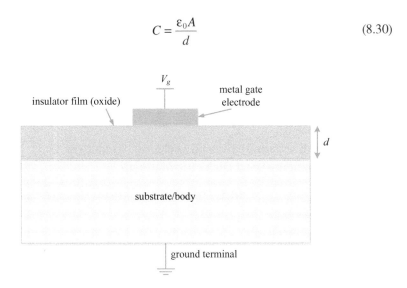

FIGURE 8.13   The side view of the physical layout of a MOS IC capacitor.

where,

    d is the separation between these plates

    ε_0 is the dielectric constant (permittivity) of the insulator layer

    A is the area of the parallel metal plates

Table 8.1 lists the most commonly used materials for the dielectric layer, including the relative permittivity and dielectric strength of these materials.

The size of the electric field across the capacitor is also a function of the separation d, where the electric field E is defined by

$$E = \frac{V}{d} \tag{8.31}$$

where V is the voltage potential across the terminals of the device. The gate voltage V_g is defined by

$$V_g = V_{fb} + \phi_s + V_{ox} \tag{8.32}$$

where,

    V_{fb} is the flat-band voltage

    ϕ_s is the surface potential

    V_{ox} is the oxide potential

The applied voltage at the flat-band condition is the difference between the Fermi levels at these two terminals, such that

$$V_{fb} = \psi_g - \psi_s \tag{8.33}$$

where,

    ψ_g is the gate work function

    ψ_s is the semiconductor work function

        at the flat-band voltage, $V_g = V_{fb}$, and $\phi_s = V_{ox} = 0$

TABLE 8.1

Relative Permittivity and Dielectric Strength of Commonly Used Dielectric Materials

Material	Relative Permittivity	Dielectric Strength (V/nm)
SiO_2 dry oxide	3.9	11
SiO_2 plasma	4.9	3–6
Si_3N_4 LPCVD[a]	6–7	10
Si_3N_4 plasma	6–9	5

[a] Low-pressure chemical vapor deposition.

The flat-band potential is where there is zero charge applied to the plates of the capacitive element and, therefore, there exists no electric field at the oxide junction. The value of the flat-band voltage depends on the concentration of the doping of the semiconductor material as well as any residual charges that exist at the semiconductor-insulator interface (typically the Si-SiO$_2$ interface). An important parameter, which defines the operating region boundaries of a MOS capacitor, is the threshold voltage. The threshold voltage, also referred to as the threshold of inversion potential, where surface inversion takes place, is defined by

$$V_t = V_{fb} + 2\varphi_B + \frac{\sqrt{qN_a 2\varepsilon_s 2\varphi_B}}{C_{ox}} \tag{8.34}$$

where,

V_{fb}	is the flat-band voltage as discussed above
φ_B	is the bulk potential
q	is the elementary electron charge ($q = 1.602 \times 10^{-19}$ C)
N_a	is the acceptor carrier concentration
ε_s	is the specific permittivity
C_{ox}	is the oxide capacitance

Each of these capacitor-related parameters is discussed in further detail in this section.

MOS CAPACITOR OPERATING REGIONS

A MOS capacitor has three primary operating regions (four, if counting the flat-band region) depending on the applied voltage between the gate and the body. These regions are surface accumulation, depletion and inversion regions. The inversion region is also characterized as weak or strong inversion, subject to the applied gate voltage with respect to the threshold voltage. These regions are summarized in Figure 8.14.

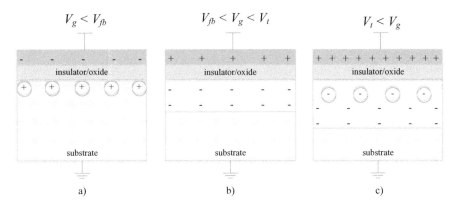

FIGURE 8.14   The charge distribution of an *n*-type MOS capacitor structure in a *p*-type substrate under (a) accumulation, (b) depletion and (c) inversion operating conditions.

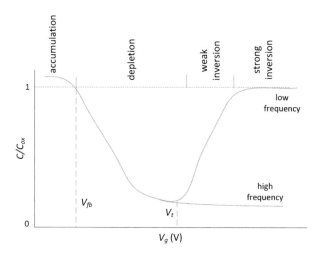

FIGURE 8.15　The operating regions of a MOS capacitor identified through CV analysis.

As shown in Figure 8.14, the device operates in the surface accumulation region if the applied gate potential is smaller than the potential of the flat-band; it operates in the depletion region if gate voltage is greater than the flat-band but lesser than the V_t. The MOS capacitor operates in inversion if the gate potential is greater than V_t. By performing a capacitance versus applied voltage sweep of a capacitor, these regions can be identified, as shown in Figure 8.15.

These operating regions are discussed in further detail in the following section. Defining the capacitor operating regions is critical for understanding the device operation, allowing more complete cognition when considering layout optimization techniques and electronic circuit enhancements to facilitate Moore's law; therefore it is briefly discussed in this chapter.

Surface Accumulation

If $V_g < V_{fb}$, then a positive charge is induced on the gate terminal and a negative charge is induced in the semiconductor substrate. As a result, the electron concentration at the surface of the capacitor is higher than at the bulk, resulting in surface accumulation. If the surface is in accumulation, a linear correlation exists among the charge per unit area at the semiconductor/insulator interface and the applied gate voltage. The slope of this relationship represents the oxide capacitance per unit area, given by

$$C_{ox} = \frac{\varepsilon_{ox}}{t_{ox}} \qquad (8.35)$$

where,

ε_{ox}　is the permittivity of the oxide material
t_{ox}　is the thickness of the oxide

The permittivity of the oxide material is a function of the carrier concentration, and its relationship is given by

$$\varepsilon_{ox} \propto N_a x_d \qquad (8.36)$$

where,

N_a is the carrier concentration

x_d is the width of the depletion region

For a p-type MOS capacitor, holes accumulate at the surface (the accumulation layer) and are referred to as accumulation-layer holes. The accumulation-layer charge of these holes is defined as Q_{acc} and is specified in units of C/cm². Similarly, for an n-type MOS capacitor, the accumulation layer would hold electrons, which are referred to as accumulation-layer electrons. With respect to (8.32), at the accumulation condition, the surface potential ϕ_s is small and can be neglected, such that the oxide potential becomes

$$V_{ox} = V_g - V_{fb}. \qquad (8.37)$$

Relating the modified version of (8.32) to the oxide capacitance, and applying the common voltage, charge and capacitance relationship $V = Q/C$, it follows that the oxide potential can be written as

$$V_{ox} = -\frac{Q_{acc}}{C_{ox}} \qquad (8.38)$$

where the negative sign appears, since for a MOS capacitor, the gate voltage is related to the charge in the substrate. Since the accumulation-layer charge Q_{acc} is zero if $V_g = V_{fb}$, (8.38) can be generalized as

$$V_{ox} = -\frac{Q_{sub}}{C_{ox}} \qquad (8.39)$$

where Q_{sub} is the charge present in the substrate at any given moment, and includes the accumulation-layer charge if the conditions of V_g and V_{fb} are met. If the MOS capacitor is operating in surface accumulation, the capacitance is independent of the gate voltage as well as the operating frequency, as long as the wave of the bulk of the carriers that add to the substrate charge remains constant with respect to the incremental velocity of the gate charge. The following paragraph reviews the oxide potential and gate voltage equations of a MOS capacitor in surface depletion.

Surface Depletion

If $V_g < V_{fb}$, then, at the interface between the gate terminal and the insulating material (oxide), a negative charge is induced. As a result, a positive charge is induced at the semiconductor/oxide interface and the surface of the semiconductor is exhausted of mobile carriers, resulting in the positive space charge region. The capacitance of the space charge region (C_D) as a function of the gate voltage is defined by

$$C_D(V_g) = \frac{\varepsilon_s}{x_d(V_g)} \qquad (8.40)$$

where x_d is the thickness of the depletion layer, variable with the applied gate potential. Increasing the positive gate voltage tends to enlarge the width of the surface depletion region, x_d; therefore, the capacitance from the gate to the substrate will decrease. From the small-signal equivalent, the capacitance of the MOS capacitor in the surface depletion condition is given by

$$C_{\text{depletion}} = \frac{1}{C_{ox}} + \frac{1}{C_D} = \frac{C_{ox}C_D}{C_{ox} + C_D} \tag{8.41}$$

where the thickness of the surface depletion increases as the potential across the gate is decreased, since additional electrons are pushed away, resulting in a thicker space charge region. The oxide potential is a function of the surface depletion charge Q_{dep} and, therefore, of the width of the depletion region, such that

$$V_{ox} = -\frac{Q_{\text{sub}}}{C_{ox}} = -\frac{Q_{\text{dep}}}{C_{ox}}, \tag{8.42}$$

which can be rewritten as

$$V_{ox} = \frac{qN_ax_d}{C_{ox}} = \frac{\sqrt{qN_a2\varepsilon_s\phi_s}}{C_{ox}} \tag{8.43}$$

where the surface potential in the surface depletion condition is given by

$$\phi_s = \frac{qN_ax_d^2}{2\varepsilon_s} \tag{8.44}$$

and, in order to determine V_{ox} and ϕ_s, x_d must first be determined as a function of V_g by rewriting (8.32) as

$$V_g = V_{fb} + \frac{qN_ax_d^2}{2\varepsilon_s} + \frac{qN_ax_d}{C_{ox}}. \tag{8.45}$$

If the potential across the gate terminal is further enlarged, the device will enter surface inversion, and depending on the frequency of operation, the capacitance will either increase (at low operating frequencies where the capacitance is subject to the oxide capacitance) or decrease (at high frequencies). The weak inversion mechanism is discussed in the following paragraph, and the threshold voltage is defined.

Weak Inversion (Threshold of Inversion Condition)

If a MOS capacitor is said to operate in inversion, the surface of the capacitor is inverted from either a *p*-type to an *n*-type material, or from an *n*-type to a *p*-type material. Therefore, at the threshold, the surface electron concentration is equivalent to that of the bulk doping. If the capacitor is at the threshold of inversion, and therefore no longer in the surface depletion condition, the bulk potential is defined as twice the surface potential at the threshold condition, such that

$$2\varphi_B = \varphi_{st} = 2\frac{kT}{q}\ln\frac{N_a}{n_i} \tag{8.46}$$

where,

φ_{st} is the surface potential at the threshold condition
k is Boltzmann's constant
T is the temperature in kelvin
n_i is the intrinsic carrier concentration

The gate voltage at the threshold condition is referred to as the threshold voltage, V_t, defined by, with reference to (8.44),

$$V_t = V_{fb} + 2\varphi_B + \frac{\sqrt{qN_a 2\varepsilon_s\phi_B}}{C_{ox}} \tag{8.47}$$

where the threshold voltage can be determined graphically by plotting the oxide thickness versus the substrate doping density. For a gate voltage higher than the threshold voltage, the MOS capacitor enters strong inversion, and the oxide potential in this region is reviewed in the following paragraph.

Strong Inversion (Beyond the Threshold Condition)

If $V_g > V_t$, then strong surface inversion occurs and an inversion layer filled with inversion electrons is created. The charge density of the inversion condition can be represented by Q_{inv}, and the surface potential, ϕ_s, does not increase much higher than φ_B, which results in the depletion region width also staying relatively constant as V_g increases. The maximum width of the depletion region is defined by

$$x_{d\max} = \frac{\sqrt{2\varepsilon_s\phi_B}}{qN_a} \tag{8.48}$$

and the gate potential for strong inversion is given by

$$V_g = V_t - \frac{Q_{inv}}{C_{ox}} \tag{8.49}$$

Equation (8.49) can be rewritten in terms of the charge across the inversion region,

$$Q_{inv} = -C_{ox}\left(V_g - V_t\right) \tag{8.50}$$

showing that, for a MOS capacitor, if $V_g = V_t$, the inversion layer charge, Q_{inv}, is zero and increases as the gate voltage is increased beyond V_t.

As the operating regions have now been defined, the following section focuses on capacitor layout optimization, as well as circuit schematic optimizations based on the operation of a MOS capacitor to reduce mismatching among devices and improve efficiency and performance, as well as to reduce the physical size required, which is typically a limiting factor introduced by passive components in IC design.

CAPACITOR LAYOUT OPTIMIZATIONS

Highly dense capacitive structures are used specifically in power applications to achieve the minimum equivalent series resistance (ESR), with proposed structures provided in Villar et al. (2003). Since all capacitors (and inductors) used in electrical circuits, which include off-chip discrete components as well as on-chip integrated solutions, are not ideal and can in most cases be approximated as a perfect (ideal) device in series with a resistive element, the ESR is typically defined as an AC resistance and measured with standardized frequencies. ESR arises from losses in the dielectric substances, electrodes or other factors in addition to the capacitive properties of the component and additional parasitic inductance, denoted as the equivalent series inductance (ESL) resulting from the inductance of electrodes, leads or tracks on ICs. A simplified two-element equivalent model of the capacitor, which includes the ESR, is shown in Figure 8.16. This model will be discussed later in this section.

The effect of the ESR in ceramic capacitors, for example, is typically very small and is often neglected. Discreet ceramic capacitors have a relatively large capacitance in the nF to μF range, and the ESR is typically in the mΩ range. However, in IC on-chip capacitors, capacitances in the fF to pF range are often used, and the ESR plays a more significant role in terms of its ratio to the capacitance. In addition, for higher frequency applications, such as in RF applications as well as in low-power applications such as energy-harvesting circuits, the ESR becomes a significant contributor that decreases RF performance and increases power consumption. RF semiconductor devices used in matching implement very low input impedances. If the input impedance of the circuit is low, in the range of the parasitic ESR, a large ratio of the power will be consumed by the ESR, decreasing output power and circuit efficiency. The figure of merit to define the dissipation of a capacitor is its quality factor (Q), defined as

$$Q = \frac{\omega C}{R_s} \tag{8.51}$$

where,

    ω is the operating frequency in rad/s (defined as $\omega = 2\pi f$, where f is the frequency in Hz)

    R_s is the ESR of the capacitor, C

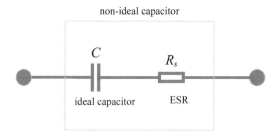

FIGURE 8.16 The two-element equivalent circuit of a non-ideal capacitor, showing the ideal capacitance, C, and the series resistance component, R_s.

The inverse of the quality factor is referred to as the dissipation factor (DF), which is a measure of the value of the affinity of dielectrics to absorb energy from an applied AC signal. The DF can be defined through

$$DF = \frac{1}{Q} = \frac{R_s}{\omega C} = \frac{\sigma}{\varepsilon \omega} \tag{8.52}$$

where,
    σ is the conductivity of the dielectric bulk material
    ε is the lossless permittivity of the dielectric

DF is mostly used as a figure of merit at lower operating frequencies, with respect to the size of the capacitor, whereas another figure of merit, the loss tangent, is used at higher frequencies. The loss tangent is the tangent of the dissimilarity between the phase angle and the capacitor potential and its current measured from the theoretically expected 90°, caused primarily by dielectric losses in the capacitor. The loss tangent, δ, is defined by

$$\tan \delta = DF = \frac{R_s}{\omega C} \tag{8.53}$$

where the product of ωC is also called the reactance (X_c) of the capacitor. ESR also affects the power consumption in the circuit, since the average power consumed in the capacitor is a function of its characteristic impedance, Z_c, given by

$$Z_c = \sqrt{R_s^2 + X_c^2}. \tag{8.54}$$

MOS capacitor models are typically more complex and include real and imaginary parasitic components of series resistance and series inductance, as well as parallel conductance. In order to extract the electrical parameters of the oxide dielectric layer accurately, a four-element equivalent model is proposed. This four-element model is represented in Figure 8.17.

In Figure 8.17, the real dielectric capacitance is represented by C, the parallel conductance due to the gate leakage current by G_p, the series resistance due to the well/substrate and contact resistance by R_s and the series inductance due to leads, probes and metal tracks by L_s. The characteristic impedance of this system, compared with the less complex representation in (8.54), is given by Baomin et al. (2009):

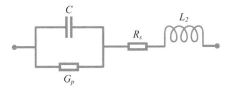

FIGURE 8.17   The four-element equivalent circuit of a non-ideal capacitor, showing the ideal capacitance C, the series resistance component, R_s, a series inductance, L_s, and the parallel conductance G_p.

$$Z_c = \frac{1}{G_p + j\omega C} + j\omega L_s + R_s \tag{8.55}$$

where the correct value of the MOS capacitance can, for example, be extracted using its characteristic impedance combined with two-frequency or multi-frequency techniques, which are not reviewed in this book. In addition, reliable capacitor measurements require a small DF (and therefore a large quality factor), given for MOS capacitors by

$$DF = \frac{R_s + R_p}{\omega C R_p^2} + \omega C R_s \tag{8.56}$$

where, if the operating frequency is high, the first term in (8.56) can be neglected and the DF follows a linear relationship with frequency, dependent on the capacitance and the ESR. The frequency dependence of DF for MOS capacitors can be found by keeping the gate voltage constant and graphing DF versus frequency, typically for various geometries. The bulk series resistance is dependent on the gate electrode diameter, or area, as studied in Iniewski et al. (1989). The effect of the bulk series resistance therefore influences the MOS capacitor high-frequency capacitance-voltage (CV) measurements, used frequently to characterize MOS capacitors, as studied in Iniewski et al. (1989). Most notably, the determination of the actual oxide thickness using CV characteristics in strong accumulation, where the capacitance is highest, will result in erroneous calculations of the oxide thickness if R_s is not considered. If measured, the capacitance displayed by the measurement equipment, C_M, is related to the actual capacitance by

$$C_M = \frac{C}{1 + (\omega R_s C)^2} \tag{8.57}$$

and, as a result of the influence of the ESR, the CV would be deformed and extracted parameters would be incorrect, unless R_s is properly accounted for (Iniewski et al. 1989). The two most relevant sources of series resistance are the resistance, R_B, of the quasi-neutral semiconductor bulk between the contact to the substrate and the depletion layer edge at the semiconductor surface, and the contact to the semiconductor wafer back. The bulk series resistance R_B for a circular-shaped MOS capacitor gate can be described as a function given by

$$R_B = \frac{\rho}{4a} f\left(\frac{l}{a}\right) \tag{8.58}$$

where,
    ρ is the semiconductor resistivity
    l is the thickness
    a is the radius of the gate electrode

The function f is typically given in tabulated form, since complex numerical three-dimensional Laplace transform calculations are otherwise required. The approximation of $f(1/a)$ is given in Iniewski et al. (1989).

The deviation from the ideal behavior, operation and capacitance of capacitors stems from various sources. Many of these deviations can be accounted for or, ideally, minimized, depending on the geometrical layout and structure of the physical component. Deviation sources and measurement errors are commonly introduced because of several factors, which include

- Matching errors (alleviated by the replication principle).
- Dielectric (oxide) gradients.
- Edge variation effects.
- Parasitic capacitances.
- Temperature and voltage dependence.

Performance deviations and inaccurate parameters of capacitors are common and depend not only on the technology variations during manufacturing but also on insufficient layout techniques of adjacent and standalone capacitors. The first deviation issue summarizes matching errors due to the fundamental layout of multiple capacitors. The effect can be quantified by investigating the geometry of, for example, two capacitors. If two capacitors of equal length but varied width, in this case varied by a factor 2, the conclusion below can be drawn. Consider the two capacitors depicted in Figure 8.18.

The capacitors in Figure 8.18 are of equal length, arbitrarily set to 5 µm, therefore

$$L_1 = L_2 = 5\,\mu m \tag{8.59}$$

where L_1 and L_2 represent the length of the capacitors C_1 and C_2, respectively. Suppose that the width of C_1 is 12 µm; this results in

$$W_2 = 2W_1 = 12\,\mu m \tag{8.60}$$

where W_1 and W_2 represent the width of the capacitors C_1 and C_2, respectively. In addition, consider a process variation on both the length and width of both capacitors of Δx µm equal to 500 nm. To find the error in matching, the ratio of the areas of both capacitors can be determined by the following scenario, where an arbitrary -1 µm deviation in the width and length is introduced; therefore

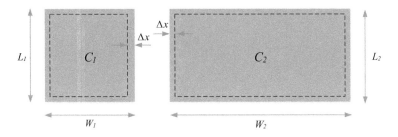

FIGURE 8.18   A representation of two semiconductor MOS capacitors, C_1 and C_2, of equal length, where the width of C_2 is twice as large as that of C_1.

$$\frac{C_2}{C_1} = \frac{(L_2 - 2\Delta x)(W_2 - 2\Delta x)}{(L_1 - 2\Delta x)(W_1 - 2\Delta x)} \tag{8.61}$$

for which, in this example, the ratio of C_2/C_1 would equate to

$$\frac{C_2}{C_1} = \frac{(5-1)(12-1)}{(5-1)(6-1)} \tag{8.62}$$

which is equated to

$$\frac{C_2}{C_1} = \frac{44}{20} = 2.2 \tag{8.63}$$

resulting in a 10% deviation from the ideal $C_2/C_1 = 2$. The aim of capacitor matching is to reduce this 10% error, in this example, to a much lower value, leading to smaller tolerances among designed components. Allen and Holberg (2002) show that the general expression of the capacitor ratios can be represented as

$$\frac{C_2}{C_1} \approx \frac{W_2}{W_1}\left(1 - \frac{2\Delta x}{W_2} + \frac{2\Delta x}{W_1}\right) \tag{8.64}$$

and therefore, as stated in Allen and Holberg (2002), the matching error can be minimized if the widths of the capacitors are equal, therefore if $W_2 = W_1$ (as opposed to $W_2 = 2W_1$ in the above example). This means that to minimize the matching error between two or more components, all components should have an equal area-to-perimeter ratio. To achieve this, consider the layout of the same two capacitances, but designed by using two capacitors in parallel to achieve the value for C_2, as shown in Figure 8.19.

For the layout in Figure 8.19, for the same deviation scenario, the C_2/C_1 ratio can be written as

$$\frac{C_2}{C_1} = \frac{2(L_2 - 2\Delta x)(W_2 - 2\Delta x)}{(L_1 - 2\Delta x)(W_1 - 2\Delta x)} \tag{8.65}$$

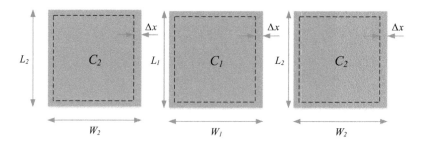

FIGURE 8.19   A representation of two semiconductor MOS capacitors, C_1 and C_2, of equal length, where the width of C_2 is achieved by placing two capacitors, of similar size to C_1, in parallel.

and, considering that for each sub-capacitor of C_2, $W_2 = W_1$, it follows that

$$\frac{C_2}{C_1} = \frac{2(5-1)(6-1)}{(5-1)(6-1)} \tag{8.66}$$

resulting in a ratio of 2, which is therefore a 0% deviation error from the expected ratio, which is also 2. This is called the replication principle, and it can be implemented for other passive components such as resistors and inductors.

The second variation factor, dielectric or oxide gradients, is introduced if there is a variation in the dielectric thickness across the wafer. Wafer thickness gradients can be presented relatively easily, depending on the equipment used to thermally grow the oxide, the physical wafer size and the variations in the even distribution of the gas flow across the wafer. Figure 8.20 represents a silicon substrate and the polysilicon gate electrode, with an uneven gate oxide grown owing to various external factors.

As seen in Figure 8.20, several mechanisms are present if an uneven gate oxide is grown on the semiconductor substrate. Leakage currents can occur in response to reliability issues due to charge injection. Non-uniformities and defects in the oxide film can cause dielectric breakdown and dopant penetration can enter the oxide through the gate electrode. In regions where the uneven oxide is thin, conduction is rather through direct tunneling, as reviewed in Chapter 7 of this book. During direct tunneling, there is no energy lost in the oxide, since there are no scattering events. The energy is released at the accumulating anode or cathode and is considerably less than in the FN regime. The damage is consequently smaller at a particular electric field and this has repercussions for the dependability of scaled components and processes. One such implication is the stresses incurred as a result of volume change during SiO_2 formation, resulting in induced stress at the semiconductor interface. The silicon and oxygen bonds are exceedingly stressed at the boundary, and less energy is required to disrupt these bonds. A hot electron or even a hole could have adequate energy to disrupt the stressed bonds and produce traps at the boundary,

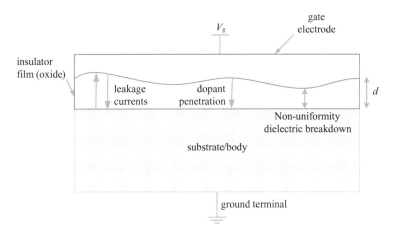

FIGURE 8.20 A silicon substrate, with an uneven gate oxide grown owing to various external factors, and the gate electrode, showing the mechanisms that decrease device performance.

leading to device degradation over time. The viscosity of the oxide layer decreases as the growth temperature increases, allowing for flow that is more viscous. At higher temperatures, it evidently takes less time for the viscous flow to ensue. To alleviate the effects of uneven oxide growth leading to stability issues, one can use the common-centroid layout technique, as discussed for MOS transistors in Chapter 7 of this book, as well as using numerous repetitions that are randomly connected to achieve a statistical error balanced over the entire area of interest. A simplified representation of the common-centroid capacitor layout is given in Figure 8.21.

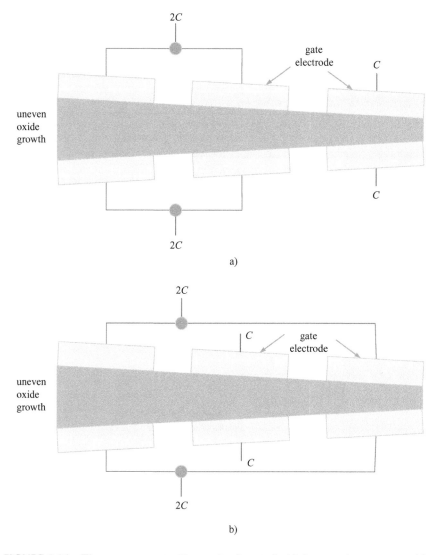

FIGURE 8.21 The common-centroid capacitor layout. In (a) the capacitors are not taking advantage of the common-centroid technique, whereas in (b) the technique is applied to alleviate the effects of uneven dielectric oxide growth on semiconductor wafers.

In the common-centroid layout technique presented in Figure 8.21, unit capacitors are connected in parallel to form a larger capacitor. Typically, the ratio between these capacitors is an important parameter based on the required geometry and capacitance. The error in a single capacitor is proportionate to the perimeter-area ratio of the larger capacitor, and dummy devices are often used to improve matching of the unit capacitors. If two unit capacitors, with capacitance C_1 and C_2, respectively, are to be matched, the following derivation can be used, where the ratio between the two capacitors is given by

$$\frac{C_1}{C_2} = \frac{C_{1,\text{ideal}}\left(1+e_1\right)}{C_{2,\text{ideal}}\left(1+e_2\right)} \tag{8.67}$$

where,

$C_{1,\text{ideal}}$ and $C_{2,\text{ideal}}$ represent the ideal required capacitance

e_1 and e_2 are the error in measured capacitance for C_1 and C_2, respectively

To minimize the total error in the ratio of the capacitances, it follows that $e_1 = e_2$, implying that the perimeter and area ratios must be exactly equal. The best technique to match capacitors effectively and completely is to design the device in a square geometry, referred to as the unit capacitor C_u, as shown in Figure 8.22.

As seen in Figure 8.22, the unit capacitor has a width of x_u and a length of y_u. Since this is a unit capacitor, both sides of the capacitor have equal dimensions, therefore

$$x_u = y_u \tag{8.68}$$

and, thus, the area of the unit capacitor, A_u, is given by

$$A_u = x_u^2 = y_u^2. \tag{8.69}$$

During the design phase, it is recommended to design the circuit, at schematic level, to operate with capacitances C_1 and C_2 being integer multiples of the unit capacitor, C_u. To achieve the total capacitance of each capacitor, these integer multiples can be used, such that the ratio of the two capacitors is

unit capacitor

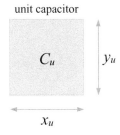

x_u

FIGURE 8.22 Matching capacitors by designing the device as a square geometry of equal width and length.

$$\frac{C_1}{C_2} = \frac{n_1 C_u}{n_2 C_u} \tag{8.70}$$

where n_1 and n_2 represent the integer multiples used for each capacitor. If an error or difference between the two capacitances exists, a separate function can be created to represent this difference. In such a case, the total capacitance of C_1 is represented by the unit capacitor and its integer multiples, such that

$$C_1 = n_1 C_u \tag{8.71}$$

and C_2 can be represented to contain the error, such that

$$C_2 = n_2 C_u + (1 + f) C_u \tag{8.72}$$

where,

f is a fraction

$(1+f)C_u$ is referred to as a non-unit capacitor

C_{nu},   which should ideally have an equal ratio of perimeter and area as the unit capacitor

In order to determine the size of C_{nu}, the parameter N can be introduced such that

$$N = 1 + f \tag{8.73}$$

which is expressed in terms of the area of the non-unit capacitor and the unit capacitor, by

$$N = \frac{A_{nu}}{A_u} = \frac{x_{nu} y_{nu}}{x_u^2} \tag{8.74}$$

where, to achieve matching, the following should hold true:

$$\frac{P_{nu}}{A_{nu}} = \frac{P_u}{A_u} \tag{8.75}$$

which can be rewritten as

$$\frac{P_{nu}}{P_u} = \frac{A_{nu}}{A_u} = N \tag{8.76}$$

which becomes

$$N = \frac{2(x_{nu} + y_{nu})}{4 x_u} \quad \therefore x_{nu} + y_{nu} = 2 N x_u. \tag{8.77}$$

From (8.77), the width of the non-unit capacitor, x_{nu}, can be expressed by

$$x_{nu} = \frac{Nx_u^2}{y_{nu}} \tag{8.78}$$

and it follows that using (8.78) in conjunction with (8.77), the length of the non-unit capacitor can be solved using

$$y_{nu} = x_u \left(N \pm \sqrt{N^2 - N} \right) \tag{8.79}$$

where the $\pm$ sign can be used as a positive sign if $y_{nu} > x_{nu}$ is required, and as a negative sign if $y_{nu} < x_{nu}$ is required.

The third potential source of deviation of the capacitance of two seemingly equal capacitors is edge variation effects that are introduced during the processing of the wafer. As for any manufacturing process, small variations, or tolerances, among components are inevitable and a well-controlled process can aim to minimize or statistically balance out these errors, but they cannot be completely removed. During the processing of semiconductor devices (this is true for all semiconductor components), variations in the edges of the devices are introduced. This typically occurs during the etching process, especially if wet etching is used, and to a lesser extent with dry etching, such as reactive ion etching (RIE). The presence or absence of adjacent structures is known to influence the definition of the edges, especially for large components such as capacitors, inductors and resistors (passive components). Figure 8.23 shows two capacitors, C_1 and C_2, at a specified distance from each other, but including a non-optimized and non-related (spatially) capacitor, C_3, in close vicinity.

As seen in Figure 8.23, the presence of C_3 disturbs the etching profile and, therefore, the definition of the edges of C_1 and C_2. This effect can lead to higher tolerances

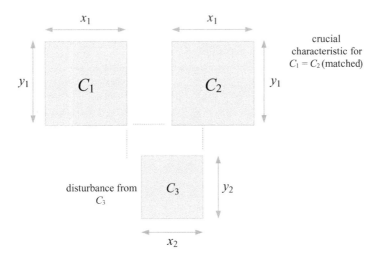

FIGURE 8.23 A simplified representation of two capacitors, C_1 and C_2, at a specified distance from each other, including a non-optimized capacitor, C_3, in close vicinity.

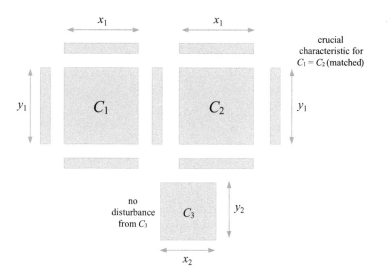

FIGURE 8.24   A simplified representation of two capacitors, C_1 and C_2, at a specified distance from each other, with their spatial surroundings matched, including a non-optimized capacitor, C_3, in close vicinity.

between C_1 and C_2, if designed to be of similar size and capacitance. To avoid edge variations and improve the matching of capacitors, the area around these capacitors should also be matched, especially in sparse layouts. The technique to match array spacing is shown in Figure 8.24.

In Figure 8.24, the immediate areas around C_1 and C_2 are exactly matched, so that each capacitor sees the same surroundings on all sides. These matching structures, and also dummy structures, should be placed in accordance with the design rules of the specific process used, depending on the layer used (metal, poly, diffusion). Capacitor C_3 now introduces a significantly smaller, ideally zero, effect on the etching profile around the edges of capacitors C_1 and C_2.

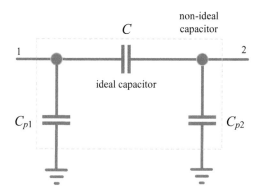

FIGURE 8.25   The simplified equivalent circuit of a MOS capacitor, showing the desired capacitance, C, and the parasitic parallel capacitances to AC ground, C_{p1} and C_{p2}.

The fourth source of deviation errors is intrinsic parasitic capacitances between the top and bottom metal (conductive) plates of the capacitor and AC ground, which is typically the semiconductor substrate. The AC equivalent circuit of the MOS capacitor, showing the parasitic capacitances, is shown in Figure 8.25.

In Figure 8.25, C_{p1} represents the parasitic capacitance between the bottom conducting metal plate and the substrate, whereas C_{p2} represents the parasitic capacitance between the top conducting metal plate and the substrate. Typically, C_{p1} can have a value ranging between 5% and 20% of the desired capacitance, C, which is relatively large. C_{p2} is typically smaller compared with the desired capacitance, ranging between 0.1% and 1% of C, and is due to the interconnections between the capacitor and other circuitry (Whitaker 1996). These parasitic capacitances have an effect on the high-frequency RF performance of the capacitor and should be minimized as far as possible. The techniques used to minimize parasitic capacitances in MOS capacitors are often application-specific and require schematic-level optimizations, geometric optimizations and, to a lesser extent, process-specific optimizations, since foundries are responsible for specifying capacitors with the lowest attainable parasitic capacitance within the limitations of the process.

The fifth potential source of capacitance is the inherent dependence on temperature and the applied bias voltage. The CV and IV characteristics of MOS capacitors degrade significantly because of higher temperature and bias voltage. At high temperatures, "an increased effective oxide thickness, oxide trapped charge density, and interfacial density of state during bias temperature stress lead to degradation of the characteristics of MOS capacitors" (Yu et al. 2013). Because of the substantial decrease in oxide thickness accompanied by aggressive scaling of semiconductors, tunneling currents, as described in Chapter 7 of this book, have become dominant sources of leakage currents in MOS devices. There is a definitive and relatively significant temperature dependence on gate tunneling leakage currents, affecting the performance of devices realized by oxide layers, as in the case of MOS capacitors. This temperature dependence can be attributed to a thermionic-type emission current at high temperatures and low electric fields (Yu et al. 2013) and Fowler-Nordheim emission models in high electric field regions. Typically, the capacitance of a MOS capacitor is decreased if the temperature is decreased, an indication of an increase in oxide thickness and a reduction in surface mobility (Lattin and DeMassa 1974). The temperature dependence of a MOS capacitor is defined as

$$T_{CC} = \frac{1}{AC_t}\frac{d}{dT}\left(AC_t\right) \qquad (8.80)$$

where,

    A is the plate area of the device

    C_t is the total MOS capacitance per unit area

    T is the temperature in kelvin (McCreary 1981)

    C_t is given by the series combination of oxide capacitance and space charge capacitance, given as

$$C_t = \left(\frac{1}{C_t} + \frac{1}{C_s} \right)^{-1} \qquad (8.81)$$

where the space charge capacitance is calculated by

$$C_s = \frac{q}{kT} \frac{dQ_s}{dU_s} \qquad (8.82)$$

where,

Q_s is the semiconductor space charge per unit area
U_s is the dimensionless surface potential
T_{CC} represents the fractional degree of variation of the overall capacitance per unit temperature

Depending on the ratio of the oxide capacitance and the space charge capacitance, McCreary (1981) furthermore resolves the temperature coefficient of MOS capacitors into three primary components, such that

$$T_{CC} = T_{CC}(th) + T_{CC}(sc) + T_{CC}(\varepsilon_{ox}) \qquad (8.83)$$

where,

$T_{CC}(th)$ represents the variation in capacitance for a plate with area A and dielectric thickness t_{ox} due to thermal expansion
$T_{CC}(sc)$ corresponds to the temperature dependence of the space charge region capacitance
$T_{CC}(\varepsilon_{ox})$ expresses the temperature dependence of the dielectric constant of the oxide, defined as k_{ox} (McCreary 1981)

In McCreary (1981), the temperature dependence due to thermal expansion and the space charge region is analyzed, and the temperature dependence of the dielectric constant is evaluated based on the results.

The following section briefly introduces the integrated inductor, with the various implementation possibilities, and the critical parameters used in designing these components for optimal performance, as well as considering the geometric aspects of these components.

INDUCTOR OPTIMIZATIONS

INTEGRATED CIRCUIT INDUCTOR LAYOUT

An integrated inductor typically consists of numerous windings of metal tracks of minimal diameter on semiconductor material. A larger diameter would reduce the ohmic resistance, but can make the inductive element physically large. On-chip inductors commonly improve the dependability and effectiveness of integrated RF elements and offer better performance if compared with discrete off-chip solutions, and additionally offer a higher level of integration (Koutsoyannopoulus and Papananos 2000). Ideal inductors have no loss of power when current flows through the inductor. Practical inductors do in fact exhibit power losses owing to several factors such as

- The skin effect causing the resistance of coil/metal tracks to increase at high frequencies.
- Eddy currents and hysteresis effects in the core that cause power loss that is dependent on both frequency and the materials.
- The coil/metal tracks having finite resistance.
- Radiation of power to the surrounding components.

The losses in an inductor due to its finite resistance can be modeled as an inductive element in series with a parasitic resistance, and include a parallel intrinsic capacitance, as shown in Figure 8.26.

In Figure 8.26, the inductance is represented by L, and the series resistance of the metal tracks is represented by R_S. The quality factor, Q, of the inductor can be determined by

$$Q = \frac{\omega L}{R_S} \tag{8.84}$$

where ω is the frequency of operation in rad/s. An intrinsic (assumed to be lossless) parallel capacitor C also exists in the equivalent model, which results in a resonant circuit around a center frequency. The resonant frequency, ω_0, as a result of the parallel capacitance, is determined by

$$\omega_0 = \frac{1}{\sqrt{LC}} \tag{8.85}$$

and the impedance, Z, of this tank circuit can be calculated by

$$Z = \frac{R_S + j\omega L}{-(\omega)^2 LC + j\omega R_S C + 1} \tag{8.86}$$

where the impedance is at its maximum at resonance. For the case of integrated inductors on semiconductor substrates (spiral inductors, multi-level spiral inductors, solenoids or bond wires), the equivalent model description of the component is more complex, with substrate impedance and capacitance and oxide capacitance having additional effects on the quality factor of the inductor. The complete equivalent model for an integrated inductor is given in Figure 8.27.

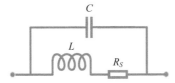

FIGURE 8.26 An integrated inductor equivalent model with series resistance and parallel capacitance representing the losses in the inductor.

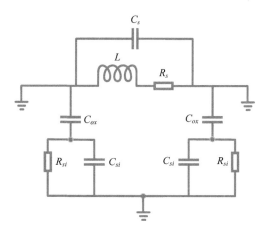

FIGURE 8.27   The complete integrated inductor equivalent model.

In Figure 8.27, L is the desired inductance, R_s is a series resistance and C_s is the intrinsic feed-through capacitance. C_{ox} represents the oxide capacitance, C_{si} are the semiconductor substrate capacitance losses and R_{si} are the semiconductor substrate ohmic losses that result in excessive eddy currents. Importantly, to optimize inductor performance and ensure efficient high-frequency operation, the three parameters to consider are the inductor quality factor, its inductance at the operating frequency and the frequency of self-resonance (Burghartz and Rejaei 2003). The primary constraints of these parameters include the substrate and conductive track losses and the oxide capacitance. Several techniques can be employed to improve the quality factor of the inductor and its inductance at the operating frequency; these techniques are not reviewed in this book. Essentially, if the aim is to maximize the quality factor (as is typically the case), then R_s and C_{ox} need to be minimized and R_{si} should be as large as possible. Burghartz and Rejaei (2003) and Lipka (2010) investigate the optimum spiral inductor designs to maximize their Q factor and provide a detailed analysis of each alternative topology. The design optimization techniques and strategies discussed in this chapter for integrated resistors and capacitors can also be applied to inductors.

CONCLUSION

Chapters 7 and 8 of this book present microelectronic circuit enhancements and design methodologies to facilitate the continuation of Moore's law with aggressively scaled semiconductor components. This chapter specifically focuses on the passive components, resistors and capacitors, as well as introducing briefly the important parameters to consider in inductor optimization.

Passive components occupy a significant amount of space on semiconductor ICs, where, alternatively, these components can be placed off-chip by using discrete components, although for superior performance, especially in low-noise applications, on-chip passive components are still preferred. Because of their large size, however, passive components undergo large variations in manufacturing parameters, relative to their smaller active counterparts. This chapter reviews the primary sources of mismatches during processing, which include systematic, random and gradient mismatches. These mismatches are either introduced by sub-optimal design strategies or by inherent manufacturing variations from equipment and processing tolerances (such as during etching, lithography or metal deposition).

This chapter therefore aims to identify the principles of passive components and elaborates on the design-specific techniques that decrease process variations, typically by addressing statistical random deviations. The first component reviewed is the integrated resistor, a component regularly used in integrated designs to limit the flow of current at specific junctions. The primary parameters of the integrated resistor are identified, and include sheet resistance, thermal resistance, current density, frequency operation and noise. A detailed description of sheet resistance and its derivation, which depends primarily on the concentration of the material and its geometry, is supplied. Three common methods used to measure sheet resistance and contact resistance (a necessary component that adds to the total resistance) are provided, with figures clarifying the concepts. These methods are the Greek cross and the bridge structure to measure sheet resistance, and the Kelvin structure to determine contact resistance. A review of the parasitic components evident in any practical integrated resistor is also provided, to be used in conjunction with design strategies, aimed at minimizing these unwanted intrinsic parameters – a precursor to a review of the noise present in integrated resistors. Through understanding and realizing these principles of integrated resistors and the methods used to measure critical parameters and to understand how the geometry of the component influences its behavior, the layout optimization techniques often used in integrated design containing resistors are presented in this chapter. The techniques/strategies include interdigitated layout topologies, common-centroid topologies and dummy elements, which are all very common techniques to reduce process variation within components and among adjacent, ideally matched components.

A similar approach is followed for integrated capacitors, another commonly used integrated component, which, if correctly designed and optimized, can reduce variations in circuit components and unwanted effects due to intrinsic parasitic components. The physical layout of the integrated capacitor is initially discussed to provide the reader with a clear understanding of its construction and limitations in semiconductor design. The operating regions of the capacitor are also reviewed, as well an effective and accurate method to analyze and quantify a capacitor, which is frequently used in the semiconductor industry at post-processing to qualify a process. These operating regions (accumulation, depletion and inversion) are discussed in detail, with specific reference to the MOS capacitor. Again, with knowledge of the physical layout of an integrated capacitor, knowledge of its operation and how semiconductor manufacturing and processing determine its performance, the regularly used strategies to match capacitors are presented in this chapter. These optimizations are supplied by first introducing the non-ideal capacitor model, which includes an equivalent series resistor, and by determining the quality of the device based on these intrinsic parameters. Deviation sources (changing the parameters from the ideal, designed and desired values) such as matching errors, oxide gradients, edge variation effects, parasitic capacitances and temperature/

voltage dependencies are furthermore reviewed in this chapter. For each source, a practical approach with supporting figures and equations is provided, enabling researchers and designers to fully optimize integrated capacitor design for high-performance circuits.

Finally, a brief introduction to integrated inductors is provided, with specific focus on the layout of typically used inductors. The most significant sources of power losses in these components are listed and a review of the non-ideal equivalent circuit model is presented. The optimization techniques of integrated inductors are not discussed; various principles and techniques used by resistors and capacitors can be applied to these components.

REFERENCES

Allen, P. E., Holberg, D. R. (2002). *CMOS Analog Circuit Design*. New York, Oxford: Oxford University Press.

Bahl, I. J. (2003). *Lumped Elements for RF and Microwave Circuits. Artech House Microwave Library*, Norwood: Artech House, ISBN 1580536611.

Baker, D. W., Herr, E. A. (1964). Parasitic effects in microelectronic circuits. *IEEE Transactions on Electron Devices,* 12(4), 161–167.

Baker, R. J. (2011). *CMOS: Circuit Design, Layout, and Simulation*. Hoboken: Wiley-IEEE Press.

Baomin, W., Guoping, R., Yulong, J., Xinping, Q., Bingzong, L., Ran, L. (2009). Capacitance-voltage characterization of fully silicided gated MOS capacitor. *Journal of Semiconductors*, 30(3), 034002–1–034002-6.

Brederlow, R., Weber, W., Dahl, C., Schmitt-Landsiedel, D., Thewes, R. (2001). Low-frequency noise of integrated polysilicon resistors. *IEEE Transactions on Electron Devices*, 48(6), 1180–1187.

Burghartz, J. N., Rejaei, B. (2003). On the design of RF spiral inductors on silicon. *IEEE Transactions on Electron Devices*, 50(3), 718–729.

Chen, C., Lee, R., Tan, G., Chen, D. C., Lei, P., Yeh, C. (2012). Equivalent sheet resistance of intrinsic noise in sub-100-nm MOSFETs. *IEEE Transactions on Electron Devices*, 59(8), 2215–2220.

Choma, J. (1985). The computation of semiconductor sheet resistance. *IEEE Transactions on Electron Devices,* 32(4), 845–847.

Dutta, P., Horn, P. M. (1981). Low-frequency fluctuations in solids: 1/f noise. *Review of Modern Physics*, 53(3), 497–516.

Enderling, S., Brown, C. L., Smith, S., Dicks, M. H., Stevenson, J. T. M., Mitkova, M., Kozicki, M. N., Walton, A. J. (2006). Sheet resistance measurement of non-standard cleanroom materials using suspended Greek cross test structures. *IEEE Transactions on Semiconductor Manufacturing*, 19(1), 2–9.

Fayed, A., Ismail, M. (2006). *Adaptive Techniques for Mixed Signal System on Chip*. New York: Springer-Verlag US.

Iniewski, K., Balasinski, A., Majkusiak, B., Beck, R. B., Jakubowski, A. (1989). Series resistance in a MOS capacitor with a thin gate oxide. *Solid-State Electronics*, 32(2), 137–140.

Jiang, S., Wu, C., Ho, T. (2012). A nonlinear optimization methodology for resistor matching in analog integrated circuits. *Proceedings of Technical Program on VLSI Design, Automation and Test*, 1–4.

Jindal, R. P. (1985). High frequency noise in fine line NMOS field effect transistors. *1985 International Electron Devices Meeting*, 31, 68–71.

Koutsoyannopoulos, Y. K., Papananos, Y. (2000). Systematic analysis and modeling of integrated inductors and transformers in RF IC design. *IEEE Transactions on Circuits and Systems-II: Analog and Digital Signal Processing,* 47(8), 699–713.

Lattin, W., DeMassa, T. E. (1974). Geometric and temperature compensating effects of the MOS enhanced capacitor. *1974 International Electron Devices Meeting (EDM),* 50–52.

Lipka, B., Zhang, Y., Kleine, U. (2010). Design of integrated matched resistors, capacitors and inductors. *Proceedings of the 17th International Conference on Mixed Design of Integrated Circuits and Systems (MIXDES 2010).* Wroclaw, Poland, 251–254, 24–26 June 2010.

Logan, M. A. (1967). Sheet resistivity measurements on rectangular surfaces -general solution for four point probe conversion factors. *The Bell System Technical Journal,* 46(10), 2277–2322.

McCreary, J. L. (1981). Matching properties, and voltage and temperature dependence of MOS capacitors. *IEEE Journal of Solid-State Circuits.* SC, 16(6), 608–616.

Murji, R., Deen, M. J. (2005). Accurate modeling and parameter extraction for meander-line N-well resistors. *IEEE Transactions on Electron Devices,* 52(7), 1364–1369.

Smits, F. M. (1958). Measurement of sheet resistivities with the four-point probe. *The Bell System Technical Journal,* 37(3), 711–718.

Srinivasan, P., Xiong, W., Zhao, S. (2010). Low-frequency noise in integrated N-WELL resistors. *IEEE Electron Device Letters,* 31(12), 1476–1478.

Vandamme, L. K. J., Casier, H. J. (2004). The 1/f noise versus sheet resistance in poly-Si is similar to poly-SiGe resistors and Au-layers. *Proceedings of the 30th European Solid-State Circuits Conference,* 365–368.

Villar, G., Alarcon, E., Guinjoan, F., Poveda, A. (2003). Optimized design of MOS capacitors in standard CMOS technology and evaluation of their equivalent series resistance for power applications. *Proceedings of the 2003 International Symposium on Circuits and Systems (ISCAS 03),* III–451–III-454.

Voldman, S. H. (2015). *ESD: Analog Circuits and Design.* Hoboken: John Wiley & Sons.

Walton, A. J. (1998). *Microelectronic Test Structures.* Kluwer.

Whitaker, J. C. (1996). *The Electronics Handbook.* CRC Press, ISBN 0849383455.

Yu, T., Jin, C. G., Dong, Y. J., Cao, D., Zhuge, L. J., Wu, X. M. (2013). Temperature dependence of electrical properties for MOS capacitor with HfO2/SiO2 gate dielectric stack. *Materials Science in Semiconductor Processing,* 16(5), 1321–1327.

9 The Evolving and Expanding Synergy Between Moore's Law and the Internet-of-Things

Wynand Lambrechts and Saurabh Sinha

INTRODUCTION

For a device to be classified in the so-called IoT, it must have (at least) three basic properties: a constant data connection, low power usage and the ability to communicate across a network (Lambrechts and Sinha 2016). CMOS semiconductor devices have, without any doubt, been a vital enabler of the IoT, both in the processing of data and in communication among nodes. CMOS devices and interconnects are constantly achieving higher processing speeds, lower power consumption, smaller sizes and larger yields. Technologies such as the FinFET transistor, high-κ dielectrics and SiGe attest to this progress, and higher levels of SoC integration lead to more powerful and more relevant IoT applications. However, with the lower voltages associated with new CMOS generations, the performance of certain analog, digital and RF applications has become degraded through lower noise margins, decreased linearity and reduced output power. III–V materials such as GaAs and GaN still play a dominant role in RF front-ends for many wireless radios used by the IoT. The IoT covers a diverse range of application realms including implementations taking advantage of wireless sensor networks, applications within the machine-to-machine (M2M) communications space, radio-frequency identification, cyber physical systems and mobile computing. The general operational phases of an IoT system consist of first acquiring, processing and handling the data, followed by data storage and, finally, transmitting the data to its required location(s). M2M communication is a testament to how wired and wireless infrastructures have affected and diversified information sharing with respect to these fundamental operational stages. The growth, projection and usefulness of the IoT are largely governed by three laws; albeit commonly a retrospective governance, it is undoubtedly associated with technology miniaturization and efficiency. These three laws are

- Moore's law.
- Metcalfe's law.
- Koomey's law.

Moore's law – the doubling of components on integrated circuits every year (adjusted historically to every 18 months) – has been established and referenced throughout this book. For the IoT, Moore's law affirms its significance by increasing the computational capacity of a constant area, and importantly, lowering the cost of computations, as summarized by the representation in Figure 9.1.

Moore's law, however, needs to hold up in four primary categories that define the IoT, namely

- Sensors that measure environmental conditions such as temperature, light or motion.
- Actuators that display data, provide acoustical outputs or control mechanical devices.
- Computational processors that run programs and perform logic operations.
- Communication interfaces that provide wired or wireless data transfer.

Sensors and actuators, however, do not necessarily follow the same scaling path as computational processors and communication interface circuits. Sensors and actuators not only interface with the environment directly but also serve, in many cases, as the human-to-machine interface. To interact with the physical environment and with humans, making these circuits smaller is not always practical; one can consider touch screens, pressure sensors or image sensors, for example. Defining technology improvements in each of the categories listed above is a difficult task because of the sheer number of potential applications of the

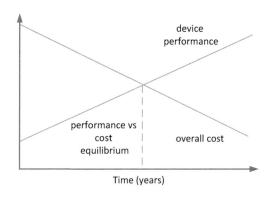

FIGURE 9.1 Simplified relationship between device performance and the overall cost of producing devices – including material cost, processing, packaging and distribution cost – as proposed by Moore's law.

IoT, and the fact that the IoT is essentially only a miniaturized electronic circuit design, borrowing concepts and designs from older technologies. Moore's law has been adapted in recent times, taking into account technology scaling not only as a holistic idea, but also as a means to achieve automated ubiquitous computing. Analysts also suggest that chipmakers are more considerate to emerging markets, but using mature designs that are miniaturized. This view includes moving away from dramatic performance leaps and instead creating incremental enhancements to chips for less complex devices that need a lesser amount of processing power. Such efforts are important in fulfilling real needs and broadening the ways in which computing can influence human lives, as opposed to pushing technology scaling without proper regard for where it is going.

Second, Metcalfe's law, not specific to integrated circuit components or semiconductor processing but relevant to the IoT, states that the *value* of a communications network is proportional to the square of the number of connected users, notwithstanding the fact that not all connections are created equal. The IoT is essentially a network of nodes or devices communicating and sharing information; therefore, Metcalfe's law is a useful figure of merit to quantify the value of such a network. Figure 9.2 represents a simplified relationship between the overall cost of implementing devices (or nodes) and the value they add.

In Figure 9.2, it is noticeable that the value added to a network follows a logarithmic model, as opposed to a linear increase in the overall cost associated with producing an increasing number of interconnected nodes. This logarithmic relationship of the value Θ of a network, as defined by Metcalfe's law, is given by

$$\Theta(n) = \frac{n(n-1)}{2} \tag{9.1}$$

where n is the number of connected nodes. A convenient and commonly used diagram, which illustrates the number of connections that can exist in a network,

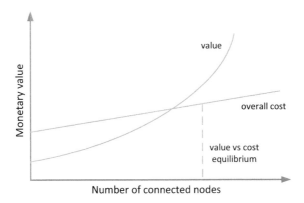

FIGURE 9.2   Simplified relationship between the overall cost of producing devices – including material cost, processing, packaging and distribution cost – and the value that connected nodes add to a network as proposed by Metcalfe's law.

therefore contributing to increasing its value, is adapted from various sources and given in Figure 9.3.

As seen in Figure 9.3, the number of possible interconnections among nodes increases exponentially as the number of nodes increases. Only a single potential connection (in both ways) exists between two nodes; for five nodes, ten potential connections exist. Continuing this trend, and not specifically shown in Figure 9.3, six nodes would result in a possible 15 connections. It has also been acknowledged by Jadoul (2016) that the IoT, specifically, needs to shift from being technology-driven to being value-driven (focusing on applications rather than only on technology advances and scaling). This is defended by Metcalfe's law tailored to network growth, value creation and customer acquisition, as opposed to purely focusing on technology innovation.

Finally, Koomey's law, which is again tailored to the semiconductor and microelectronic industry, states that the energy efficiency of computation doubles roughly every 18 months. In the IoT, Koomey's law is often used to predict the power requirements of computations of low-energy (ideally, energy harvesting from renewable sources) devices or nodes. These governing laws of the IoT and its relationship with increasing

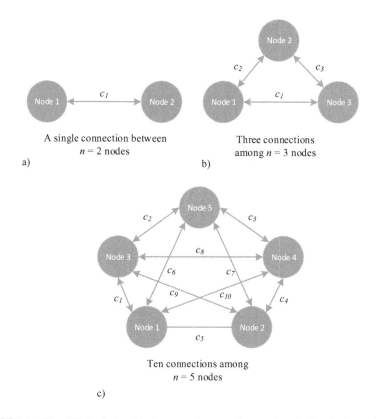

FIGURE 9.3   Simplified relationship between the overall cost of producing devices – including material cost, processing, packaging and distribution cost – and the value that connected nodes add to a network as proposed by Metcalfe's law.

performance at lower cost and smaller footprint, essentially translating to lower power consumption and less expensive implementations, are shown in Figure 9.4.

Figure 9.4 represents the relationship among the governing laws of the IoT and shows the primary drivers of IoT integration and development, namely size, cost, value, performance and power consumption. This chapter further investigates the performance of these drivers as a result of technology miniaturization, advancements in the semiconductor industry and the ways in which information is gathered, stored, analyzed and transmitted in modern networks. An analysis of the commonly used circuit implementations in IoT systems is presented in this chapter.

As with any emerging technology, the IoT is not without its share of challenges. A fast-growing and rapidly evolving industry, with potentially billions of devices, inevitably leads to a drop in standards, proprietary protocols and security. Moving away from the notion of how information is presented to users, traditionally through cloud-based or back-end servers, the IoT enables, and in some ways defines, how information is presented by the interaction of data, devices, applications and users. This is, again, a testament to the laws formulated by Gordon Moore, Jonathan Koomey and Robert Metcalfe. However, technology advances and a form of control over how IoT devices need to work in unison are needed for the IoT to reach its full potential. As the internal processing capacity of pervasive devices increases, the flexibility to innovate, creating diverse, energy-efficient and ubiquitous devices, also increases. A growing IoT market leads to technology innovation and, similarly, technology innovation results in a growing IoT market. For this reason, a comparison between the digital transformation of the IoT and the semiconductor industry, particularly the technologies driven by Moore's law (and inevitably Koomey's and Metcalfe's laws), is drawn in this chapter.

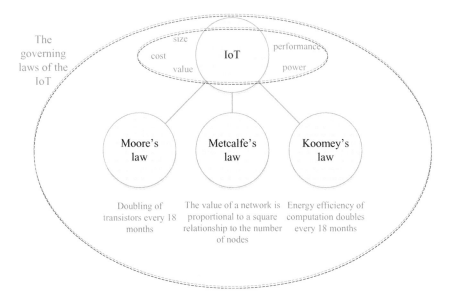

FIGURE 9.4 Relationship among the governing laws of the IoT – Moore's law, Metcalfe's law and Koomey's law.

Since the IoT is a multidisciplinary integration with a multitude of devices, interfaces, protocols, architectures and energy sources combined, the diversity as well as the sheer variety of IoT applications and technology makes it challenging to propose an inclusive and comprehensive assertion for the hardware and software requirements of an IoT system. In a relatively general manner, serving as a good guideline for IoT development, an IoT ecosystem can be defined by a seven-layer reference model. This model contains layers for services, applications, analytics, integration, interconnection, acquisition and market relevance. The seven-layer reference model of an IoT ecosystem is given in Figure 9.5.

Essentially, proposing a hardware and software requirement for an IoT ecosystem, and determining if current technology solutions can satisfy these requirements (or if predictions based on Moore's law envisage short-term realization of these requirements), requires the identification of several key characteristics of the necessities of such a system. These key characteristics include

- Definition of the technical properties of the devices and the applications.
- Hardware and software architectures for the IoT system and the relevant sub-systems.
- Identification of the required EDA tools.
- The necessary connectivity (Samie et al. 2016).

Each of these key characteristics is well defined and thoroughly reviewed in Samie et al. (2016), and certain highlights of each key characteristic are adapted and presented in this section.

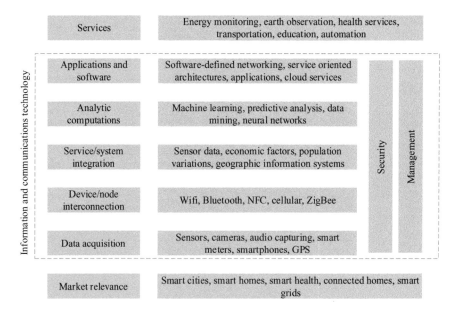

FIGURE 9.5   Seven-layer reference model of an IoT ecosystem.

In defining the physical properties of the devices and the applications, Samie et al. (2016) divide the characteristic into two primary categories, namely application areas and applications versus devices. Application areas determine the impact of the IoT system by either providing new services or by improving existing services, such as enhancements to traditional M2M networks, for instance. Application areas cover a variety of environments where IoT can offer expansion. The most common applications are in healthcare, assisted living, smart buildings, smart homes, smart cities and smart industries. Samie et al. (2016) divide the applications versus devices category into four classifications, depending on the amount of IoT devices and the amount of services offered. These classifications are one-to-one (single device – single application), one-to-many (single device – multiple applications), many-to-one (multiple devices – single application) and many-to-many (multiple devices – multiple applications) services (Samie et al. 2016).

The second key characteristic to be identified is the defining of the hardware and software architectures of the IoT system. The main goal is efficiency at each of the operational stages (acquisition, processing, storage and transmission), where efficiency can be optimized at both the hardware and software level, as described in further detail in Samie et al. (2016). To process large amounts of data, typically required when streaming audio, video or raw data, a relocation of computation and processing to different essential layers in the IoT chain relieves the computational overload during instantaneous high-bandwidth operations. These various computing layers and platforms include device-centric computation, gateway-centric heterogeneity between networks, cloud-centric computing and hybrid approaches (Samie et al. 2016). Computations can also be *approximate*, or *exact*, and each of these adds to the complexity, required resources and, ultimately, the bandwidth limitations and latency of networks.

The increasing complexity of traditionally simplistic SoC solutions and a high demand for designing and implementing embedded devices require sufficient and powerful EDA tools, driven historically by personal computers, mobile computers, smartphones and, more recently, IoT embedded design. The substantial market value of the IoT and future predictions of the industry have led to multiple design houses, focused on designing new application-specific SoC solutions for the IoT. EDA tools are at the center of realizing application-specific designs, often combining abilities to design analog and digital integration, energy harvesting through MEMS devices and wireless or wired connectivity. Designer expertise and the intellectual property of core libraries are rapidly becoming increasingly valuable and essential globally. For these reasons, companies providing EDA tools are increasingly focusing on supplying IoT-specific core libraries and expanding current solutions traditionally tailored to larger corporations.

The final key characteristic, as described in Samie et al. (2016), is the choice of connectivity for an IoT system, based on device and application requirements. Connectivity of IoT systems encompasses not only wireless (or wired, in some circumstances) communication technologies, but also incorporates consideration of the timing of communication (continuous, sporadic or event-driven), and the bandwidth, modulation and data rate of IoT sensors. Several communication protocols are implemented specifically for IoT systems. The most commonly used RF protocols used in the IoT are

- Bluetooth/Bluetooth low energy, operating at 2.4 GHz.
- Proprietary standards such as ZigBee at 2.4 GHz.
- Z-wave communications at approximately 900 MHz, typically used for home automation, HVAC, security systems, home cinema and access control.
- Wifi, also operating at either 2.4 or 5 GHz.
- Cellular RF technology at 900/1800/1900/2100 MHz.
- A drive toward ultra-low-power RF communication protocols.

To scale up wireless connectivity to the level required to drive the IoT requires frequency allocation and management to avoid interference among devices. The typically used industrial, scientific and medical (ISM) radio bands used in IoT systems are summarized in Table 9.1.

As seen in Table 9.1, the ISM frequency bands used for IoT are not specifically designated to the IoT, but are rather a set of unlicensed frequencies used for short-range wireless communication to limit interference from nearby devices specifically because of the large number of IoT devices that could exist in an area. These lower frequencies (315–928 MHz) are most often used for networked devices in an IoT architecture to communicate and transfer data among nodes. Higher RF frequencies are typically used between a central node, responsible for capturing data from a variety of sensors, and the internet. The bandwidth and data rate depend specifically on the IoT application, where processing power requirements are estimated based on the type and regularity of data transmission. Before reviewing the technical specifications associated with various subsystems of a general IoT system, certain economic and technical limitations of the IoT are discussed in the following section.

THE LIMITATIONS OF THE IOT

The early generations of the IoT concentrated on framework solutions that allow communication with and between sensors classified as smart. As the IoT evolved,

---

TABLE 9.1

ISM Frequency Bands used for IoT Systems

IoT ISM Frequency	Typical Applications
315 MHz	Rolling-code (cope-hopping) garage door remote controls
433 MHz (LPD433)	Short-interval hobbyist remote control, short-range voice communications
902–928 MHz	Amateur radio continuous wave or single-sideband communications
863–870 MHz	Short-range device telecommunication
2.4 GHz	Wifi, Bluetooth
5.8 GHz	Amateur fixed-satellite and radiolocation services

newer systems focused on integrations that enabled back-end competences of connected devices through the same channel, the internet. There are, however, various complications when taking such an approach, predominantly a lack of differentiated platforms, outdated or non-existent regulatory statures, vague business models and a lack of economically viable and market-penetrable applications. The challenges of the IoT can be divided into four categories, namely

- Platform.
- Connectivity.
- Business models.
- Applications.

The platform category includes lack of clarity and intuitiveness of the form and design of user interfaces, analytic tools that process large amounts of gathered data and scalability. Connectivity concerns are becoming less prevalent, with many manufacturers of smart devices integrating compatibility with technologies such as Bluetooth and NFC. The two more pressing matters, both from a less technical perspective, are the lack of comprehensive and universal business models and applications able to satisfy the requirements for electronic commerce and consumer markets, which include sub-standard methods of presenting relevant information to improve the quality of life. Apart from concerns about privacy, safety, compatibility, complexity and independence from human intervention (which includes factors such as a reliable and constant internet connection), many factors that decrease the potential and forecast growth of the IoT are at play, but from a purely technological perspective, there are additional challenges and limitations with similar consequences.

One such technical limitation stems from the low-power nature of IoT devices, placing stringent design requirements on the transmitter and receiver of these components. With respect to the receiver, a commonly used figure of merit is the noise figure (NF), which defines the sensitivity of the receiver. The NF is typically defined as

$$NF = P_{min} + 174\text{dBm} / \text{Hz} - 10\log(BW) - MG - SNR \qquad (9.2)$$

where,

P_{min} is the reference sensitivity of the receiver in dBm

BW is the transmission bandwidth in Hz

MG is the design margin in dB and SNR is the signal-to-noise ratio of the receiver for error-free demodulation, in dB (Song et al. 2017)

In addition, the energy per bit (E_{bit}) and complexity of transmitters and receivers should be optimal for low-power applications, including accuracy in crowded frequency spectrums in lower ISM bands. This often leads to the use of external crystals or large antennas and batteries, in cases where energy harvesting cannot produce the required energy quantities. For these reasons, a shift in terms of low-power radio transceivers with simplistic design and comparable performance is required for the IoT (Shehhi et al. 2016). Modern RF receivers have explicit require-ments driving power consumption, a direct function of baseband bandwidth, sensi-tivity, carrier frequency, modulation and interference (Nilsson and Svensson 2014). Before this chapter reviews these requirements, a brief analysis of the differences between current-generation IoT systems and traditional M2M systems is presented. M2M systems paved the way for establishing appropriate protocols and architectures that enable efficient IoT systems, specifically in terms of the requirements of modern RF receivers.

M2M COMMUNICATION AND THE IOT

M2M is the forerunner of the IoT. Although closely related and, in certain circum-stances, seemingly similar, there are fundamental differences between an M2M communications system and the holistic idea of the IoT. In Cao et al. (2016), a con-cise description of the primary differences is presented. The IoT focuses on end-point information presentation and on the connection of physical objects to each other – including to human operators. An M2M system is connectivity-centric, refer-ring principally to automated systems that involve devices capable of automatically collecting data from a remote source, exchanging this information and acting on the immediate environment through a public network infrastructure. A simplified depic-tion of the primary differences between an M2M communication system and the IoT is given in Figure 9.6.

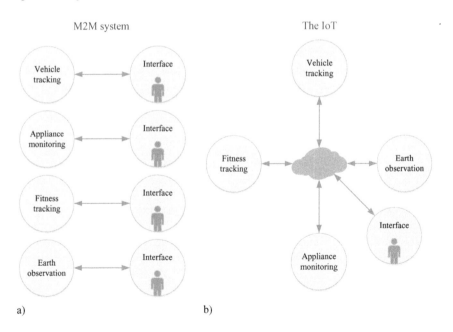

FIGURE 9.6   Simplified depiction of the primary difference between (a) an M2M commu-nication system and (b) the IoT.

As shown in Figure 9.6, the primary difference between an M2M communication system and the IoT is the method through which devices (or machines or *things*) are interconnected to an interface where data are either displayed or processed. In an M2M system, each machine or thing is directly connected to an interface, displaying the accumulated sensor data for a human either to intervene or to adjust the environment automatically based on a set of rules. The IoT connects numerous devices through a cloud-based or central server where the data are relayed to an interface that can be used to monitor or control the environment. A deliverable common to both systems is remote device access and how this is achieved.

Traditional M2M systems depend on embedded hardware with point-to-point communication. This communication can be wired or wireless, depending on the application and other limitations. These solutions offer "remote access to machine data, traditionally directed at point solutions in service management applications" (Polsonetti 2014). These data are rarely integrated with enterprise applications to improve business or data-gathering performance. This is partly why M2M market sustainability and its inability to forecast future growth of this sector has caused the apparent demise of M2M as the leading format of information-gathering infrastructure, especially since access to remote devices provides the fundamental value proposition for networked solutions (Metcalfe's law). In contrast, IoT solutions depend on internet-protocol (IP) networks to interface various device data to a cloud or middleware platform (Polsonetti 2014). Its high potential for enterprise integration, the capacity to accommodate a broader variety of devices and its reliance on software rather than hardware add to the value benefits and visibility of the IoT and its success. Hassel (2015) summarizes the differences between M2M and the IoT, adapted here and presented in Table 9.2.

M2M, although it has been around for many years, has recently been presenting new innovations and opportunities for enterprise, even while the IoT is enjoying

---

TABLE 9.2

Fundamental Differences between M2M Communication Systems and the IoT

M2M Communication Systems	The IoT
Point-to-point infrastructure typically embedded in hardware at the client site	Networked infrastructure using IP networks and varying communication protocols (such as Bluetooth)
Most devices use wired or wireless connections	Data delivery through a middle-layer hosted central service (cloud)
Connections are not bound to an internet connection	Devices typically require a connection to the internet
Limited device integration due to corresponding standard and protocol requirements	Unlimited integration options that require a central management solution

most of the attention. A key contributor to the heralding of M2M and new solutions to deal with high-value assets and logistics management is Moore's law. The price of purchasing sensors has decreased dramatically and low-cost wireless implementations of networks not connected to the IoT are becoming less complex and more cost-effective and are providing dedicated solutions for industries. The communication of M2M systems with the IoT is also becoming more commonplace, as standards and protocols are enabling seamless integration with the internet, at a reduced cost compared with historic implementations. M2M is now able to open up new ways of conducting business, including dynamically priced user-specific solutions that allow new clients and customers to engage with users. The indirect effects of Moore's law not only reduce implementation prices; the computation capabilities of these specific solutions also increase dramatically, enabling the data manipulation and analysis of large amounts of sensor-gathered data. Database management systems such as SAP HANA permit the storage and retrieval of data locally (when not connected to the IoT) and can perform analytics on the data.

For both infrastructures, however, there are several challenges, both economical and technical, that must be overcome as each new generation of devices is presented. These challenges are reviewed in the following section.

REMOTE DEVICE MANAGEMENT CHALLENGES

Since the introduction of the Apple iPhone in 2006, smartphones have been the overwhelming driver of innovation in the technology and miniaturization industry. This technology includes the driving forces for development in cameras, connectivity, batteries, sensors, processors and memory. Chipmakers have been under constant pressure to produce smaller and more powerful components for each generation of smartphones. It is clear, however, with smartphone innovation still at the forefront of technology improvements, that the IoT is heralded as the new industry driving technology miniaturization and developing denser, faster and more cost-effective chips for low-power ubiquitous computing. This hardware and software industry will, in the near future, run on billions of sensors in a multitude of environments and applications. Several manufacturing companies are investing in these smart products, which will be imperative in driving the future of the IoT and the continuation of Moore's law. Investing companies include Texas Instruments, Intel, Qualcomm, ARM, Microchip and Freescale. Figure 9.7 lists the companies driving the IoT through the system-on-chip (SoC) integration of high-performance and low-power components, showing the respective market cap of each company as of June 2017.

As shown in Figure 9.7, the companies that currently have the largest market cap and investments in driving the IoT have market caps ranging from US $3.43 billion (Atmel) upward to US $165.08 billion (Intel). The IoT and other low-power mobile device application industries have created a significant market interest, and many smaller companies are being established to take advantage of the growing potential. Semiconductor manufacturing companies, as well as *fabless* design-oriented companies, already provide integrated solutions and portfolios that enable IoT designs and

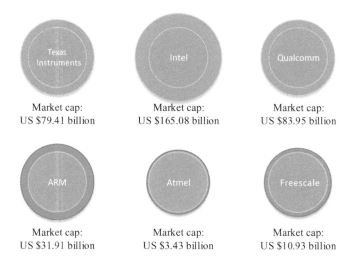

FIGURE 9.7   Semiconductor manufacturing companies and their respective market caps (June 2017) – the driving forces of the IoT. Green circular border color and larger thickness represent higher market cap.

applications in various industries. These solutions include low-power IoT-tailored semiconductor chips optimized for

- Wireless and wired connectivity.
- Scalable and unified processors.
- High-performance and/or low-power microcontrollers.
- Area- and power-efficient sensing devices.
- Power management hardware and software tools.
- Efficient battery management implementations.
- AC and DC controllers and converters.
- LED lighting solutions.
- Low-power amplifiers and data converters.
- Resourceful signal conditioning.
- Signal isolation.

To implement these solutions successfully, especially in a competitive market, there are four fundamental device management requirements for IoT device deployment, as indicated by Weber (2016). These fundamental device management requirements are listed and shown in Figure 9.8.

As described in Weber (2016), the four fundamental device management requirements for an IoT system include

- Provisioning and authentication.
- Configuration and control.
- Monitoring and diagnostics.
- Software updates and maintenance (Weber 2016).

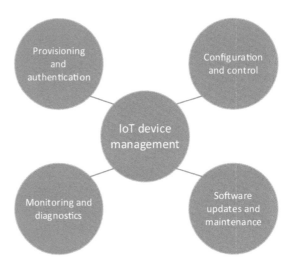

FIGURE 9.8   The four fundamental device management requirements for IoT device deployment. (Adapted from Weber, J., 2016, Fundamentals of IoT device management. Retrieved 9 June 2017 from http://iotdesign.embedded-computing.com.)

Provisioning and authentication enroll a device into a system and are related to device security, securely establishing the identity of the device to ensure that it is a trusted device or service. Configuration and control involve registering parameters and settings such as unique identification numbers between the client software and the device, as well as establishing control to configure, activate or deactivate the device remotely. Monitoring and diagnostics, as the terms suggest, are crucial for the secure and uninterrupted operation of (thousands of) devices in an interconnected architecture, providing capabilities of monitoring network and device statistics and indicating potential security breaches and diagnostics of software or hardware device management. Finally, software maintenance and updates are required if any incorrect behavior or bugs are identified in device operation. Software maintenance of remote devices is a long-term and continuous process and requires persistent connection to the remote device. These connections are not always reliable, especially if the devices are moving or are in remote areas, whereas in such cases, periodically scheduled software maintenance is performed.

As seen from the above discussion on IoT device management, wired or wireless connections to the *things* are crucial from the deployment phase through to maintenance and repair. The sheer number of connected devices in an IoT topology requires efficient, effective and low-complexity configuration, control, monitoring and maintenance. The universal limitations of Moore's law, as presented in Chapter 1 of this book, are all relevant in the IoT industry, and it remains important to understand how these limitations affect the growth of the IoT. Significant obstacles to device scaling have been identified in Chapter 1, ranging from

- Challenges in manufacturability.
- Reduced reliability.

- Reduced yield.
- Limitations to power supply and threshold voltage.
- Effective thermal management.
- Scaling performance.

All these obstacles can be compared with the fundamental device requirements for an IoT system required to manage and use these architectures to their full potential. The principal characteristics of technologies that must be overcome to realize a fully automated (capable of energy harvesting) and ubiquitous topology of an IoT share the same caveat compared with earlier technology adoption, evident in technologies such as personal computing, notebooks, smartphones and tablets. The IoT will also share similar challenges. These characteristics of miniaturization limitations are given in Figure 9.9.

As seen from Figure 9.9, there are three major challenges that not only slow down Moore's law based on traditional computing systems but also affect the fabrication of the IoT and M2M facilities, owing to the strong drive toward device miniaturization. In Figure 9.9, photolithography refers to the fidelity of patterns transferred from radiation in the imaging system, as well as the physical properties of the photoresist. Minimum feature sizes are directly proportional to the wavelength of the light incident on the photoresist. As feature size decreases, it becomes more difficult to reliably produce features with minimal defects. In Figure 9.9, interconnects between devices and components are also becoming shorter and thinner as the device technology size decreases, and interconnect density also increases with increased circuit complexity. As more electrical currents are now converted to heat, extracting dissipated heat in these small form factors is also becoming increasingly difficult to achieve. IoT devices are rapidly becoming powerful enough to dissipate high amounts of energy that need to be extracted from the system to improve device reliability.

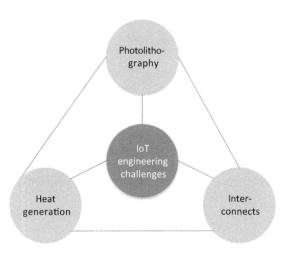

FIGURE 9.9 Major challenges that slow down Moore's law and presents limitations in miniaturization.

Heat dissipation during resource-intensive tasks can be an effective technique to avoid thermal damage, typically caused by thermal runaway. However, researchers and designers are increasingly interested in improving the core efficiency of IoT devices, not only to limit heat generation but also to extend the lifetime of devices powered by either batteries or renewable sources. The following section reviews IoT core energy efficiency.

IOT CORE ENERGY EFFICIENCY

Energy leakage, especially in IoT systems and applications, is a primary design consideration and constraint, since these devices are designed to operate using the minimum amount of available energy, often using energy harvesting to ensure constant operation in remote and difficult-to-reach environments. The total power consumed in an IoT system is the sum of the power consumed by all sub-components at a specific time, and depends on several parameters, conditions and design factors (both in hardware and software). These parameters include the

- Process technology (scaling) node.
- Architecture of the processing core.
- Power modes of the peripherals.
- Voltage and frequency requirements.
- Temperature of the processing core and certain peripherals.
- Complexity and number of instructions that are carried out (Henkel et al. 2016).

The most effective method to reduce energy consumption of a task or instruction in a microprocessor core is to lower the operating voltage, although there are several limitations to this approach. To improve efficiency, and therefore reduce energy leakage in a processing core, clock frequency scaling should be considered, since increasing the clock frequency reduces the time to complete a task, and vice versa, depending on the instantaneous requirements of the circuit or system (Pinckney et al. 2012). IoT processing cores can be equipped with dynamic voltage and frequency scaling (DVFS) capabilities to enhance their efficiency and dynamically scale their performance based on immediate requirements (Henkel et al. 2017). For a core to support a specific operating frequency and DVFS, its supply voltage must be adjusted above a predetermined minimum value. A higher frequency requirement translates to a higher minimum voltage; therefore, at a constant operating voltage, a core should be operated below a specified maximum frequency. This relationship occurs since the amount of energy stored in the tank circuit of an oscillator, E_{stored}, typically a combination of inductance, resistance and capacitance, is given by

$$E_{\text{stored}} = \frac{CV_{pk}^2}{2} \tag{9.3}$$

where,

$\quad C \quad$ is the capacitance of the tank circuit
$\quad V_{pk} \quad$ is the peak voltage across this capacitor

To achieve a specific oscillating frequency proportional to an input voltage, a voltage controlled oscillator (VCO) is commonly used in various applications, including low-power IoT devices requiring frequency control. The instantaneous frequency $f(t)$ of a VCO is also defined by a linear relationship with its instantaneous control voltage, such that

$$f(t) = f_0 + K_0 v_{in}(t)$$ (9.4)

where,

f_0 is the quiescent frequency of the oscillator
K_0 is the gain of the oscillator, specified in Hz/V
$v_{in}(t)$ is the instantaneous voltage applied to the VCO

The relationship between the supply voltage and the core frequency for stable execution, therefore f_{stable}, is given in Henkel et al. (2016) as

$$f_{stable} = k \frac{(V_{dd} - V_{th})^2}{V_{dd}}$$ (9.5)

where,

k is a fitting parameter depending on the oscillator architecture/topology
V_{dd} is the supply voltage
V_{th} is the threshold voltage for a given CMOS technology

For a specified supply voltage, operating a device below its stable frequency will produce a stable output; however, the device efficiency will be lowered. For a device with an operating voltage equal to the threshold voltage, the stable frequency in (9.5) is zero, therefore the circuit cannot be stable at any given frequency. The device should, therefore, as with most CMOS circuits, be biased above the threshold voltage of the process. The normalized stable frequency as a function of supply (biasing) voltage with a constant threshold voltage is modeled in Figure 9.10.

As seen in Figure 9.10, the normalized stable frequency given in (9.5) is zero if the bias voltage is equal to the threshold voltage. With increasing bias voltage, the normalized stable frequency increases linearly, and is dependent on the maximum process technology supply voltage (5 V in this example).

Low-power IoT systems require low-power subsystems to prolong the lifetime of the device and decrease the footprint required for energy excitation. Voltage references are commonly used in low-power systems, devices producing a constant voltage irrespective of the load, temperature variations or power supply perturbations. In IoT systems, voltage references are the most common circuits used as power supplies. Analog-to-digital converters that convert an analog signal from a sensor to a digital representation thereof and the converse, digital-to-analog converters as well as various subsystems responsible for measurement and control, are

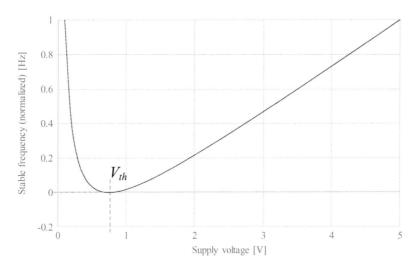

FIGURE 9.10   The normalized stable frequency as a function of supply (biasing) voltage with a constant threshold voltage.

key building blocks of IoT systems. Some of the key requirements for a voltage reference are to

- Produce an output that is voltage- and temperature-independent.
- Generate an output that is power-supply-independent.
- Provide operation across a wide range of supply voltages.
- Have the ability to easily scale the output voltage (Sanborn et al. 2007).

Although band gap voltage references are among the most common types of voltage references, and are representative of the characteristics listed above, their standby power consumption can be relatively high in terms of IoT devices and they are used less often in IoT systems, especially when relying on energy harvesting. An on-chip, and therefore not externally realized using discrete components but integrated in the IC, voltage reference is an essential circuit in IoT and low-cost SoC applications. It provides the biasing current and compensation in output voltage resulting from changes in process parameters, supply voltage and temperature (Maddikatla and Jandhyala 2016); these factors are typically associated with remote and ubiquitous devices using energy harvesting and operating in harsh environments. In addition, traditional band gap voltage references usually require a supply voltage that is greater than 1.2 V (Sanborn et al. 2007), a specification that was not a limiting factor in previous-generation technologies. Even in the proposed work by Wang et al. (2015), which utilizes the silicon band gap narrowing effects of bipolar transistors, the voltage reference still requires an operating voltage of 1.3 V. In the work presented in Sanborn et al. (2007), the circuit utilizes an operating voltage of 1 V by using a reverse band gap voltage principle.

To grasp the principle operation and design of the band gap voltage reference, the basic principle of operation of a CMOS band gap voltage reference, adapted from Vittoz and Neyroud (1979), is given in Figure 9.11.

In Figure 9.11, the CMOS band gap voltage reference output voltage is denoted by V_{REF}. The band gap/diode potential is acquired over the base-emitter junction of a bipolar transistor, T_S, denoted by V_J. This transistor is accessible in most CMOS technologies by implementing an n-type substrate as the collector, a p-type well as the base and a highly-doped n-type diffusion region used as the emitter. Transistor T_S is biased at its emitter by a current source, I_E, which has a relatively relaxed requirement with respect to its accuracy, typically requiring only $\pm 10\%$ precision (Vittoz and Neyroud 1979). A voltage that is proportionate to absolute temperature ($PTAT$) is added to V_J to compensate exactly for the variance between V_J and the gate voltage of T_S. $PTAT$ supplies a voltage V_2, and is compared with the voltage V_1 through a differential amplifier A; the value of V_1 is

$$V_1 = \frac{R_2}{R_1 + R_2} V_{REF} - V_J \tag{9.6}$$

where R_1 and R_2 is a resistive voltage divider circuit at V_{REF} (Vittoz and Neyroud 1979). The voltage output of A, denoted by V_A, drives the gate of a p-channel regulating transistor, T_R, and the loop is in equilibrium, therefore supplying a constant voltage reference, at

$$V_{REF} = \left(1 + \frac{R_1}{R_2}\right) + \left(V_J + V_2\right) \tag{9.7}$$

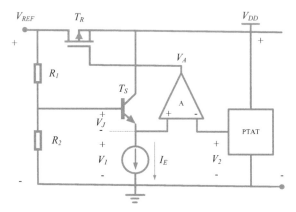

FIGURE 9.11 Simplified block diagram explaining the principle of the CMOS bandgap voltage reference. (Adapted from Vittoz, E. A. and Neyroud, O., *IEEE Journal of Solid-State Circuits*, SC-14(3), 573–577, 1979).

which is independent of temperature if

$$V_{G0} = V_J + V_2 \tag{9.8}$$

where V_{G0} is the gate voltage of T_R. V_{REF} can be adjusted to practically any value above V_{G0} by varying the ratio of R_1/R_2 (Vittoz and Neyroud 1979). In the work presented by Maddikatla and Jandhyala (2016), an integrated temperature sensor combined with a CMOS voltage reference biased in the sub-threshold region, specifically for use in IoT applications, is presented. The circuit has a precision of 65 ppm/°C over a 3σ deviation in process and ±10% discrepancy in the supply, within a temperature range of −40°C to +100°C. The offered work also attains a precise *PTAT* voltage and the supply sensitivity of the output voltage is 4000 ppm/V, consuming approximately 12 µW with an output voltage of 266 mV (Maddikatla and Jandhyala 2016). The device is operated by a 1.8 V supply voltage.

As an alternative to band gap voltage references, the subthreshold voltage references are low standby power consumption voltage references, in the picowatt range, and are thus more reliable and suitable for IoT systems (Lee et al. 2017), although their output voltage is relatively low compared with band gap voltage references. This low power consumption is achieved by subtracting threshold voltages, V_{th}, of stacked NMOS and PMOS transistors. In Lee et al. (2017), a subthreshold voltage reference, where the reference voltage, V_{REF}, is similar to that of a band gap voltage reference (approximately 1.2 V for silicon), is proposed. This is achieved by increasing the number of stacked PMOS transistors, and a prototype is implemented in a standard 180 nm CMOS process. The subthreshold voltage reference operates on the principle of biasing a MOSFET in its subthreshold region, therefore by applying a gate-source voltage, V_{GS}, smaller than V_{th} of the transistor. In this region, the drain current through the transistor is small, but is a non-zero value, and exhibits an exponential dependence on V_{GS}. The drain current I_D in the subthreshold region is described by

$$I_D = I_0 \exp \frac{V_{GS}}{\alpha V_T} \tag{9.9}$$

where α is a non-ideality factor, typically $\alpha > 1$, and

$$V_T = \frac{kT}{q} \tag{9.10}$$

where,

    k is Boltzmann's constant
    T is the temperature in kelvin
    q is the elementary electron charge

The equation for I_D can be rewritten such that (assuming a constant drain-source voltage, V_{DS})

$$V_{GS} = \alpha V_T \ln\left(\frac{I_D}{I_0}\right).$$

(9.11)

The subthreshold current is an exponential function of the gate-source potential, albeit very small (in the nA range). The subthreshold current is also dependent on variations in process parameters and temperature fluctuations. Process variations are primarily categorized into two groupings: within-die or intra-die variations caused by mismatches among transistors within a chip, and die-to-die, or inter-die variations caused by mismatches of components among samples. The process dependence of the subthreshold current is determined by

$$\frac{\Delta I_D}{I_D} = \frac{\Delta\mu}{\mu} - \frac{\Delta V_{TH}}{\eta V_T}$$

(9.12)

where the variation in the mobility ($\Delta\mu$) is typically less than the variation in the threshold voltage (ΔV_{TH}), and the subthreshold current variation primarily depends on ΔV_{TH}. In Cheng and Wu (2005), a subthreshold voltage reference using a peaking current mirror is proposed, adapted and presented to clarify the fundamental operation of the subthreshold voltage reference. The proposed circuit in Cheng and Wu (2005) is presented in Figure 9.12.

The subthreshold voltage reference in Figure 9.12 operates with all transistors biased in the subthreshold region. The bias current, I_B, is found by following the closed-loop derivation,

$$I_B = \frac{V_{GS2} - V_{GS1}}{R_2}$$

(9.13)

where,

V_{GS1} and V_{GS2} are the gate-source voltages of transistors M_1 and M_2, respectively
R_2 is the biasing resistor as shown in Figure 9.11

The bias current can also be given by

$$I_B = \frac{\eta V_T \ln\left(\frac{K_1}{K_2}\right)}{R_2}$$

(9.14)

where K_1 and K_2 are the geometry characteristics (W/L aspect ratios) of M_1 and M_2, respectively. The output voltage of the circuit, therefore the reference voltage V_{REF}, can be described by

$$V_{REF} = V_{GS2} + I_B R_1$$

(9.15)

and can consequently be rewritten such that

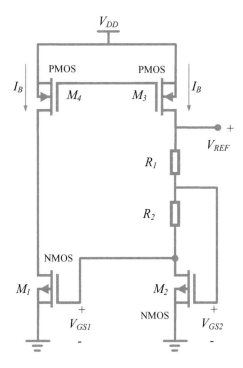

FIGURE 9.12 A subthreshold voltage reference circuit using a peaking current mirror, as proposed. (From Cheng, M. H. and Wu, Z. W., *Electronics Letters*, 41(10), 572–573, 2005 and Frey, D. R., *IEEE Transactions on Circuits and Systems – I: Regular Papers*, 60(12), 3267–3278, 2013.)

$$V_{\mathrm{REF}} = V_{GS2} + \frac{R_1}{R_2} \eta V_T \ln\left(\frac{K_1}{K_2}\right) \qquad (9.16)$$

and since V_{GS} has a negative temperature coefficient and V_T has a positive temperature coefficient, the reference voltage can be adjusted to have an ideally zero temperature coefficient by adjusting the ratio of the resistors R_1 and R_2.

To its disadvantage, the proposed circuit uses resistors to achieve a temperature-independent reference voltage, a passive component that increases the required die area on-chip and adds thermal noise in electronic circuits. Circuit topologies and architectures using resistor-less designs are also used in various applications and implement MOSFET transistors at the output of the internal current mirrors, typically operated in the strong inversion region where they behave similarly to a resistor. Nilsson and Svensson (2014) provide a systematic study of power bounds for RF systems and offer a synopsis of bandwidth and sensitivity requirements in low-power systems and their correlation with power consumption. The bandwidth and sensitivity of an electronic circuit are typically associated with the receiver architecture, efficiency, modulation and topology. In the IoT, specific receiver technologies are preferred, primarily because of the low-power operational requirements. A review of energy-efficient IoT receivers is presented in the following section.

ENERGY-EFFICIENT IOT RECEIVERS

The full networked topology of IoT systems ideally comprises simplistic nodes able to collect and transmit data to a central controller or gateway that provides connectivity to the internet. In the case of M2M systems, the performance-limited nodes transmit data to individual computing centers that are able to receive, analyze and display the gathered data. Both the nodes and the gateways must be designed to minimize power consumption, provide reliable and robust network connections and extend wireless connectivity as far as possible (Char 2015). The core of an IoT system is its microcontroller unit, a device that processes data, runs software stacks and interfaces with connectivity protocols of the system. The specifications of the microcontroller unit are application-specific, and several off-the-shelf alternatives are available based on the system specifications. Apart from the industry leaders in IC design and manufacturing, many smaller companies are investing in designing and producing their own application-specific integrated circuits (ASICs) for the purposes of their IoT requirements. Power efficiency and physical area are both dependent on a multitude of factors, a topic that is more appropriate in a dedicated mixed-signal design optimization guide, but not applicable to the scope of this book.

Supporting subsystems, such as sensors and communication circuits, do, however, implement a range of fundamental energy-efficient techniques, where the transceiver (transmitter and receiver) architecture specifically allows for a wide variety of optimization. This section focuses specifically on the receiver fundamentals used in IoT systems, and identifies and reviews the most common implementations for low-power and low-noise receivers. According to Nilsson and Svensson (2014), the three most effective low-power transmission systems used in IoT systems are the

- Tuned RF receiver, where a prior low-noise amplifier (LNA) can either be implemented or omitted, depending on the application requirements.
- Superheterodyne receiver.
- Super-regenerative receivers, which have recently elicited renewed interest.

The basic principles and motivations for why these low-power receivers are used in IoT systems, with renewed interest in recent years, are described in the following paragraphs. The first receiver that is reviewed is the Tuned RF Receiver.

TUNED RF RECEIVER ARCHITECTURE

Tuned RF (TRF) receivers are known to have the simplest architecture, comprising an antenna, an LNA and an envelope detector (Moazzeni et al. 2012; Otis et al. 2004). An incoming signal is received by the antenna, amplified and analyzed to source the data from the signal. A simplified block diagram of the TRF receiver is adapted from Nilsson and Svensson (2014) and presented in Figure 9.13.

The simplified TRF receiver, as shown in Figure 9.13, consists of a resonant antenna that receives and converts the incoming signal to a usable voltage, followed by a bandpass filter, B_{RF}, to remove any unwanted frequency components outside

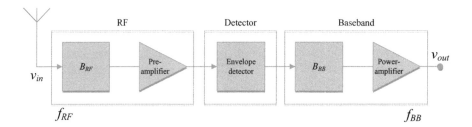

FIGURE 9.13 Block diagram of a traditional TRF receiver architecture. (Adapted from Nilsson, E. and Svensson, C., *IEEE Journal on Emerging and Selected Topics in Circuits and Systems*, 4(3), 273–283, 2014.)

the desired spectrum, defined within the RF frequency, f_{RF}. The incoming signal is amplified by a pre-amplifier before detection, and this step has a significant influence on the power budget of the circuit as a result of RF amplification. The envelope detector then transforms the signal from its modulated state to baseband, and the simplified schematic representation of a CMOS implementation of the envelope detector is presented in Figure 9.14, adapted from Nilsson and Svensson (2011).

The CMOS implementation of an envelope detector circuit, as shown in Figure 9.14, uses transistor M_1, biased in the subthreshold region, and M_2, a constant bias current source, to charge the output capacitor C_{out} through the output current, i_o. The detection of the RF signal is achieved by the non-linear transfer function from a voltage to a current in M_1. The conversion gain from a peak RF input to a baseband output potential is defined in Nilsson and Svensson (2014) by

$$G_D = \frac{V_{BB}}{V_{RF}} \tag{9.17}$$

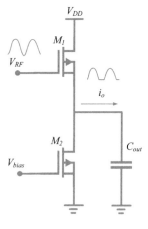

FIGURE 9.14 A typical CMOS envelope detector circuit used in TRF receivers. (Aadapted from Nilsson, E. and Svensson, C., *Proceedings of the 20th European Conference on Circuit Theory Design*, 773–776, 2011.)

where,

V_{BB} is the output potential of the detector

V_{RF} is the peak RF input voltage, where this expression can be rewritten such that

$$G_D = \frac{V_{RF}}{4V_0} \tag{9.18}$$

where V_0 is the effective output voltage, defined in Nilsson and Svensson (2011). Harmonics formed by the detector are filtered by the baseband filter, B_{BB}, and the signal is amplified, at baseband, and presented at the output of the TRF receiver. Partly because of its simplicity, and since the TRF receiver does not have a local oscillator (LO), the power consumption of the TRF receiver is lower compared with heterodyne or superheterodyne receivers (Moazzeni et al. 2012). The filtering and channel selection of TRF are more complex and typically lead to degraded performance compared with superheterodyne receivers, since they have to be done at RF frequencies. This also increases the complexity of these sub-circuits, and modern technology, where high-GHz receivers are commonplace, has led to TRF receivers being replaced by heterodyne and superheterodyne receivers that perform most tasks at an intermediate frequency (IF). In addition, to their detriment, TRF receivers are incompatible with frequency modulated signals and have low sensitivity owing to the use of envelope detection circuits with relatively small conversion gain (Moazzeni et al. 2012). Envelope detection is traditionally used with amplitude modulation schemes; although more power-efficient, these schemes are spectrally inferior to frequency modulated schemes, especially with respect to the bandwidth requirements of modern electronic circuits. Amplification of the signal prior to envelope detection can proliferate the sensitivity of the TRF receiver, although it has a sizeable influence on power consumption, as gain at RF frequencies is resource-intensive owing to the inherent trade-offs between sensitivity and power consumption.

Introduced in 1920, and an improvement on the TRF receiver, is the superheterodyne receiver, implemented using a local oscillator to produce the sum and difference of frequency components, where detection takes place at lower frequencies (baseband). This receiver is discussed in the following section.

THE SUPERHETERODYNE RF RECEIVER ARCHITECTURE

The superheterodyne (supersonic heterodyne) RF receiver is the most widely used radio receiver architecture, and is commonly used for low-power and low-noise receivers used in IoT applications. IoT applications typically have low bandwidth (low data-rate) requirements for the transmission of data, presuming that audio or video is not transmitted, and therefore place less stringent performance specifications on these receivers. This complements the low-noise and high-sensitivity requirements of such systems. The superheterodyne receiver can be tuned to different RF frequencies through the LO used in the system. The receiver is consequently more sensitive to instabilities in the LO frequency; nevertheless, there are techniques to mitigate its effects with more tolerant IF stages. A simplified representation of the building blocks of a superheterodyne RF receiver architecture is adapted from Neu (2015) and Nilsson and Svensson (2014) and presented in Figure 9.15.

The block diagram of the traditional heterodyne RF receiver architecture shows the relevant subsystems for wireless communication (excluding the transmission protocols of over-the-air (OTA) transmission). These subsystems consist of bandpass filters (BPFs), an LNA, a mixer and LO, an IF amplifier and an analog-to-digital anti-aliasing filter (AAF). The superheterodyne receiver is different from the TRF receiver, since it has a down-converting stage that decreases the frequency of the signal before detection. As a result, the superheterodyne receiver offers

- Higher sensitivity.
- Superior channel filtering, as it is easier to filter out unwanted signals at the lower IF frequency compared with the RF frequency.
- Lower power requirements to amplify at the IF as opposed to RF.
- Lower cost of components and lower complexity of filtering circuits.

To its detriment, the LO requires constant power and therefore increases the active and passive power consumption of the receiver; the power consumption is also proportional to the operating frequency. High-quality factor and efficient local oscillators are among the primary design considerations in extremely low-power superheterodyne receivers. Two types of LO architectures that can be used in superheterodyne receivers, also compatible with IoT receivers, depending on the frequency, are resistor-capacitor (RC) oscillators and inductor-capacitor (LC) oscillators. The RC and LC oscillators are reviewed in Nilsson and Svensson (2014), and are adapted and briefly presented in Figure 9.16.

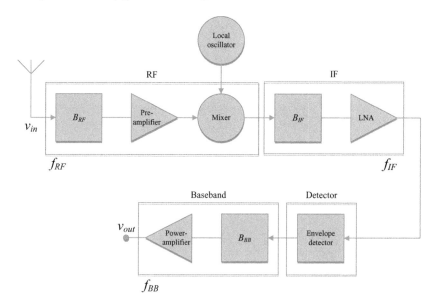

FIGURE 9.15 Block diagram of a traditional heterodyne RF receiver architecture. (Adapted from Neu, T., Direct RF conversion: From vision to reality. *Texas Instruments*, 2015.)

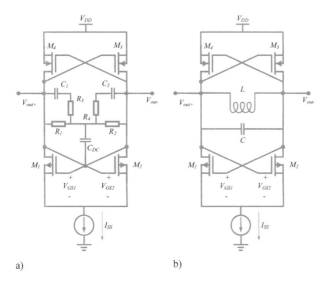

a)   b)

FIGURE 9.16   The (a) RC oscillator and (b) LC oscillator typically used in superheterodyne receiver circuits.

In the RC oscillator in Figure 9.16a, the oscillating frequency is determined by varying/shifting the phase of the signal in the loop that provides feedback to the circuit. Setting the capacitances C_1 and C_2 as equal and at their minimum requirement, the resistor values can be obtained with respect to the LO angular frequency ω_{LO} such that

$$R_1 = R_2 = R_3 = R_4 = \frac{1}{\omega_{LO}C_{min}}$$ (9.19)

where R_1, R_2, R_3 and R_4 are shown in Figure 9.16a and

$$C_{min} = C_1 = C_2.$$ (9.20)

The required transconductance (g_m) to ensure oscillation is determined by

$$g_m = \frac{3}{R_1} = 3\omega_{LO}C_{min}$$ (9.21)

and is delivered by two complementary transistors with equal bias current (Nilsson and Svensson 2014). As for the LC oscillator in Figure 9.16b, a load resistor with value equal to

$$R_L = \frac{Q}{\omega_{LO}C_{min}}$$ (9.22)

is typically connected to the output of the circuit, where Q is the quality factor of the tank circuit and C is the capacitance of the capacitor among the output stages. The required transconductance for oscillation to occur in an LC oscillator is determined by

$$g_m = \frac{\omega_{LO} C}{Q} \qquad (9.23)$$

and is also delivered to the circuit by two complementary transistors, as shown in Figure 9.16b (Nilsson and Svensson 2014). As seen in these derivations, the required transconductance of the RC oscillator, therefore the required bias current through the transistors, is a function of the resistive elements in the circuits. Although the RC oscillator is more cost-effective to integrate on-chip in the absence of large-area inductors, and because of the inverse proportionality to the resistance, these circuits are lossy and imply higher power consumption, a lower quality factor and, therefore, increased phase noise when compared with the LC oscillator (Nilsson and Svensson 2014). The LC oscillator can yield superior performance, especially in view of its lower power consumption, if off-chip inductors are used with a higher quality factor and larger attainable inductance. An off-chip inductor, however, leads to increased integration complexity and a larger size, but adds a degree of modularity by enabling interchangeable inductors. The power budget of the superheterodyne receiver can be further enhanced by adding a preceding LNA circuit, which also improves the sensitivity of the receiver as reviewed and analyzed by Nilsson and Svensson (2014).

To reduce the power budget of IoT receivers further, the super-regenerative amplifier (SRA) receiver, invented in 1922 by Edwin H. Armstrong (Armstrong 1922), has high gain and sensitivity and consumes minimal DC power (Lee and Mercier 2017). A review of the super-regenerative receiver is given in the following section.

SUPER-REGENERATIVE RECEIVERS

The modern demand for cost-effective wireless communication operating with low-power requirements, driven essentially by the IoT, has reinvigorated SRA-based receivers for reasons similar to why it was sought after in the early 1900s – the capacity to produce "high gain and high sensitivity at low power" (Lee and Mercier 2017). These systems are typically used as high-gain RF amplifiers implementing modulation schemes such as on-off keying (OOK). SRA receivers can also be used in receivers that demodulate frequency shift keying or phase shift keying signals (Lee and Mercier 2017). In recent low-power technologies, this has been an important characteristic of SRA receivers, as it provides higher-order modulation schemes. The time-varying and often non-linear behavior of these systems allows a large degree of approximation and intuition when designing SRA receivers (Frey 2013), making their implementation relatively complex, although recent developments in applying classical mathematical techniques to model and predict their behavior completely have alleviated these challenges.

The SRA receiver is essentially an oscillator, intermittently switched on and off amid a high-gain startup phase and a steady-state oscillatory phase (Lee and Mercier 2017), owing to its lower power consumption if compared with superheterodyne

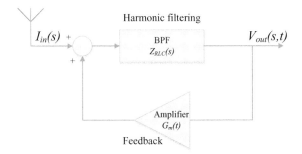

$I_{in}(s)$ +

Harmonic filtering

BPF
$Z_{RLC}(s)$

$V_{out}(s,t)$

Amplifier
$G_m(t)$

Feedback

FIGURE 9.17   Simplified block diagram of an SRA receiver. (Adapted from Frey, D. R., *IEEE Transactions on Circuits and Systems – I: Regular Papers*, 60(12), 3267–3278, 2013 and Thotla, V., et al., *IEEE Transactions on Instrumentation and Measurement*, 62(11), 3006–3014.)

receivers. The SRA receiver can be described by a feedback network model with gain and a bandpass filter, as shown in Figure 9.17.

As shown in Figure 9.17, the SRA receiver consists of a fixed bandpass filter with a positive feedback applied via a time-varying network (Frey 2013) that acts as a quench oscillator. Modulation leakages from the quench oscillator result in unintentional emanations and the harmonics within these received emissions govern the quench of the SRA receiver (Thotla et al. 2013). The quench tones need to be filtered from the data signal to circumvent corruption of the baseband data and, therefore, the oversampling ratio originates from the complexity of the filter. The quench frequency should ideally be above the limit of human hearing, typically approximately 25 kHz, but otherwise as low as possible. Excessively high quench frequencies reduce the maximum sensitivity of the receiver. The quench frequency must, however, be higher than two times the highest frequency component of the baseband signal, following the Nyquist criterion for modulation and demodulation. The rate at which the oscillator is switched off and on is henceforth called the quench frequency. To its advantage, and a motivation for the SRA receiver gaining popularity, is its ability to combine blocks of this architecture with small intrinsic gain, leading to a significant combined total gain and high selectivity. A simplified circuit model of the SRA receiver is presented in Figure 9.18, adapted from Bohorquez et al. (2009).

As shown in Figure 9.18, the parameters of interest of the SRA receiver are derived from the internal RLC components. These parameters include the resonant frequency ω_0, the characteristic impedance Z_0, the quality factor Q_0, and the quiescent damping factor γ_0. The relationships among these parameters, with respect to the discrete components in Figure 9.18, are given by

$$\omega_0 = \frac{1}{\sqrt{LC}} \qquad (9.24)$$

$$Z_0 = \sqrt{\frac{L}{C}} = \omega_0 L = \frac{1}{\omega_0 C} \qquad (9.25)$$

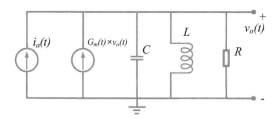

FIGURE 9.18 Simplified circuit model of an SRA receiver (Adapted from Bohorquez et al., 2009.)

$$\gamma_0 = \frac{1}{2RC\omega_0} = \frac{1}{2Q_0} = \frac{Z_0}{2R} \tag{9.26}$$

where,
  R is the impedance of the resistor
  L is the inductance of the inductor
  C is the capacitance of the capacitor in Figure 9.18

The total impedance of the parallel RLC resonant circuit is defined as the Laplace transformation

$$Z_{RLC} = \frac{Z_0\omega_0 s}{s^2 + 2\gamma_0\omega_0 s + \omega_0^2} \tag{9.27}$$

and the instantaneous damping factor, also referred to as the damping function of the system, is defined as

$$\gamma(t) = \gamma_0\left(1 - G_m(t)R\right) \tag{9.28}$$

where $G_m(t)$ is the "positive feedback transconductance and can be modeled as a negative resistance that only affects the damping factor of the second-order system" (Bohorquez et al. 2009). The gain of the SRA receiver is presented by Otis and Rabaey (2007), and briefly revised in this section. In the logarithmic mode of operation (non-linear), the startup time of the oscillator, defined as the time from when the oscillator is enabled until it reaches its saturation voltage, V_{osc}, can be derived as

$$\tau_{\text{rise, thermal}} = \tau_{\text{rise}} \ln\left[\frac{V_{osc}}{\sqrt{v_n^2}}\right] \tag{9.29}$$

where,
  $\tau_{\text{rise,thermal}}$ is the thermal noise signal of the oscillator tank device that causes the initial oscillation
  v_n is the noise voltage source in the tank circuit

The relevance of this is that the oscillator saturation is reached at a different time than its startup time. If the RF signal is much larger than the noise voltage, the thermal rise time of the oscillator can be written as

$$\tau_{rise,\ thermal} = \tau_{rise} \ln \left[\frac{V_{osc}}{V_{RF}} \right] \tag{9.30}$$

where V_{RF} is the RF oscillator signal. Furthermore, to "extract the modulated data from the RF signal, a running average of the oscillator envelope is taken" (Otis and Rabaey 2007), yielding the demodulated output voltage, assuming a high signal-to-noise ratio, derived such that

$$V_{BB} = V_{osc} \left[Q_{dc} - \tau f_q \ln \left(\frac{V_{osc}}{2V_{RF}} \right) \right] \tag{9.31}$$

where,

Q_{dc} is the quench waveform duty cycle
τ is the oscillator startup time constant
f_q is the quench frequency

The gain, G, of the SRA receiver can therefore be determined by

$$G = \frac{V_{BB}}{V_{RF}} \tag{9.32}$$

which results in

$$G = \frac{V_{osc}}{V_{RF}} \left[Q_{dc} - \tau f_q \ln \left(\frac{V_{osc}}{2V_{RF}} \right) \right] \tag{9.33}$$

and is nearly a linear function of the amplitude of the input signal (Otis and Rabaey 2007). The SRA receiver can therefore be demonstrated as a linear amplifier trailed by a logarithmic detector, as further described and analyzed by Otis and Rabaey (2007). The SRA is tuned with a bandpass filter, as depicted in Figure 9.17. This bandpass filter effectively sets the free-running oscillation frequency, and through positive feedback, the effective quality factor is also increased, leading to gain enhancement and increased bandwidth as a result of the improved Q. This is called the regenerative mode of operation and it is prone to unwanted oscillations, but periodically allowing oscillation effectively improves the robustness of the system, called super-regenerative mode. According to Otis and Rabaey (2007), the bandwidth of the SRA receiver is also derived from the basic principles that define an oscillator bandwidth, being

$$f_{3dB} = \frac{f_0}{Q_{tank}} \tag{9.34}$$

where,

f_0 is the resonant frequency of the oscillator

Q_{tank} is the tank circuit quality factor

Recent interest and advances in SRA receivers have provided high-quality and low-power receivers specifically designed for the IoT. Shehhi and Sandalunea (2016), for example, presented a K-band (24 GHz) SRA logarithmic-mode receiver, using OOK modulation and demodulation and achieving a peak energy efficiency of 200 pJ/bit at 4 Mb/s at a bit-error rate (BER) of 10^{-6}. Its peak and average power consumption are 800 and 8 μW, respectively, and the sensitivity of the proposed receiver is −60 dBm at the same BER and data rate. The receiver was realized in a standard 65 nm CMOS process and occupies an area of 740×670 μm^2, which includes the VCO and LNA on-chip. This is a testament to the performance abilities of SRA receivers used in low-power systems such as the IoT, where, traditionally, bandwidth-limited applications are replaced by low-cost and small-area implementations with much improved bandwidth capabilities.

The following section concludes the reviews and discussions presented in this chapter.

CONCLUSION

One of the main considerations in IoT systems is the power consumption and energy efficiency of the integrated solution. Typical IoT applications are operated either by batteries or by energy-harvesting circuits, and are expected to operate autonomously for extended periods of time, often exceeding 20 years. The microcontroller unit at the heart of the IoT system can be optimally designed with ASIC design strategies, and switched to low-power or standby modes if not used. The dynamic power consumption, applicable when the device is operating at its rated capacity, is a function of the operating frequency and the topologies used in each subsystem, a value that can be designed for, adjusted and optimized, and more importantly, estimated and measured. Static power consumption, however, although measurable, accounts for energy draining of the power source, which does not contribute to the application and should be minimized or, ideally, eliminated. However, with a decrease in device size, as driven by Moore's law, and an inherent increase in leakage currents coupled with an increasing number of active devices on a single integrated chip, IoT systems require complex and considerable amounts of design, simulation and measurement. This chapter aims to highlight the portion of the IoT system that can benefit from identifying the primary culprits typically associated with high power consumption in receivers, which are an integral component in any IoT system. This chapter also reviews the main considerations of IoT core efficiency, without necessarily simply reducing the operating voltage of the core. This method has become more difficult, with operating voltages reaching inherent and inevitable limitations as they encroach on the threshold voltage of transistors.

REFERENCES

Armstrong, E. H. (1922). Some recent developments of regenerative circuits. *Proceedings of the Institute of Radio Engineers,* 10(4), 244–260.

Bohorquez, J. L., Chandrakasan, A. P., Dawson, J. L. (2009). Frequency-domain analysis of super-regenerative amplifiers. *IEEE Transactions on Microwave Theory and Techniques*, 57(12), 2882–2894.

Cao, Y., Jiang, T., Han, Z. (2016). A survey of emerging M2M systems: Context, task, and objective. *IEEE Internet of Things Journal*, 3(6), 1246–1258.

Char, K. (2015). Internet of things system design with integrated wireless MCUs. Retrieved 15 June 2017 from http://semanticscholar.com

Cheng, M. H., Wu, Z. W. (2005). Low-power low-voltage reference using peaking current mirror circuit. *Electronics Letters*, 41(10), 572–573.

Frey, D. R. (2013). Improved super-regenerative receiver theory. *IEEE Transactions on Circuits and Systems – I: Regular Papers*, 60(12), 3267–3278.

Hassel, M. (2015). IoT and M2M, What's the difference? Retrieved 17 June 2017 from http://www.incognito.com.

Henkel, J., Pagani, S., Amrouch, H., Bauer, L., Samie, F. (2017). Ultra-low power and dependability for IoT devices. *2017 Design, Automation and Test in Europe Conference and Exhibition (DATE)*. 954–959, Lausanne, Switzerland, 27–31 March 2017.

Jadoul, M. (2016). About Moore's law, Metcalfe's law and the IoT. Retrieved 8 June 2017 from http://blog.networks.nokia.com

Lambrechts, J. W., Sinha, S. (2016). *Microsensing Networks for Sustainable Cities.* Switzerland: Springer International Publishing.

Lee, D., Mercier, P. P. (2017). Noise analysis of phase-demodulating receivers employing super-regenerative amplification. *IEEE Transactions on Microwave Theory and Techniques*, PP(99), 1–13.

Lee, I., Sylvester, D., Blaauw, D. (2017). A subthreshold voltage reference with scalable output voltage for low-power IoT systems. *IEEE Journal of Solid-State Circuits*, 52(5), 1443–1449.

Maddikatla, S. K., Jandhyala, S. (2016). An accurate all CMOS bandgap reference voltage with integrated temperature sensor for IoT applications. *IEEE Computer Society Annual Symposium on VLSI*, 524–528.

Moazzeni, S., Cowan, G. E. R., Sawan, M. (2012). A comprehensive study on the power-sensitivity trade-off in TRF receivers. *10th IEEE International NEWCAS Conference*, 401–404.

Neu, T. (2015). Direct RF conversion: From vision to reality. Retrieved 5 June 2017 from http://www.ti.com.

Nilsson, E., Svensson, C. (2011). Envelope detector sensitivity and blocking characteristics. *Proceedings of the 20th European Conference on Circuit Theory Design,* 773–776.

Nilsson, E., Svensson, C. (2014). Power consumption of integrated low-power receivers. *IEEE Journal on Emerging and Selected Topics in Circuits and Systems*, 4(3), 273–283.

Otis, B., Chee, Y., Lu, R., Pletcher, N., Rabaey, J. (2004). An ultra-low power MEMS-based two-channel transceiver for wireless sensor networks. *Proceedings of the VLSI Circuits Symposium*, 20–23.

Otis, B., Rabaey, J. (2007). *Ultra-low Power Wireless Technologies for Sensor Networks.* Springer Science and Business Media.

Pinckney, N., Sewell, K., Dreslinski, G., Fick, D., Mudge, T., Sylvester, D., Blaauw, D. (2012). Assessing the performance limits of parallelized near-threshold computing. *Design Automation Conference (DAC)*, 1143–1148, 2012.

Polsonetti, C. (2014). Know the difference between IoT and M2M. Retrieved 8 June 2017 from http://www.automationworld.com

Samie, F., Bauer, L., Henkel, J. (2016). IoT technologies for embedded computing: A survey. *2016 International Conference on Hardware/Software Codesign and System Synthesis (CODES+ISSS)*, 2–7 October 2016.

Sanborn, K., Ma, D., Ivanov, V. (2007). A sub-1-V low-noise bandgap voltage reference. *IEEE Journal of Solid-State Circuits*, 42(11), 2466–2481.

Shehhi, B. A., Gadhafi, R., Sanduleanu, M. (2016). Ultra-low-power, small footprint, 120 GHz, CMOS radio transmitter with on-chip antenna for Internet of Things (IoT). *5th International Conference on Electronic Devices, Systems and Applications (ICEDSA)*, 1–4.

Song, Z., Liu, X., Zhao, X., Liu, Q., Jin, Z., Chi, B. (2017). A low-power NB-IoT transceiver with digital-polar transmitter in 180-nm CMOS. *IEEE Transactions on Circuits and Systems I: Regular Papers*, PP(99), 1–13.

Thotla, V., Tayeb, M., Ghasr, A., Zawodniok, M. J., Jagannathan, S., Agarwal, S. (2013). Detection of super-regenerative receivers using Hurst parameter. *IEEE Transactions on Instrumentation and Measurement*, 62(11), 3006–3014.

Vittoz, E. A., Neyroud, O. (1979). A low-voltage CMOS bandgap reference. *IEEE Journal of Solid-State Circuits*. SC-14(3), 573–577.

Wang, B., Law, M., Bermak, A. (2015). A precision CMOS voltage reference exploiting silicon bandgap narrowing effect. *IEEE Transactions on Electron Devices*, 62(7), 2128–2135.

Weber, J. (2016). Fundamentals of IoT device management. Retrieved 9 June 2017 from http://iotdesign.embedded-computing.com

10 Case Studies
Technology Innovations Driven by Moore's Law

Wynand Lambrechts and Saurabh Sinha

INTRODUCTION

The amount of computing power that can be harnessed from fairly small devices has become a remarkable accomplishment in recent years, compared with what has been achieved historically – not necessarily referring to centuries ago but, more fittingly, decades and, in certain circumstances, years ago, depending on the figure of merit. Currently, the most relevant is the processing power of mobile smartphones when compared with desktop computing of a few years ago. At the heart of this progress is still Moore's law of 1965, albeit indirectly, but a clear driving force for semiconductor manufacturers to reduce the size of their components such that more components can fit on a given chip area. Notwithstanding the improvements made to these devices in terms of manufacturing techniques to increase yield and performance and reduce cost, the reason for indicating that Moore's law is indirectly responsible for computing progress is that it is a law of economics, not of physics – an informal opinion formed around the seeming evolution of what was, in 1965, an early chip industry. It is (was) only a prediction, and not a set of rules, that dictates how and when improvements in manufacturing should take place, and there is no guide book that prescribes how to overcome challenges and keep meeting this predicted law. It only serves as a caveat to conventional thinking, a 'poster-on-the-wall' that highlights the wish list of manufacturers. It has, however, inspired great change and innovation in the computing industry, and looking back in history, many advances have contributed to sustaining, or surpassing, Moore's law.

Technology at the time of writing is one of the largest industries in the world, and the internet and its accompanied software and services accounted for 25% of the net profit margin of all industries in 2016, as reported by Forbes (Chen 2016). Among the largest industries in the world are

- Retail and food.
- Alcohol/tobacco.
- Oil and gas.
- Pharmaceuticals.
- Telecommunications.

Importantly, it should be noted that all these industries are in some way driven, controlled, maintained and standardized by technology. Intellectual property forms the basis of these industries, but information is processed in almost all forms through technology, making it an integral part of all these industries.

There are many institutions worldwide that design, manufacture, research or consult on semiconductor devices. Because of their integral role in the everyday lives of humans, their vast market reach, their prevalence and popularity and the profitable future of the technology, these devices are driving industries to adapt to take advantage of their growth. In terms of the scope and focus of this book, the industry leaders of semiconductor research, specifically those who are into reducing the size of transistors, are discussed in this chapter. These industry leaders are either producing state-of-the-art microelectronic components, or are at the brink of introducing advances that aim to steer Moore's law through the next decade, or in some instances, achieving both.

In Figure 10.1, the technologies and companies that are implementing these technologies are summarized. The list in Figure 10.1 is arranged by the physical gate length achievements of each case study, from smallest to largest in a top-down approach. This chapter reviews each of these cases in a bottom-up approach, therefore leading up to the smallest attainable gate length to date (2018).

From Figure 10.1, starting from the bottom, the commercially available graphics processing unit (GPU) from Nvidia, utilizing the 12 nm FinFET technology, is briefly discussed in the following paragraph. This processing technology encompasses a mature technology that is implemented in the latest generation (announced in 2017) of Nvidia commercial products. A record-setting 21 billion transistors are implemented in the GPU chips, with a noteworthy improvement in core efficiency.

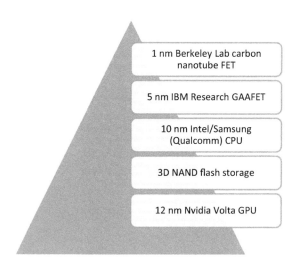

FIGURE 10.1 The most prominent transistor technologies and innovations that are currently (2017) either under development and being researched or used in high-volume commercial manufacturing. The technologies are *stacked* from the smallest gate length (top) and discussed in this chapter from the bottom up.

The second topic to be discussed, with reference to Figure 10.1, is that of 3D NAND storage, used by various manufacturers, including Samsung and Western Digital. 3D NAND storage is used as a non-volatile storage mechanism in solid-state hard drives (SSDs), a momentous improvement over traditional mechanical hard disks. Flash storage, albeit much faster than conventional magnetic disks, is still lagging behind volatile RAM, but is implementing high-density transistor technology that utilizes 3D stacking of transistors to increase overall density, without compromising the performance (in terms of speed) of storage.

A recent implementation of state-of-the-art transistor technology is that of the 10 nm gate length implementation used by Intel and Qualcomm in their latest central processing unit (CPU) generations. The 10 nm transistor technology, using various advances in gate structuring methodology and self-aligned patterning, allows for the use of gate lengths that are currently considered the smallest in the commercial market. The section delves into the current generation of processors offered to the mobile as well as the desktop computing market and reviews the improvements that allow for the mass-market use of this 10 nm technology.

As for technologies that are not yet fully commercially mature, smaller transistor gate lengths of 5 nm and below (down to 1 nm) are discussed in the final sections of this chapter. The 5 nm gate length, as proposed by IBM Research, a gate-all-around FinFET implementation with III–V materials resulting in extremely low off-state current, has been researched, prototyped and published in various journals and publications in recent years (Bu 2017). Although still a fair amount of time away from commercial integration, the 5 nm transistor technology shows the promise of being the technology to ensure that Moore's law continuation is assured in the foreseeable future. Finally, and currently in prototyping, another technology that aims to warrant the future of Moore's law is the 1 nm carbon nanotube gate length proposed by Berkeley Lab. The 1 nm gate length transistor uses transition metal dichalcogenide materials to realize a monolayer gate channel that can be grown without being limited by current photolithography restrictions.

The following paragraph briefly reviews the 12 nm FinFET technology implemented by current-generation GPUs offered by the manufacturer, Nvidia.

NVIDIA VOLTA 12 NM GRAPHICS PROCESSING UNITS

In 2017, at the annual GPU Technology Conference, the chief executive officer of Nvidia, Jen-Hsun Huang, stated in his keynote address that "we need to find a path forward for life after Moore's law" (Nvidia 2017). Nvidia is one of the largest corporations manufacturing GPUs, and is continuously aware and confronted by limitations that could derail Moore's law; however, the company maintains its capacity to produce processing units that meet and, in certain circumstances, surpass Moore's law.

Its latest offering, announced in 2017 and planned to be mass-produced in 2018, is the Volta line of GPUs, a processing unit based on a 12 nm FinFET technology node, housing 21 billion transistors in its main processing core on a die size of 815 mm^2 (Nvidia 2017). Its predecessor, the Pascal line of GPUs, was built on a 14 nm technology and contained 15 billion transistors on a die size of 610 mm^2 on its primary processing core. Huang additionally said, during this keynote address, that the Volta

GPU is at the limits of photolithography, therefore acknowledging that this process step (photolithography) is the primary challenge to adhering to Moore's law. The Volta line of GPUs has a redesigned microprocessor architecture with respect to the Pascal line, and is able to operate 50% more efficiently than its predecessor. In addition, these GPUs implement high bandwidth memory (HBM) for their video RAM (VRAM), as opposed to the traditional, albeit less costly, double data rate (DDR) memory – currently type 5, GDDR5 (Nvidia 2017). HBM uses vertically stacked dynamic RAM (DRAM) memory chips interconnected by through-silicon vias, shortening the path between individual memory chips, effectively reducing power consumption, reducing the required area and allowing for higher bandwidth at lower clock speeds.

With respect to storage, flash or long-term data storage, transistors have become a much more intricate means to store data, especially when compared with traditional magnetic spinning disk technology. Their density and operating speed are crucial to their becoming a viable replacement for older technologies. 3D NAND storage is used in modern hard drives and is reviewed in the following section.

3D NAND Storage

Flash memory, as used in SSDs, smartphones and secure digital cards, has a series of floating gate transistors, assigned to a value of either *on* (1) or *off* (0). These memory blocks are then arranged in a 2D layout, and as technology improves and transistors scale to smaller nodes, more memory blocks can be placed on a single die, increasing the memory capacity of these devices. Ideally, if technology exactly followed Moore's law, the number of transistors, and hence the memory capacity, would double every 18 months to two years. However, as the number of memory cells on a die increases, physical problems start to arise. Transistors used for memory applications demonstrate the same issues as high-speed processors, where if the walls among the cells become smaller, electrons leak out and a memory cell can become volatile. NAND flash, as opposed to DRAM, does not need to be refreshed to retain its charge state, and is able to retain its charge state even when no power is connected to the component – assuming that very little to no leakage current occurs from these devices.

A single bit per cell describes a single-level cell (SLC) flash storage mechanism, where each cell only contains a single bit, but allows for higher speed and long-term endurance. Cells can also be divided up into four levels of charge, for two bits per cell, known as a multi-level cell (MLC), and in many cases to eight levels of charge, allowing for three bits of data per cell, referred to as a triple-level cell (TLC). MLCs and TLCs consequently increase storage density but decrease the read and write speed and durability, since there are multiple levels of charge that must be differentiated. SSDs, implemented using NAND flash storage, are not as fast as volatile DRAM memory, but can operate at speeds far higher than conventional mechanical spinning hard drives (HDDs). A summary of the quantity of bits for each data cell for SLC, MLC and TLC is given in Figure 10.2.

As shown in Figure 10.2, the quantity of bits for each data cell increases from one, to two, to three, for SLC, MLC and TLC, respectively. As a result, the number of discrete voltage levels in each implementation also increases by 2^n, where n represents

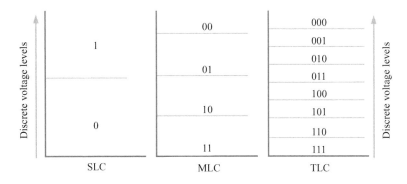

FIGURE 10.2   The number of bits per data cell for SLC, MLC and TLC.

the number of bits per data cell. This results in higher storage density as n increases, but typically leads to slower performance (latency). The typical access latencies for storage mediums are summarized in Table 10.1, adapted from Hruska (2017).

As seen in Table 10.1, the program/erase (P/E) cycles for SLC, MLC and TLC storage decrease significantly as the number of data bits per cell increases, a metric that is not applicable to conventional hard drives or RAM. Seek latency is a metric specific to HDDs, defined as the time required for the mechanical head (needle) to move from its current position to the position where the applicable data are located. Also evident from Table 10.1 is the read, write and erase latency that increases as the quantity of bits for each cell of NAND flash memory increases, albeit much lower compared with conventional HDDs, but far from the performance reached by RAM (Hruska 2017).

One of the practical restrictions of SSDs (NAND flash storage) is that, although these devices can read and write data rapidly to a blank drive, the overwriting of data takes significantly more time. SSDs read and write data at individual rows within the NAND memory grid (page level), supposing that the adjacent cells are vacant. However, this storage medium can simply delete data at the block level, since the erase operation requires a higher voltage compared with reading or writing.

TABLE 10.1
Typical Access Latency for Storage Mediums

	SLC	MLC	TLC	HDD	RAM
P/E cycles	100 k	10 k	5 k	–	–
Bits per cell	1	2	3	–	–
Seek latency (µs)	–	–	–	9000	-
Read latency (µs)	25	50	100	2000–7000	0.04–0.1
Write latency (µs)	250	900	1500	2000–7000	0.04–0.1
Erase latency	1500	3000	5000	–	–

Source: Adapted from Hruska, J., 2017, How do SSDs work? Retrieved 14 July 2017 from http://www. extremetech.com.

Theoretically, NAND flash memory can be erased at page level, but this operation at the higher voltage places additional stress on the individual cells. "For an SSD to update an existing page, it must copy to the contents of the entire block into memory, erase the block, and then write the contents of the old block as well as the updated page" (Hruska 2017). If the drive is at full capacity, it is required that the SSD first scans for blocks that are demarcated for erasure, erases these blocks and then writes the data to this page – which significantly increases the latency, as summarized in Table 10.1.

NAND flash memory is rapidly evolving from 2D to 3D structures, primarily since 2D NAND flash has reached various physical limitations in terms of size and density. The non-volatile flash memory market is largely occupied by NAND flash memory owing to its high integration density, low production cost per bit and high reliability. In order to overcome the scaling limitations of 2D planar cell NAND memory structures, multiple types of 3D NAND flash memory have been introduced from approximately 2006 (Kim et al. 2017) with a double-stacked planar cell structure. Architectures such as the terabit cell array transistor (TCAT), for example, increase chip density by increasing the number of stacking gate structures, instead of the conventional shrinking of unit cells, as Moore's law proposes. Since the start of mass production of 3D NAND flash with TCAT structures, the generations of TCAT have evolved to offer from 24 up to 64 stacked planar cells, with TLC applied to the second generation of TCAT flash memories. With each generation of 3D NAND technology, for the past four generations, the bit areal density has approximately doubled for each generation (Kim et al. 2017). When compared with 2D NAND flash, 3D NAND flash also has more reliable data retention characteristics (Mizoguchi 2017).

The integral component in NAND memory is the floating gate (FG) transistor (Spinelli et al. 2017), in which a polysilicon area is inserted between a gate (also referred to as the tunnel) oxide and an inter-poly dielectric that is attached to a control gate (Spinelli et al. 2017). The cross-sectional view of a floating gate transistor, if compared with a traditional MOSFET, is shown in Figure 10.3.

As seen in Figure 10.3b, the floating gate of the FG transistor is flanked by the gate/tunnel oxide and the inter-poly dielectric that is attached to the control gate. The storage of a charge is achieved within this floating gate. If a charge is stored in the FG, elementary electrostatics indicate that the V_t, the threshold voltage of the cell, befits

$$V_t = V_{t0} - \frac{Q}{C_{pp}} \tag{10.1}$$

where,

V_t is the threshold voltage
V_{t0} is the neutral V_t that corresponds to zero charge on the FG
Q is the charge on the device
C_{pp} is the inter-poly dielectric capacitance

Therefore, the threshold voltage changes depending on the amount of FG charge. Electrons can be injected or ejected into and out of the FG through the tunnel oxide by applying an electric field across it. Through its intrinsic characteristics, whereby

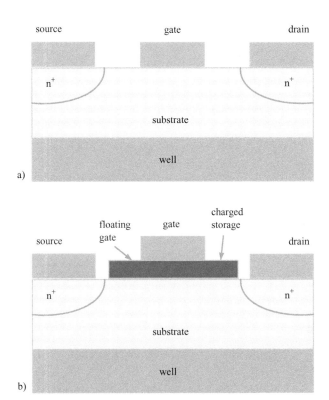

FIGURE 10.3 Simplified cross-sectional view of a (a) a traditional MOSFET transistor and (b) a floating gate transistor.

the charge cannot simply (ideally) leak from the isolated FG, the non-volatile memory characteristic is realized. In a typical (planar) NAND memory array structure, FG transistors are interconnected in series with a sequence, and its control gates are connected in an altered series sequence. This institutes the word lines (WLs), each forming a page of the memory cell. Sequences are connected to bit lines (BLs) and the mutual source line through certain transistors powered by a drain select line (DSL) or a source select line (SSL), depending on the architecture. Such an array creates a block and its size is a vital characteristic of NAND array unification. The simplified schematic representation of a NAND memory array structure is given in Figure 10.4, adapted from Spinelli et al. (2017).

The simplified schematic illustration of a NAND flash memory array, as depicted in Figure 10.4, shows the architecture with respect to the individual bit lines, drain or source select lines, WLs and the common source line. The number of WLs depends on the technology and architecture of the NAND cell. The top view of such an array is depicted in Figure 10.5 – with reference to the single transistor cross-section in Figure 10.3.

Figure 10.5 illustrates the arrangement of a flash memory array, a structure that can become extremely compact based on the chosen technology node. Given a feature size, F, of a specific technology/process, the elementary memory cell, as shown

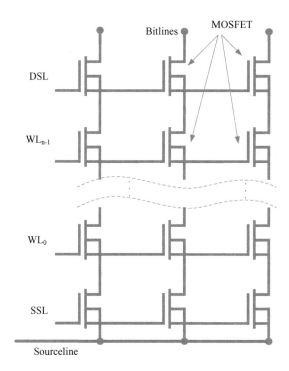

FIGURE 10.4 Simplified schematic illustration of a NAND flash memory array. (Amended from Spinelli, A. S., et al., *Computers*, 6(2), 16, 2017.)

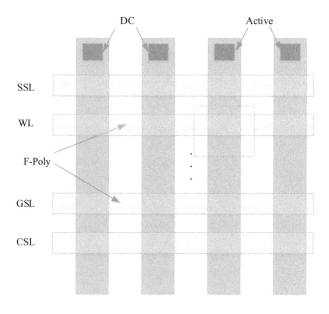

FIGURE 10.5 The top view of a NAND flash memory array depicting the individual WLs of the cell.

in Figure 10.5, uses an area on the substrate of, ideally, $4F^2$. Conversely, the actual cell area, therefore "the total chip area divided by the number of cells", is typically slightly bigger due to overhead supporting circuitry, as well as required components on the die such as string contacts, layout-optimization dummy cells and control transistors. The array operation of NAND memory structures is well-documented and described in papers such as that of Spinelli et al. (2017), and is not discussed in this chapter as it falls outside the scope of these case studies.

Substantial challenges are presented by 3D NAND flash memory if it is to continue not only surviving, but also flourishing in the non-volatile memory market. One such challenge is the increasing aspect ratio with increasing stacks of planar cells, which increases the manufacturing complexity of these devices, especially in maintaining such a large aspect ratio during etching of the channel holes. These high-aspect ratios also increase the mechanical stress within the device and complicate uniform dielectric layer deposition (Kim et al. 2017).

Samsung has also made various architectural changes to planar NAND memory structures and uses 3D NAND technology (termed V-NAND) in its current generation of non-volatile flash storage. V-NAND effectively triples the number of pages and the block sizes of traditional planar memory. Table 10.2 is a summary of the key differences between planar NAND memory and Samsung's V-NAND technology, based on 3D NAND flash.

As shown in Table 10.2, the second-generation V-NAND flash memory storage of Samsung improves considerably on its traditional planar storage. Noticeable from Table 10.2 is that real-world performance does not increase substantially (random read time from 52 to 35 μs and data transfer rates from

---

TABLE 10.2

Key Features of Samsung's Planar Flash Memory versus its Second-Generation 3D NAND Implementation, Termed V-NAND (V2)

Feature	21 nm Planar NAND	V-NAND (V2)
V_{DD} (V)	3.3	3.3
Page size within a block (bytes)	8 K + 640	16 K + 1536
# of pages per block	128	384
Block size (bytes)	(1 M + 80 K)	4 M + 384 K
# of planes	4	2
Page program per page (bytes)	8 K + 640	2 × (16 K + 1536)
Page program time (ms)	2	0.39
Page read per page (bytes)	8 K + 640	16 K + 1536
Random read time (μs)	52	35
Data transfer rate (Mbps)	400	667
Block erase time (ms)	5	4

Source: Adapted from Samsung, 2013, White Paper: Samsung V-NAND technology. Retrieved 11 July 2017 from http://www.samsung.com.

400 to 667 Mbps – approximately a 60% increase), but the storage density in V-NAND technology shows a substantial increase (block size from 1 MB + 80 kB to 4 MB to 384 kB – approximately a 400% increase).

In the determination to shrink cells to increase the density in 2D NAND flash, processing and optimizations become challenging and sometimes impossible, especially for UV light to breach a mask to transfer a geometric shape or pattern to the light-sensitive photoresist. The decrease in light output power limits the patterning procedure and can completely prevent the process from taking place. Samsung's 3D V-NAND design stacks cells vertically, leading to a broader gap among cells, which overcomes some of these patterning restrictions and limitations. The cell-to-cell dimensions in a planar process are normally around 15 nm, whereas with its V-NAND flash technology, this space increases to between 30 and 40 nm. Continually shrinking planar 2D NAND technology can also lead to interference among cells, which can lead to the corruption of data (Samsung 2013). Below a cell dimension of approximately 20 nm, the possibility of interference increases dramatically, making these cells unreliable. In Samsung's V-NAND technology, this issue is moderated by implementing charge trap flash (CTF) technology, resulting in a structure that is virtually free of interference between cells. CTF is based on a non-conductive film of silicon nitride that briefly captures and traps electrical charges to preserve cell reliability. This film is altered into a 3D arrangement to confine the control gate of the specific cell, operating as a padding that holds charges and prevents data corruption caused by interference between cells (Samsung 2013). Furthermore, from a programming perspective, planar NAND memory needs to isolate the sequencer by the least and most significant bits, therefore requiring two operations. V-NAND flash memory is capable of dual-page programming, enabling the memory to sequence the bits all at one time, leading to improved performance in terms of bandwidth usage.

INTEL CORPORATION 10 NM PROCESS TECHNOLOGY

Intel Corporation is determined to keep implementing Moore's law as a figure of merit for current and future-generation processors. Since the introduction of the Core central processing unit (CPU) range in 2006, many critics have viewed Intel's progress in architecture and speed gains as lacking, claiming that improvements have been incremental. Intel is committed to addressing these claims, and at the Intel Technology and Manufacturing Day 2017 in San Francisco, Intel presented leading figures that covered a range of areas, from Moore's law to how the company achieved its 10 nm manufacturing process, and how it believes that it is not just meeting Moore's law, but exceeding it. Intel did also address the fact that, in modern times, the time it takes to move from one node to another is taking longer and is much more expensive. Companies such as Intel must therefore recover the costs and/or ensure that the yield is high enough to merit the additional cost. Also, in order to meet Moore's law, companies should ensure that, if it takes a longer time to move to smaller nodes, there is a significant increase in transistor density to account for the longer time. Intel claims that it overcame both these factors, at least for now, by using hyper-scaling manufacturing in its 10 nm process, effectively increasing the number of passes during patterning to increase the yield, albeit at a higher tooling cost. Hyper scaling is a form of multiple patterning derived from the self-aligned double patterning technology used in 14 nm processing that allows features smaller than the limits incurred by the 193 nm wavelength lithography, without yet moving to EUV. Intel, however, claims that the yield improvements, leading to lower cost per transistor, are enough to cancel out the higher tooling costs, and eventually lead to lower cost when compared with its predecessor.

As a result, Intel claims that its hyper-scaling technology – in this case, the self-aligned quad patterning – allows for a transistor density of 100 million transistors per square millimeter. Hyper scaling in 10 nm technology offers improvements (a reduction) in the pitch among transistor fins, the pitch among metal conductors, cell height and transistor gate pitch, and it reduces the requirement for a double dummy gate to a single dummy gate, which further increases area scaling. From Intel's Technology and Manufacturing Day 2017 brochure, the reductions are specified as

- Fin pitch reduction from 42 to 34 nm, a 19% reduction in size,
- Minimum metal layer pitch reduction from 52 to 36 nm, a 31% reduction
- Cell height reduced from 399 to 272 nm, a 32% reduction in height
- Transistor gate pitch reduced from 70 to 54 nm, a 22% reduction.

Furthermore, Intel introduced the contact over active gate (COAG) feature, where contact to the gate can be placed on top of the transistor, leading to a further 10% reduction in area scaling. A representation of the COAG technique is presented in Figure 10.6.

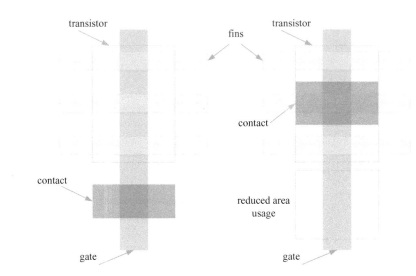

FIGURE 10.6 Intel's COAG feature to reduce the area usage in 10 nm nodes by approximately 10%, adapted from the Intel Technology and Manufacturing Day 2017 brochure. (Available at Intel, 2017, Intel Technology and Manufacturing Day 2017. Retrieved 11 July 2017 from http://www.newsroom.intel.com.)

All these features combined effectively nearly triple the transistor density when compared with its predecessor, the 14 nm process used in products such as the Kaby Lake processors, which reached 37.5 million transistors per square millimeter. Figure 10.7 is a representation of Intel's achievements with respect to transistors per square millimeter for its last five process manufacturing generations.

As shown in Figure 10.7, the progress of transistor density since the introduction of the 45 nm node approximately ten years ago has seen significant increases, following the predictions of Moore's law. The increases are not only attributed to a decrease in node size, but are also a result of processing advances, such as the hyper-scaling technology introduced in the 10 nm node. Figure 10.7, however, shows that even though technology size reduction is slowing down, especially noticeable from the time it took to move from 14 to 10 nm, Moore's law is still being observed with respect to transistor density, with a 2.7 times increase in density by moving from the 14 nm to the 10 nm node. Intel additionally claims that the 10 nm transistors have the capacity for greater speed and superior energy efficiency compared with their predecessors. The higher than 2× transistor density increase, although taking more than two years, effectively means that Intel is still meeting (and in a sense exceeding) the traditional Moore's law scaling. Adapted from the Intel Technology and Manufacturing Day 2017 brochure, Figure 10.8 presents the scaling trends of Intel's process technologies.

As seen in Figure 10.8, a relative die scaling of 0.62× was achieved by reducing the node size from 45 to 32 nm and to 22 nm, where the total area for a similar function was reduced from 100 to 62 to 38.4 mm². However, since the introduction of the 14 nm node, as well as for the 10 nm one, reductions of 0.46× and 0.43× have been achieved, respectively. The figures posted by Intel have also received a fair amount of criticism, especially from rival chipmakers such as TSMC, who questions the validity of the figures presented in Figures 10.7 and 10.8. An important issue

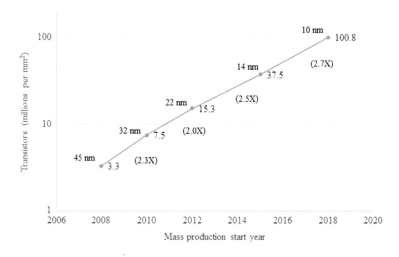

FIGURE 10.7 Intel's achievements with respect to transistors per square millimeter for its last five process manufacturing generations (manufacturing process nodes of 45, 32, 22, 14 and 10 nm).

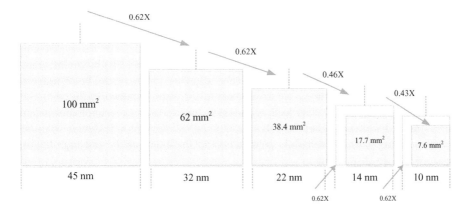

FIGURE 10.8 Intel's achievements with respect to die scaling for its last five process manufacturing generations (manufacturing process nodes of 45, 32, 22, 14 and 10 nm).

that is raised when companies announce such accomplishments is the taking into account of the actual number of 'usable' transistors per square millimeter. Although the maximum number of transistors per area can theoretically reach upward of 100 million transistors per mm², design decisions such as cache sizes and performance targets can lead to significant variations in these values.

Samsung/Qualcomm 10 nm Transistors

Leading mobile chip manufacturer, Qualcomm, announced in late 2016/early 2017 that its new mobile processor, the Snapdragon 835, utilized Samsung's 10 nm second-generation FinFET manufacturing process, allowing for 3 billion transistors per chip. The 10 nm manufacturing process allows for further improvements and advances in mobile device battery life, computing immersion, video connection, mobile connectivity, machine learning and security – characteristics deemed crucial in the mobile device market. Its previous processor, the Snapdragon 820, was based on a 14 nm process, and with the new generation of chips, these companies are aiming to stay ahead of competitors such as Intel, also motivated to keep up with Moore's law as technology scaling becomes increasingly cumbersome. As part of Qualcomm's marketing campaign, the company advertises that smaller and more efficient transistors lead to, again with emphasis on the mobile device market,

- More feature integration.
- Smaller die sizes and packaging.
- Thinner mobile phones.
- More space to increase battery size.
- Lower power consumption leading to better battery life.

The Snapdragon 835 is the first chip to utilize the second-generation FinFET technology, where the width among circuits in the chip is miniaturized to 10 nm, compared with its previous-generation 14 nm spacing. Practically, this leads to a die size reduction of 35% and power consumption reduction of 25%, according to Qualcomm.

The mobile market in 2017 is also clearly focusing its efforts on the virtual reality (VR) market, as well as augmented reality (AR) – both technologies becoming more viable, prevalent and commercially available. Mobile chipmakers such as Qualcomm are also now able to increase their graphics chips from 8-bit to 10-bit architectures as a result of miniaturization improvements, allowing for increased color usage in VR and AR implementations. Improvements in software frameworks due to higher processing performances also lead to enhancements in camera technologies, where focusing and shutter lag can be software-defined. Modern mobile device solutions are increasingly used for mobile payments, and for high-quality and high-fidelity encryption of sensitive data.

IBM RESEARCH 5 NM TRANSISTORS

Theoretically, as a result of continuous and aggressive node scaling, a limitless decrease in the pitch of the contacted gate is suggested, resulting in a compromise between the length of the gate, the thickness of the source/drain spacer and the contact size of the source/drain in a classic lateral device (Yakimets et al. 2015). Transistor gates should ideally be relatively lengthy to preserve and minimize short-channel effects. The minimum space thickness of the source/drain is defined by dependability necessities and/or by the capacitance amid the gate and the source/drain terminals, while maintaining low contact resistance of narrow source/drain contacts. FinFETs have provided a means to achieve these features in sub-20 nm devices; however, to maintain short-channel effect control, "fin thickness should be scaled together with gate length, which can result in too strong threshold voltage variability" (Yakimets et al. 2015).

Lateral gate-all-around FETs (GAAFETs), designed using nanowires, permit comfortable channel dimensions with regard to fin thickness while facilitating adequate short-channel effect control. Vertical GAAFETs are constrained to a lesser extent by the length of the gate and the thickness of the spacer, since they have a vertical orientation and exhibit superior scalability (Yakimets et al. 2015). The persistent assumption is that GAAFETs with vertical or lateral nanowire channels will replace FinFETs as the primary active device from approximately the 7 nm node (Huang et al. 2017). The more complex and therefore costly processing and manufacturing requirements of GAAFETs could be prohibitive to their immediate success, but their advantages and potential would possibly overcome this limitation.

In June 2017, IBM Research announced that it had created a new type of transistor for semiconductor devices that enables 5 nm transistor node construction, currently the smallest feature size for silicon processors (Etherington 2017). By IBM's own admission, this technology will only be ready for mass-market commercialization in 10–15 years from its research prototype stage (2017), but it is a promising candidate for the future of semiconductor processing. IBM Research collaborated with GlobalFoundries and Samsung to achieve this technology, and aided in breaking barriers that essentially signaled the end of Moore's law. This new method of manufacturing transistors makes it possible to construct silicon processors with upward of 30 billion transistors on a chip approximately equal to the size of a human fingernail. Among various processing improvements, the primary technique employed to achieve this feat uses GAAFET transistors, horizontal layers of stacked silicon, to effectively create a fourth gate on the transistors on these chips. A known phenomenon in FET transistors nearing the 5 nm feature size is channel width variations that can cause undesirable variability and mobility loss in the semiconductor devices. In GAA CMOS transistors, the gate is placed on all four sides of the active channel, and in some cases, GAAFETs could have III–V materials such as InGaAs in the channel for mobility improvements. These devices are characteristic of silicon nanowires with a gate circumventing them. In IBM's GAA manufacturing process, dual separate landing pads are made on a substrate and the nanowires are independently made and suspended horizontally on the landing pads. Then, vertical gates are patterned over the suspended nanowires, and multiple gates are thus formed over a common suspended region. After this process is completed, a spacer is formed and the silicon nanowires are cut externally from the region around the gate terminal. Silicon epitaxy is grown (doped in-situ) from the uncovered cross-sections of the silicon nanowires at the brink of the spacer. Classical self-aligned nickel-based silicide contacts and copper interconnects are used to finish off the device, as per IBM's process description. The cross-sections of a cylindrical and rectangular junction-less GAA FET are shown in Figure 10.9, adapted from Hur et al. (2015).

As shown in Figure 10.9, the (a) cylindrical GAA and (b) rectangular GAAFETs are formed by surrounding the transistor with a thin gate dielectric, where t_{ox} is the thickness of the oxide layer and R_{si} and W_{si} are the channel radius and channel width, respectively.

As a result, GAA leads to an estimated 40% performance improvement over current 10 nm chip designs, while consuming the same amount of power. By means of Si/SiGe superlattice epitaxy and a doping procedure for stacked wires that is performed in-situ, researchers have developed a stacked, four-wire GAA FET with gate lengths of 10 nm, constructed on an electrostatic scale length of 3.3 nm. These devices have exhibited an *off*-state current of less than 100 nA/μm and supply voltage V_{dd} of 0.68 V, and from the combination of steep subthreshold slope and low drain-induced

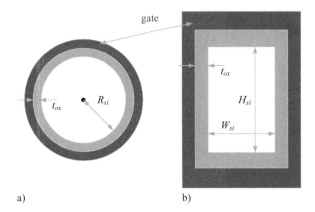

FIGURE 10.9   Cross-sectional representation of junction-less (a) cylindrical GAA and (b) rectangular GAAFETs. (Adapted from Hur, J. et al., *IEEE Transactions on Electron Devices*, 62(7), 2285–2291, 2015.)

barrier lowering, these devices can also yield a relatively high on-state current and possibly faster performance compared with conventional CMOS transistors.

WIND RIVER/INTEL CORPORATION INDUSTRIAL IoT INFRASTRUCTURE

Intel announced in March 2017 that it is advancing the industrial IoT with the availability of the software platform, Wind River Titanium Control, based on open standards and an on-premise cloud-scale infrastructure. The control software enables critical infrastructure companies to cost-effectively evolve legacy control systems that were not specifically designed to support the connected nature of the IoT. The Wind River Titanium Control software platform is optimized for the Intel Xeon line of processors and pre-validated on hardware from leading providers of Intel-based servers.

In April 2016, however, Intel underwent restructuring in its IoT department, cutting 12,000 jobs to refocus on IoT and its data centers, which make up a significant portion of its total revenue. As a result, in October 2016, Intel announced two lines of low-power IoT-dedicated processors, the E3900 series for connected devices and wearables and the A3900 series, specifically for smart automobile applications. The E3900 series is based on the Intel Apollo Lake line of processors, a 14 nm technology targeted at IoT endpoint applications. Processor cores range between two and four, with between 6 and 12 watts of power consumption under load. The E3900 series additionally supports up to 8 GB of RAM and features error correction code capabilities, as well as high-definition image capturing. The A3900 series of processors, also part of the Apollo Lake family, shares a similar infrastructure, but with the focus on digital instrument clusters and vehicle infotainment systems.

By June 2017, Intel had terminated a large number of its efforts to infiltrate the IoT market, shelving several of its processors tasked with giving them a foothold in the ARM-dominated market (powering the likes of the Raspberry Pi, for example).

BERKELEY LAB 1 NM CARBON NANOTUBE

A team of scientists at Berkeley Lab has fashioned a transistor with an operational 1 nm carbon nanotube-based gate. In a statement by Desai et al. (2016), it is mentioned that the choice of proper materials can lead to more room to shrink electronic components, far more than previously considered (the physical limitation is around 5 nm). The key to achieving a working 1 nm transistor gate in Desai et al. (2016) was to use carbon nanotubes and molybdenum disulfide (MoS_2), a universally known solution used as vehicle engine coolant, on silicon. MoS_2 is a transition metal dichalcogenide (TMD), an atomically thin semiconductor of the type MX_2, where M is the transition metal atom, such as Mo or W, and X is a chalcogenide atom, such as S, Se or Te.

As per its chemical composition, a single layer of transition metals is inserted between two films of chalcogenide atoms. This material is a layered 2D structured semiconductor and has been identified as a possible alternative to silicon for channel material, where each material displays altered band structures and characteristics. Even thickness regulation with atomic-level accuracy in TMD structures is possible because of its layered nature, a highly desirable characteristic for highly controllable electrostatics in ultrashort gate length transistors. As opposed to graphene, which has a characteristically high carrier mobility, low losses accompanied by the Joule-Thompson effect and a band gap of 0 eV, TMD monolayers are structurally stable, have carrier mobility comparable with silicon and have a finite band gap, making these materials usable to manufacture transistors. A typical monolayer MoS_2 FET is shown in Figure 10.10.

As shown in Figure 10.10, a typical monolayer MoS_2 FET is constructed as epilayers on a traditional silicon substrate. The source and drain terminals are separated from the substrate by a SiO_2 layer (ranging between approximately 200 and 300 nm in thickness in Radisavljevic et al. 2011), acting as a pedestal for these terminals. The monolayer MoS_2 is situated within a hafnium oxide (HfO_2) gate dielectric, whereas the gate terminal is placed on top of the gate dielectric above the monolayer MoS_2 channel. HfO_2 is a relatively inert electrical insulator with a band gap of between 5.3 and 5.7 eV, typically seen in optical coatings, used as a high-κ dielectric material in DRAM capacitors and in advanced CMOS processes, for example TMD devices.

The first FET manufactured by TMD materials in 2004, using a bulk of WSe_2, had a reported carrier mobility of 500 $cm^2V^{-1}s^{-1}$ for p-type conductivity at 300 K. This is approximately half the carrier mobility obtained with silicon FETs, but a low *on/off* current ratio was achieved. In 2011, reported in Radisavljevic et al. (2011), a HfO_2 gate dielectric was used on a single-layer MoS_2 substrate to demonstrate a mobility of 200 $cm^2V^{-1}s^{-1}$ at room temperature, a significant improvement from a previously reported mobility in the 0.5–3 $cm^2V^{-1}s^{-1}$ range. This improved mobility is comparable with that of graphene nanoribbons, but with the added advantage of a finite and direct band gap (due to the monolayer) and a high *on/off* current ratio of approximately 10^8 – leading to an ultralow standby power for these devices. Radisavljevic et al. (2011) additionally mention that monolayer MoS_2-based

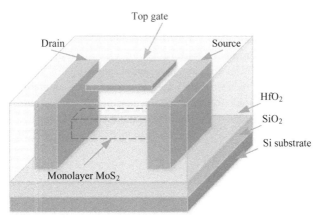

FIGURE 10.10   A typical monolayer MoS_2 FET where the gate length is defined by the single-walled carbon nanotube. (Adapted from Desai, S. B. et al., *Science*, 354(6308), 99–102, 2016.)

transistors can supplement graphene in uses that need thin and translucent semiconductors, for example in optoelectronics and energy harvesting. The optical properties of a monolayer TMD device have been taken advantage of to manufacture phototransistors and highly sensitive photodetectors.

Both MoS_2 and silicon have a crystalline lattice arrangement; however, electrons moving through silicon have a reduced mass related to MoS_2. Below 5 nm, the quantum mechanical tunneling mechanism, and therefore short-channel effects, is initiated and the gate blockade or barrier is no longer capable of preventing the electrons from rushing over and through from the source terminal toward the drain of the transistor. Direct source-to-drain tunneling and the losing of gate electrostatic regulation on the channel rigorously worsen the *off* state leakage currents (therefore increasing leakage currents), consequently restricting the scaling of silicon transistors (Desai et al. 2016). For an effective mass that is heavier, a greater band gap and a lesser in-plane dielectric constant lead to smaller direct source-to-drain tunneling currents (Desai et al. 2016).

Since electrons moving through MoS_2 have a greater real (effective) mass, their movement can be regulated with reduced gate lengths. Is it also possible to scale MoS_2 down to sheets that are 0.65 nm (650 pm) thick with a comparatively lesser dielectric constant? As per photolithography limitations, even considering EUV, the gate of the transistor in Desai et al. (2016) was constructed using 1 nm thin cylindrical carbon nanotubes. This combination of technical innovations enables a 1 nm thin transistor capable of controlling the flow of electrons.

Desai et al. (2016) demonstrated a 1D gated, 2D semiconductor FET (1D2D-FET) with a single-walled carbon nanotube gate, a MoS_2 channel where the 1D2D-FET exhibited switching features with an *on/off* current quotient of 10^6 and a subthreshold swing of approximately 65 mV per decade at 300 K and a DIBL of approximately 290 mV/V. The gate dielectric in Desai et al. (2016) is manufactured using ZrO_2 (as opposed to HfO_2) on a 50 nm SiO_2/Si substrate. As proposed by Desai et al. (2016), "TMDs offer the ultimate scaling of thickness with atomic-level control, and the 1D2D-FET structure enables the study of their physics and electrostatics at short channel lengths by using the natural dimensions of carbon nanotubes, removing the need for any lithography or patterning processes that are challenging at these scale lengths". For TMD devices, extensive manufacturing and processing on a large scale, especially down to such small gate lengths (1 nm), is challenging, and still requires future innovations. Research on developing "process-stable, low-resistance ohmic contacts to TMDs, and scaling of the gate dielectric by using high-κ 2D insulators are essential to enhance device performance" (Desai et al. 2016) and improve manufacturability.

CONCLUSION

This is the concluding chapter of the book titled *Extending Moore's Law Through Advanced Semiconductor Design and Processing Techniques*, developed to summarize trends, limitations and technology advances that endure, or threaten to derail, Moore's law. The chapters in this book take a detailed look at technologies that are currently, in 2018, deemed the current and future generations to sustain Moore's law. The chapters of this book are summarized below.

The Driving Forces Behind Moore's Law and its Impact on Technology deals with the generalized technical and economic factors that are driving Moore's law, from past accomplishments, since it was introduced in 1965, up to more recent advances and restrictions that are driving miniaturization in the foreseeable future.

The Economics of Semiconductor Geometry Scaling delves deeper into the economic characteristics that have formed an integral part of Moore's law and have motivated and encouraged manufacturers to push technologies to higher limits, from an advantageous economic perspective, especially with mobile smartphones and IoT devices becoming more powerful and effectively smaller.

The Significance of Photolithography for Moore's Law dissects photolithography as one of the most important processing steps during semiconductor manufacturing. The reason that this step is considered one of the most crucial is that semiconductor patterning of geometries below a certain threshold requires various alterations to traditional UV lithography, or for future devices, exposure at much lower wavelengths. Optical photolithography equipment is an expensive commodity for most foundries, and must be able to produce high yields to overcome initial capital expenditure.

Photolithography Enhancements to Facilitate Moore's Law focuses on various enhancement techniques to improve photolithography yield, resolution enhancements and proximity correction.

Future Semiconductor Devices: Exotic Materials, Alternative Architectures and Prospects is a chapter specifically aimed at reviewing methods, materials and breakthroughs that are being reported on or are openly accessible on the internet. These technologies (graphene, optoelectronic waveguides and photonic crystals, molecular electronics, spintronics and solid-state quantum computing) are discussed in various degrees of detail.

Microelectronic Circuit Thermal Constrictions Emanating From Moore's Law reviews the thermal considerations that must be taken into account as miniaturization leads to large amounts of transistors densely integrated onto a single chip. The thermal conductivity of materials alone cannot be used to ensure heat is transferred from active devices; circuit design considerations should also be adhered to.

Microelectronic Circuit Enhancements and Design Methodologies to Facilitate Moore's Law (1) and *(2)* take a deeper look at both physical placement (layout) and enhancements in schematic design, and the incorporation of circuit-level improvements to mitigate challenges such as high thermal energy generation, thus improving performance, reducing overall chip size and reducing the cost of microelectronic designs, and to design for ease of manufacturability and high yield.

The Evolving and Expanding Synergy Between Moore's Law and the IoT combines the predictions of Moore's law with two relevant and related laws, Metcalfe's law and Koomey's law, and reviews how the IoT has benefited from aggressively scaled semiconductor components. Through energy harvesting and small but powerful computing, a new era in environmental sensing and information delivery is rapidly growing as applications that benefit the population evolve.

Case Studies: Technology Innovations Driven by Moore's Law, the concluding chapter of this book, looks at the current generation of semiconductor technologies that dominate commercial computing products, especially those used in devices that have led to significant paradigms shifts in how information is captured, stored, analyzed and transferred. Desktop computing, mobile computing, smartphones and the IoT are all among the technologies driving semiconductor manufacturers to further reduce the size of components while maintaining high efficiency and yield. In addition, this chapter looks at recent advances in sub-5 nm transistor technology and technically reviews publicly available knowledge on its progress.

REFERENCES

Bu, H. (2017). 5 nanometer transistors inching their way into chips. Retrieved 28 June 2017 from http://www.ibm.com.

Chen, L. (2016). The most profitable industries in 2016. Retrieved 9 July 2017 from http://www.forbes.com.

Desai, S. B., Madhvapathy, S. R., Sachid, A. B., Llinas, J. P., Wang, Q., Ahn, G. H., Pitner, G., Kim, M. J., Bokor, J., Hu, C., Wong, H. S. P., Javey, A. (2016). MoS_2 transistors with 1-nanometer gate lengths. *Science*, 354(6308), 99–102.

Etherington, D. (2017). IBM creates a new transistor type for 5 nm silicon chips. Retrieved 9 July 2017 from http://techcrunch.com

Hruska, J. (2017). How do SSDs work? Retrieved 14 July 2017 from http://www.extremetech.com

Huang, Y., Chiang, M., Wang, S., Fossum, J. G. (2017). GAAFET versus pragmatic FinFET at the 5 nm Si-based CMOS technology node. *IEEE Journal of the Electron Devices Society*, 5(3), 164–169.

Hur, J., Moon, D., Choi, J., Seol, M., Jeong, U., Jeon, C., Choi, Y. (2015). A core compact model for multiple-gate junctionless FETs. *IEEE Transactions on Electron Devices*, 62(7), 2285–2291.

Intel. (2017). Intel Technology and Manufacturing Day 2017. Retrieved 11 July 2017 from http://www.newsroom.intel.com

Kim, H., Ahn, S., Shin, Y., Lee, K., Jung, E. (2017). Evolution of NAND flash memory: From 2D to 3D as a storage market leader. *2017 IEEE International Memory Workshop (IMW)*, 1–4.

Mizoguchi, K., Takakashi, T., Aritome, S., Takeuchi, K. (2017). Data-retention characteristics comparison of 2D and 3D TLC NAND flash memories. *2017 IEEE International Memory Workshop (IMW)*, 1–4.

Nvidia. (2017). Nvidia Volta. Retrieved 13 November 2016 from http://www.nvidia.com.

Radisavljevic, B., Radenovic, A., Brivio, J., Giacometti, V., Kis, A. (2011). Single-layer MoS_2 transistors. *Nature Nanotechnology*, 6, 147–150.

Samsung (2013). White Paper: Samsung V-NAND technology. Retrieved 11 July 2017 from http://www.samsung.com

Spinelli, A. S., Compagnoni, C. M., Lacaita, A. L. (2017). Reliability of NAND flash memories: Planar cells and emerging issues in 3D devices. *Computers*, 6(2): 16.

Yakimets, D., Eneman, G., Schuddinck, P., Bao, T. H., Bardon, M. G., Raghavan, P., Veloso, A., Collaert, N., Mercha, A., Verkest, D., Thean, A. V., De Meyer, K. (2015). Vertical GAAFETs for ultimate CMOS scaling. *IEEE Transactions on Electron Devices*, 62(5), 1433–1439.

Index